The World the Game Theorists Made

The World the Game Theorists Made

PAUL ERICKSON

THE UNIVERSITY OF CHICAGO PRESS   CHICAGO AND LONDON

PAUL ERICKSON is assistant professor of history and science in society at Wesleyan University.

The University of Chicago Press, Chicago 60637
The University of Chicago Press, Ltd., London
© 2015 by The University of Chicago
All rights reserved. Published 2015.
Printed in the United States of America

24 23 22 21 20 19 18 17 16 15   1 2 3 4 5

ISBN-13: 978-0-226-09703-9 (cloth)
ISBN-13: 978-0-226-09717-6 (paper)
ISBN-13: 978-0-226-09720-6 (e-book)
DOI: 10.7208/chicago/9780226097206.001.0001

Library of Congress Cataloging-in-Publication Data

Erickson, Paul, 1976– author.
    The world the game theorists made / Paul Erickson.
      pages ;   cm
    Includes bibliographical references and index.
    ISBN 978-0-226-09703-9 (cloth : alk. paper)
    ISBN 978-0-226-09717-6 (pbk. : alk. paper)
    ISBN 978-0-226-09720-6 (ebook)
    1. Game theory.   2. Science—Methodology   3. Von Neumann, John, 1903–1957.
Theory of games and economic behavior.   I. Title.
    HB144.E75 2015
    330.01′51932—dc23
                                                           2015010760

♾ This paper meets the requirements of ANSI/NISO Z39.48-1992 (Permanence of Paper).

Contents

The Game Theory Phenomenon

This book is about one of the great intellectual conversations to unfold in the latter half of the twentieth century: a great sequence of debates, now lingering into their seventh decade, about the prospects for building a mathematical theory of rational decision-making that might not only revolutionize the study of human behavior and social interaction, but that had the potential to rationalize decision-making in every area of human affairs, from business to foreign policy. At the heart of these debates lay game theory—a theory of strategic interaction between rational individuals—which was launched in 1944 by the publication of mathematician John von Neumann and economist Oskar Morgenstern's *Theory of Games and Economic Behavior*. In the years that followed, buoyed by postwar military funding and the grand ambitions of the game theorists themselves, the theory (and its close intellectual relatives, from mathematical programming to social choice theory) established footholds in an array of fields from mathematics and operations research to social psychology, political science, economics, philosophy, and evolutionary biology. Game theory also found itself at the heart of discussions of some of the crucial problems of the Cold War era, from the design of effective weapons systems to the resolution of the nuclear arms race. Indeed, many familiar problems and concepts—from "mutual assured destruction" to "the tragedy of the commons"—have become difficult to conceptualize except in game-theoretic language.

Certainly, the ambitions of game theory's proponents have at times been sweeping. Von Neumann and Morgenstern introduced their ideas by placing them next to the great triumphs of mathematical science in ages past, and envisioned a future in which their work might bring the study of society out of its present dark age into a truly scientific future. And toward the end of his career, Morgenstern would seek to write a history of game theory that located its roots in enlightenment debates over the calculus of probabilities—thereby tying game theory's distinctively modern formulation of rational behavior to the philosophical luminaries of the age of reason.[1] Yet quite apart from the size of the historical stage upon which game theory's creators imagined themselves, it is hard not to be struck by the sense of glamor, power, and prestige that has attached itself to game theory during the high years of the Cold War and beyond. Echoing von Neumann and Morgenstern's aspirations, game theory rapidly became the province of a new fraternity of intellectuals—men like von Neumann and Morgenstern, Thomas Schelling, or John Harsanyi, to name a few—who blended old and new, classical and midcentury modern. They imagined themselves in part as revivers of the great enlightenment tradition of moral philosophy, diving with flair into the great problems of ethics and philosophy of science, into the study of how one ought to think and act that existed well before modern "social science" took form in the nineteenth century. Yet they typically were also imbued with the ethos of an era that glorified science, sporting an ultramodern mathematical idiom, fluent in the language of computers, man-machine systems, ICBMs, and other manifestations of Cold War high technology. They moved at will from the traditional settings of academia—the ivied walls of Princeton or Harvard—to the new corridors of power in midcentury America: the Pentagon, military-funded think tanks like the RAND Corporation (resplendent in its modernist glass and steel Santa Monica headquarters), or the research campuses of Bell Labs and IBM.

Such was the promise. Yet looking beyond the glamor, the assumed lineage, and the breathtaking ambitions, game theory's allure was (and remains) enigmatic and contested. Its widespread use in the social and behavioral sciences might be understandable if it received high marks according to some commonly cited standards for assessing the usefulness of a "theory"—for example, its predictive power or its ability to describe the way individuals make decisions. But many of the same scholars who have most raised the profile of game theory in the social and behavioral sciences have simultaneously served as the strongest critics of the theory's empirical adequacy. This is true, for example, whether we look to Herbert Simon's notion of "bounded rationality," which explicitly assumes that humans lack the computational power to be true rational actors in the sense of game theory; or to the extensive literature, following the lead of experimental psychologists Amos Tversky and Daniel Kahneman, which claims that in many plausible situations humans routinely violate the principles of probability theory and decision theory that underlie the larger theory of games. Moreover, to this day, there remains little agreement among game theorists as to what constitutes a "solution" for many games: as a recent textbook of game theory for economists notes, "One of the intriguing and frustrating aspects of game theory is that there does not yet exist a universally accepted solution concept that can be applied to every game."[2] Indeed, when it comes to many of the particular "games" we care the most about—nations trying to abide by an arms control agreement, corporations deciding whether or not to enter a market, or children deciding how to share a slice of cake—we are typically left not with one predicted outcome or a definite set of rules of rationality to follow, but many.

From those standing at a greater distance from the brand of rational choice theorizing embodied in game theory, the criticisms have been still more vocal. New Left intellectuals of the 1960s—not to mention their more activist allies in the streets—assailed game theory for its ties to military funding agencies, for its presumed complicity in rationalizing an aggressive national security policy, and for its ethical inadequacy. In their eyes, the game theorists' pretensions to scientific neutrality coexisted uneasily with the notion that game theory might somehow tell us how we *should* make decisions in social situations, not to mention in the realm of nuclear policy. Such critiques on ethical grounds have often been accompanied by disciplinary struggles: from political science to economics to philosophy, rational choice modeling has coincided with significant transformations of these fields that saw the displacement of historicist or interpretive styles of inquiry across a broad swath of their former terrain. Corresponding to this intellectual shift has been a remaking of the academic disciplines and working practices in ways that many commentators have found malign: the rise of "economic imperialism," and a shift in power toward mathematically trained experts who hold sway in policy debates.[3]

This book seeks to make sense of what one might call the "game theory phenomenon"—the broad if contested spread of a mathematical tradition—between the publication of von Neumann and Morgenstern's seminal 1944 book and the theory's firm establishment in economics during the 1980s. It does so not by rendering a magisterial history of ideas. The intellectual corpus that has grown out of von Neumann and

Morgenstern's work, it is safe to say, is by now far too varied and diverse to imagine that any single narrative of development or growth over time could capture all its intricacies. Rather, this book explores the contours of the world that the game theorists made during this period: the kinds of intellectual and institutional connections that they catalyzed, the sheer variety of research programs their theory enabled and inspired, and the conversations and controversies that the new mathematics of rational choice facilitated. In light of the critical assessments of the theory outlined above, this kind of world-building has arguably been game theory's most impressive contribution to our intellectual culture over the past seven decades. But at the least, examining the game theory phenomenon sheds much light on a number of key features of the Cold War sciences: the vogue for mathematical techniques, especially models of rational choice; the connection between these new techniques and disciplinary alliances and patterns of collaboration spurred by new forms of patronage during this period; and, above all, the significance of a far-flung set of debates over the nature of rationality and choice in an age of scientific and technological enthusiasm.

Outlining a History

If there is a canonical history of game theory extant today, it reflects the theory's high-profile appearances in economics since the 1980s and a resulting growth of interest in its history among practitioners and historians of economics. Ostensibly pitched at economists, von Neumann and Morgenstern's *Theory of Games and Economic Behavior* continues to be cited as a founding work of the game-theoretic tradition in economics, even if the research programs it inspired subsequently came to overshadow the original work and even substantially subvert its message. In particular, the award of a Nobel Prize in economics to a trio of game theorists in 1994—John Nash, John Harsanyi, and Reinhard Selten—focused attention on the history of a particular approach to solving games that was not present in von Neumann and Morgenstern's work—the "Nash program" of reducing all games to noncooperative games (in which players could not form binding agreements) and identifying their "Nash equilibrium" points. Nash—working against a colorful if only vaguely relevant backdrop of US military support for research in game theory, as depicted in the 2001 film *A Beautiful Mind*—proved the fundamental theorems about such games in his papers of the early 1950s. Harsanyi subsequently generalized Nash's analysis to situations involving incomplete information, and Selten provided powerful economic applications of the theory, especially in the field of industrial organization.[4]

More recent Nobel prizes in economics for individuals known for their contributions to game theory—Thomas Schelling and Robert Aumann in 2005, and Leonid Hurwicz, Eric Maskin, and Roger Myerson two years later—added additional datapoints to the history, either by honoring applications of the theory to problems of "mechanism design," or to the game-theoretic analysis of "bargaining" in economics and elsewhere. As a result, historians and practitioners of economics have tended to outline a history of game theory that starts with von Neumann and Morgenstern, but that rapidly moves away from their problematic "cooperative" theory of multiplayer games and gravitates toward Nash's program, which proved to be a much more natural fit with economics' interest in the effect of incentives on rational individuals. Indeed, some have gone so far as to suggest that von Neumann and Morgenstern's book retarded the development of game theory in the social sciences by overshadowing the work of the young John Nash and those who took inspiration from it.[5]

The history I relate here certainly features some of these high-profile individuals and their celebrated achievements, and to an extent their work structures the narrative arc of the book. Some of the *dramatis personae* will be familiar (in name at least) to modern practitioners of game theory: von Neumann and Morgenstern, Nash, Harsanyi, Schelling, Selten, and Aumann, among others. But these figures will often be overshadowed by others whose careers and research programs wielded significant influence on game theory's earlier development, even if the resulting account will seem to a contemporary practitioner of game theory to dwell rather far from the action "at the front." As we shall see, this relative shift in emphasis reflects two realities of game theory's history prior to the 1980s, the decade in which noncooperative game theory celebrated its ascendancy within economics. First, it was not clear that Nash's "noncooperative" theory of multiplayer games might attain primacy among the many alternative "solution concepts" that gained attention in the first years after 1944 (including von Neumann and Morgenstern's own "cooperative" solution concept). Nor was it clear that the most important constituency for game theory would prove to be economists, even if an economist (Morgenstern) played a key role in producing

Theory of Games and Economic Behavior, and even if some individual economists were from the start deeply engaged in the conversations that flowed from the publication of von Neumann and Morgenstern's book.

Decentering the rise of noncooperative game theory within economics since the 1980s brings with it both historiographical promise and peril. This earlier period in game theory's history was marked by the emergence of a wide variety of research programs representing tantalizing possible futures for what the theory might have—and might yet—become. Consequently it is difficult to hang the story of game theory between the 1940s and the 1980s on any of the traditional objects of analysis in the history of the sciences: an individual life, or even a manageable collection of individuals with common life stories and intellectual pedigrees; a well-defined discipline, research program, or "paradigm" with well-developed borders or trajectory; or a particular institution. While there were a number of particular patrons who supported research involving game theory, there were no readily identified "departments of game theory" on university campuses during this period; aside from a relatively small number of research centers identified with particular prominent scholars, it is equally difficult to identify any today. The first regular academic journals specifically devoted to game theory (most notably, the *International Journal of Game Theory*) only began to appear in the 1970s. And while there was a small cadre of individuals who might have self-identified as "game theorists" working during this period, they often lived wandering lives, moving between departments of economics, the social and behavioral sciences, mathematics, industry, various regions of the defense establishment, and fields like operations research, industrial organization, and others that remain relatively difficult to classify.[6]

In particular, histories focused on the traditional academic disciplines couple only loosely with the rise of game theory during this period. The theory's deepest intellectual roots lay in mathematics, and individuals trained in mathematics would prove game theory's strongest constituency for much of its early history; yet there is evidence that, at least initially, the theory's status in the mainstream of "pure" mathematical research was precarious. Even labeling game theory a branch of "applied mathematics" provides it with little disciplinary foundation, given the institutionally nebulous position of applied mathematics in early twentieth-century American academia. Instead, game theory—much like cybernetics and information theory, which appeared at roughly the

same time—was closely associated with broader cross-disciplinary research agendas and patterns of funding that a growing body of literature on the postwar human sciences has identified as a key feature of
the period between World War II and the 1970s. During this period, existing university departments, pedagogy, academic journals, scholarly
societies, and patrons did not disappear, but they were increasingly com-
plemented (and possibly dwarfed) by new networks of funding, connections, seminars, conferences, and research institutes bound together less
by traditional disciplinary expertise than by common practical endeavors and problems, modeling idioms, and methodologies. New areas of
inquiry and expertise like operations research, "management science,"
"general systems," or "behavioral science," all associated with the circulation of game theory during this period, typified the postwar growth
of what Peter Galison has termed the "interdisciplines."[7] Reflecting this
fact, scholars interested in the postwar human sciences broadly speaking
have increasingly crafted intellectual histories of this era that leave traditional disciplinary boundaries behind. Steve J. Heims and Paul N. Edwards have explored the impacts of cybernetics and computing in psychology and the social sciences; Philip Mirowski has called our attention
to the legacy of "operations research" for postwar economic thought;
S. M. Amadae has followed the trail of rational choice theory across the
behavioral sciences, political science, and economics; Deborah Hammond has explored the legacy of general systems theory; and more recently, much work by Hunter Heyck, Jamie Cohen-Cole, and Joel Isaac
has shed light on this kind of intellectual fusion in other areas of the
postwar human sciences.[8]

So if game theory's history between the 1940s and 1980s cannot be
traced to a particular discipline or research program, or even a single
"interdiscipline," what then can explain the theory's wide appeal during this period? To date, the Cold War context has loomed large in the
historiography as a possible explanation. S. M. Amadae's work has connected the vogue for theories of rational choice (including game theory)
with a set of high-level Cold War ideological and intellectual struggles
to "rationalize capitalist democracy," to create a liberal vision of economics and politics that might neutralize the appeal of socialism and
statist calls for a planned society to further the common good. Such an
account elegantly explains the emergence of several key theoretical traditions in political theory and economics, but it is less readily adapted
to explain other episodes in the history of Cold War–era rational choice

mathematics—for example, game theory's appearance in the work of Marxist evolutionary biologists in the 1960s. Another approach follows the lead of an extensive literature in the history of science that has examined the impact of Cold War military patronage on the conduct of the physical and biological sciences.[9] Certainly, for much of its early history, game theory circulated in close association with military money, via organizations like the Air Force–funded RAND Corporation and the Office of Naval Research, among others. And although military patronage would become less central to the theory's history over time, it is hard to deny that it helped the theory survive at a time when it might otherwise have disappeared into obscurity.

It is tempting to draw connections between military funding and imperatives on the one hand, and the intellectual content of game theory on the other. One possible historiographical approach thus focuses on tracing the emergence of theories, models, and metaphors, and associating their spread with the influence of dominant discursive frameworks or modeling idioms that were characteristic of the Cold War era, or the demands of powerful patrons (especially military patrons) of science. Information theory, cybernetics, and systems modeling thus spread across fields like psychology, ecology, and molecular biology in part through their resonance with military-promoted discourses of command-and-control, man-machine systems, and the like.[10] In a similar vein, depending on which history you consult, game theory survived in the postwar era in part through its apparent ability to capture the "image of conflict," the paranoid us-or-them mindset of the military establishment and strategic community; it provided a natural framework for thinking about "the problem of the bomb"; or it resonated with a Manichean "ontology of the enemy" that emerged out of World War II military operations.[11]

Such approaches to interpreting the history of game theory and related models, metaphors, and theoretical frameworks have great appeal. It is hard to dispute that the activity of theorizing is essentially discursive, concerned with verbalizing, notating, and articulating. And the identification of some semblance of analytic or stylistic unity in the welter of ideas circulating across operations research, cybernetics, systems theory, and the like could to some extent overcome the historiographical challenges posed by game theory's peculiar institutional and disciplinary position between the 1940s and the 1980s. Yet looking for unity in the face of diversity runs the risk of essentializing game theory, or focusing on one aspect of the theory (such as the Nash equilibrium concept or von Neumann and Morgenstern's theory of the two-person zero-sum game) at the expense of others. Tying game theory to the interests of any particular patron or to some overarching mentality, discourse, or *zeitgeist* also risks providing an unnecessarily reductive explanation for its influence, one that cannot fully account for the striking diversity of homes and uses game theory has found, from operations research to conflict resolution, economics, and evolutionary biology. Rather, it seems at least as plausible that the widespread circulation of game theory was substantially the result of the theory's interpretive flexibility and disunity, its ability to be selectively appropriated and reinterpreted for use in service of many different disciplines and agendas.

Indeed, when viewed up close, game theory comes across less as a coherent entity with a well-defined character, message, or ideological content, and more as a cluster of conceptually pregnant metaphors, notations, axioms, and techniques that hang together not because of their internal structure but as a result of accidents of history.[12] Consider the sheer scope of the mathematical theory that von Neumann and Morgenstern's book brought together: theorems on ordering, preferences, and utility; elements of graph theory and the extensive form of games; fixed-point theorems, the minimax theorem, and related results; the theory of subadditive measures; the characteristic function form of the game; and many others. While these component parts of game theory became linked with the publication of von Neumann and Morgenstern's book, they were not *a priori* part of a common conversation. Nor did they fully move together in lockstep after 1944, even if after this date they were widely recognized as being component parts of the larger edifice of "game theory."

This book therefore adopts a slightly different—yet ultimately, complementary—approach to integrating the intellectual history of game theory with the context of the postwar era: by characterizing the theory as a heterogeneous collection of what one historian of the physical sciences has called "theoretical tools" that could be selectively appropriated, interpreted, and used to do work in a variety of contexts.[13] Emphasizing the internal *diversity* of the game-theoretic corpus avoids the temptation to essentialize game theory, equating it with the Nash equilibrium concept or granting it some essential character, and embracing the fact that game theory's identity was a constant point of contestation throughout this era and afterward. Focusing on the *use* of elements within the game theoretic corpus moves beyond charting the disembod-

ied circulation of metaphors and concepts, or high-level clashes of ideologies and ideas, and toward grounding game theory's history in the
economy and working practices of research in parts of mathematics and
the human sciences in the postwar era, thereby more fully integrating
the cultivation of game theory with its context. By focusing on the use of
game theory by varied groups of researchers, this approach helps us understand how the theory could travel as it did along the contours of the
cross-disciplinary research programs that featured so prominently in the
Cold War social and behavioral sciences and acted as a key point of connection between academic research agendas and the needs of state patrons of science. Further, it helps us to understand what I have called
"the game theory phenomenon" as a product of game theory's interpretive flexibility and the inclination of multiple groups to enroll themselves
in the game-theoretic tradition. What ties this history together is less
the theory's analytic or stylistic coherence, but rather these tools' common genealogy: their presence together in von Neumann and Morgenstern's 1944 book, of course, but also their subsequent invocation by different communities of researchers in an enduring sequence of debates
about the theory's potential to serve as a model of human rationality, as
a guide to decision-making, and as the foundation for theory-building in
the human sciences.

Theories and Tools

The approach I take in this book is inspired, at least in part, by a diverse
literature in the history of science, sociology, and anthropology that has
emerged in recent decades to provide a novel way of approaching the
histories of mathematics and theory-building activities in the sciences.
Historians of physics, in particular Andrew Warwick and David Kaiser,
have begun to focus especially on the pedagogy of theory: how theory
is learned via the mastery of theoretical tools and techniques, and how
theorists learn the rich sets of associations that constitute the "application" of their theories to the world.[14] Anthropological literature focusing
on the observation of the cognitive practices involved in counting, measuring, and calculating has begun to explore the diversity of ways that
humans conceptualize and perform these seemingly straightforward and
universal procedures, from shoppers managing their food budgets to Yo-
ruba schoolchildren learning to measure distance.[15] And, finally, an ex-

ceptionally diverse and rich literature has focused on the practices surrounding the use of "models" in the sciences, from the "paper tools"
used by chemists to visualize complex molecules, to computer models
of social systems or atmospheric processes, or indeed to the "model organisms," from lab rats to the tobacco mosaic virus, whose characteristics make them convenient investigative stand-ins for wide classes of life
forms. In each case, scientists construct and think with simplified systems that help mediate between their limited manipulative and perceptual capabilities and a highly complex external reality.[16]

Taken as a whole, this literature directs our attention away from histories of disembodied ideas, or narratives of the gradual discovery of time-
less mathematical truths, and toward viewing theorizing and calculating
as a kind of practice, a form of work that must be learned and carried
out in real time. And by permitting a provocative set of comparisons between extensive studies of *physical* working practices in the sciences and
daily life on the one hand, and approaches to the history of mathematics
and of theory on the other, it suggests several novel possibilities for the
latter. Theoretical tools initially designed for one purpose prove capable
of being repurposed and put to work in many different ways. The operations of mathematical theorizing are not inherently rule-bound and universal; instead, there may be quite varied ways of calculating with and
manipulating mathematical objects whose use otherwise seems straightforward. And finally, like laboratory scientists, for whom replication of
experiments involving a given piece of equipment is often dependent on
a certain amount of relatively inscrutable "tacit knowledge," theorists
likewise may draw upon a common stock of notations, metaphors, and
patterns of inference and manipulation, yet require additional elements
of skill, judgment, or perhaps "taste" in using these in pursuit of their research agenda.

As a concrete example, consider one of the simplest games to analyze, a standby of game-theoretic pedagogy almost from the creation of
the subject: the game of rock-paper-scissors. Two players simultaneously
name one of the three objects; paper beats rock, scissors beat paper,
rock beats scissors, and naming the same object results in a tie, to be
overcome via additional rounds of the game. A mid-twentieth-century
mathematician would develop this game via a very particular set of representations, derivations, and discursive practices. First, the game itself
would be notated via a matrix of numbers that captures (in the mathematician's view) all the essential features of the game itself: the choices

Player I: \ Player II:	Rock	Paper	Scissors
Rock	0	−1	+1
Paper	+1	0	−1
Scissors	−1	+1	0

FIGURE I.I. Rock-paper-scissors.

available to the two players, and the rules governing who wins and who loses given the choices those players make. What goes up on the blackboard is depicted in figure 1.1. The outcomes "winning" and "losing" have been replaced by numbers in the matrix (or "payoffs")—in this case, with +1 representing a win for Player I (and loss for Player II), −1 a loss, and 0 a tie.

The introduction of these numbers constitutes a first step toward relating the problem to a principle of rational choice: each player, the game theorist tells us, will (or perhaps, should) choose a *strategy* for playing the game—a set of rules for how to act in every contingency—that maximizes the numerical payoff she can expect. This assertion then generates a series of questions and answers that follow, seemingly inexorably. Do any strategies exist that allow the players to simultaneously maximize the expected value of their payoffs? By a set of further (quite complicated) symbolic manipulations, one can prove that the answer is yes, but only by permitting the players the possibility of choosing an object at random: for example, rock with probability 1/2, paper with probability 1/3, and scissors with probability 1/6. Next, can we find out what this best strategy is? More calculations. Yes: the players will (or should) choose rock, paper, and scissors completely at random, with equal probabilities. Finally, how can one actually select that random strategy? People are actually not very good randomizers, or so the statisticians tell us. An electronic computer programmed with an algorithm for generating random numbers can perhaps do the trick, or perhaps a set of calculations based on a table of statistically certified random digits.[17]

Of course, the mathematician's game is substantially unlike the original rock-paper-scissors contest. To most players the matrices, proofs, and random-number generators of the mathematicians would be thoughts yet unthought and objects unimagined. As a child I had recourse instead to the usual stock of playground techniques for winning at this game:

practical knowledge on how to expose your hand to the view of your opponent at the last permissible moment; ways to read the subtle movements of your opponent's wrist that suggest what object she will name;
ways to confuse your opponent by appropriate verbal taunts and distrac-
tions, or to discern and recall her tendencies and style of play. But since
then, via a combination of learning and socialization, I came to imagine that these two quite distinct sets of activities—playing a child's game,
and scratching symbols on a blackboard—are related, to the point that
today I can hardly think of one without the other. This intellectual on-
togeny has recapitulated a historical phylogeny: over the past seventy
years, a similar process has unfolded across the numerous fields in which
game theory has become embedded. Economic exchange and business
decision-making, international diplomacy, the Darwinian "struggle for
existence"—all have become "games" in this peculiar sense.

It may be tempting to view this process of building connections between the game theorist's somewhat idiosyncratic analysis and a wide
variety of situations as episodes of the "application" of mathematics to
the world, with separate and distinct histories. In this vein Martin Shubik, an economist who was part of the first generation of postwar practitioners of game theory, has suggested a possible division of game theory
into three parts. "High church" game theory is essentially a branch of
pure mathematics, concerned with working out the logical consequences
of collections of axioms, while "low church" game theory is essentially
a branch of industrial organization theory focused on the application of
existing game-theoretic concepts to economic problems. Finally, what
Shubik dubbed "conversational game theory" is anything that "deals
loosely with preformalized vague but strategic problems."[18] Yet even if
the categories of "high church" and "low church" fairly characterize the
recent relationships between game theorists and industrial organization
theorists, for example, we should be cautious about projecting them back
onto earlier historical eras. As historians of the sciences have noted, the
language of "pure mathematics" and "applied mathematics," like the
distinctions between "pure science," "basic research," and "applied science," is often part of a politically charged boundary-drawing exercise
aimed at insulating certain areas of research from external influence and
control.[19] And indeed, during much of the history under consideration
here, the interaction between what one might consider "high church"
game theorists and various game-theoretic research programs in the social, behavioral, and biological sciences was never unidirectional, but

was characterized by a constant dialogue, a looping feedback that makes it difficult to discern divisions between the two.

More broadly, the "pure-applied" (or theory-data) distinction does not quite capture the dynamics of game theory's spread during the immediate postwar period, as the existence of the third category of "conversational" game theory in Shubik's taxonomy might suggest. This is where the conception of game theory outlined above provides its greatest insights. Consider the game matrix, one of the signature features of game theory since 1944. Postwar mathematicians tended to view it through the lens of topology and geometry, the logical-deductive investigation of curves and regions in abstract space. By contrast, social psychologists studying team behavior largely discarded the mathematicians' calculations, viewing the matrix principally as a record-keeping device for concisely notating the laboratory situations encountered by their experimental subjects. Operations researchers looked to game theory's axiomatic structures and calculational techniques for guidance on how one ideally ought to structure complex decision-making processes, even if the ideal was seldom met in everyday life, while others viewed the theory as the basis for models to predict collective human behavior. In each case, the significance of the matrix—the meaning attributed to the numbers within it, and the repertoire of calculations and derivations that grew from this highly stylized representation of a game—varied substantially, as did the purpose and potential of a "theory" of games. The mathematicians' use of the game matrix was not simply imported wholesale into other contexts: rather, the tools that game theory provided were significantly reworked in terms of the existing practices and needs of a number of different groups. Game theory's world-building ability to further (and remake) many agendas at once while knitting them together into a common conversation explains much of its tenacity and power.[20]

Rules and Rationality

As the forgoing suggests, one feature of game theory that was interpreted very differently across disciplinary, intellectual, and institutional divides concerns the theory's relationship to the "rational" behavior that was its ostensible object of study. Are the reasoning processes of the mathematician at the blackboard a reflection of how most human beings reason and reckon in daily life? If not, can the theory's characterization of human decision-making nevertheless serve as the foundation for building a theory of social behavior, much as the postulates of Newtonian mechanics describe the behavior of the mute and inscrutable bodies of the physical world? Or does the theory seek to provide guidance for how humans *should* think and act and, if so, what gives this particular guidance moral force, setting it above other collections of precepts, rules, or paradigms? Differences over how to answer these questions were a frequent point of contention and confusion in the postwar era, and the resulting debates and controversies have strongly shaped the history of the subject down to the present day.

To some extent, these debates echo themes present in earlier attempts to cast reasoning in mathematical form, most notably, the development of the calculus of probabilities in the seventeenth and eighteenth centuries.[21] As Lorraine Daston has shown, the enlightenment probabilists tended to view the calculus of expectations, probabilities, and inference as both a characterization of and a guide to rational thought and action. Or to use a vocabulary that would become current in the wake of arguments over how to interpret game theory in the mid-twentieth century, the classical theory of probabilities was to some extent both prescriptive and descriptive in nature.[22] The probabilists were able to bridge the gap between "is" and "ought," actual and potential behavior, because their final point of appeal in developing their calculus was the common sense of an idealized, enlightened "reasonable man," for whom the rules of logic and the calculus of probabilities were second nature. Within the broader framework of moral philosophy, the calculus of probabilities served as a guide both to individual enlightenment and to institutional reform, from the writing of fair contracts to the constitution of juries.[23] However, the classical understanding of probability declined in the late eighteenth and early nineteenth centuries, as the French Revolution fractured the community of self-appointed reasonable men, and as moral philosophy was overshadowed by the new "social sciences" that concerned themselves with uncovering the laws governing masses of potentially a-rational social atoms.[24] And indeed, recent work has reinforced the intellectual distance between the nineteenth-century social sciences—for example, as embodied in neoclassical economics—and both earlier and later attempts to mathematize rational behavior. To the extent that game theorists imagined for themselves a heritage encompassing both the enlightenment calculus of probabilities and the social sciences of the nineteenth century, they had to smooth over a set of tensions between these intellectual programs.[25]

Like the enlightenment philosophers, Cold War America's scientific and policy elites proclaimed themselves the standard-bearers of reason. And if the French Revolution had an equivalent in their time, it was the Vietnam War and the subsequent cultural backlash against "the best and the brightest" and the scientistic ethos that they represented, and a broad "critique of rationality" leveled by New Left intellectuals and the campus counterculture. Variations of this storyline abound especially in popular accounts touching on game theory's history during the Cold War, emphasizing as they do the hubris of the era's "defense intellectuals" and government planners, and the limits of logic in the management of human affairs.[26] The "reason" of the enlightenment and the peculiar brand of "rationality" embraced by the Cold War elites were, however, substantially different in spirit. Cold War rationalists—a group that included many practitioners of game theory—were drawn to a conception of rationality defined by rules that could be followed mindlessly, potentially by a machine, and certainly by a bureaucrat in this age of the "organization man." The enlightenment philosophers never went this far in their quest to mathematize rationality, maintaining a role for mindfulness even in following the rules of simple arithmetic calculation.[27] Nor was the ideal of Cold War rationality necessarily met in practice, with even some of the most vocal exponents of the virtues of rational calculation reserving a place in their working practices for the exercise of intuition, imagination, and improvisation.[28] Yet as an ideal, and as a yardstick against which to measure thoughts, actions, and interventions in the world, this conception of rationality remained a constant point of reference.

A number of intertwined trends, impulses, and fears drove the emergence of this conception of rationality. One was the increasing bureaucratization and scale of research in the human sciences, which went hand-in-hand with a drive for methodological rigor and theoretical sophistication that might bring order to the sprawling research efforts that were the products of this scaling-up of research. Game theory's "rules" were first and foremost rules for practitioners of the human sciences themselves: rules for what they could say and what they could not, rules for notating and reasoning with the kinds of experiments and simulations that they produced. This new Cold War conception of rationality also sprang from worries about the unreliability of the unaided human intellect in staving off destruction in an age of nuclear weapons, ICBMs, and complex command-and-control systems. In the struggle for security in a world overshadowed by arms races and the possibility of war, the problem of choosing rationally would ideally be removed from human discretion and judgment altogether. The years of the high Cold War thus saw a great sequence of debates over the possibility of reducing reasoning to rules, even if this goal (and the possibility of realizing it) was constantly questioned. And then, starting in the 1970s, the debate began to unravel along disciplinary lines. The Cold War–inspired problems that had knit various experts together around problems of practical reasoning—how best to guide convoys across the North Atlantic, or how to design, staff, and manage the air defense systems that would guard America against Soviet attack—lost their urgency, and related streams of funding dried to relative trickles.

This skein of debates—and the Cold War context that enabled them and gave them their particular urgency—provides further structure to this book. Might the game theorists' new mathematics accomplish what the enlightenment philosophers had not: a reduction of common sense to a calculus, a set of precepts for how we should reasonably act or think in a wide class of situations, from poker to superpower rivalries? And could this spare and elegant conception of rationality also somehow form the foundation for some kind of unified theory of individual and social behavior? It was possible to glimpse such ambitions in the new theory of games when it debuted in 1944, and they would bubble up repeatedly throughout the theory's subsequent history. But such expansive hopes were repeatedly dashed in a great series of booms and busts, as initial optimistic expectations gave way to clarification of the theory's limitations in a range of different contexts.

Given this fact, the history of game theory related here is not a tale of cumulative development and spreading application, or of a theory progressively conquering problem after problem and spinning off insights and applications. It is instead a story of a great sequence of arguments, ever shifting and rarely conclusive, and drawing in disciplines from mathematics to psychology to evolutionary biology. In them, the pregnant collection of metaphors, notations, and results gathered together in von Neumann and Morgenstern's book often provided the central inspiration and idiom. They functioned at once as a code, a *lingua Franca*, a way of speaking and writing that both conjured new worlds into being and connected existing ones. A collection of tools, adopted by diverse

research programs for strikingly different reasons having to do with the
local ergonomics of their use, they changed the disciplines with which
they interacted and were in turn remade by them. But, through their
general recognition, they simultaneously helped draw together a remarkable collection of researchers around the contemplation of "rationality,"
"decision-making," and "choice" during the high years of the Cold War.

Mapping a Narrative

Chapter 2, "Acts of Mathematical Creation" is largely expository. It examines the work commonly acknowledged as contemporary game theory's founding text: mathematician John von Neumann and economist
Oskar Morgenstern's 1944 *Theory of Games and Economic Behavior*.
Nevertheless, the book looks strikingly dissimilar from texts of game
theory studied today. This reflects the fact that, while *Theory of Games*
came into existence at the confluence of von Neumann and Morgenstern's intellectual concerns and influences in the 1920s and 1930s, these
have been gradually stripped away in subsequent waves of the theory's
appropriation and exposition. Hence, the chapter presents a reading of
Theory of Games and Economic Behavior. While von Neumann and
Morgenstern did try to provide the book with a coherent gloss and interpretation, game theory as it stood in 1944 proved less a unified axiomatic
approach to the study of rational behavior and more a diverse collection
of terminology, notational representations of games, and results, coupled with more completely fleshed-out theories of utility and of the two-
person zero-sum game. Likewise, while the work makes an apparently
reasonable set of assumptions about what constitutes rational conduct,
these are nevertheless somewhat idiosyncratic and internally diverse, a
reflection of the theory's roots in von Neumann's varied mathematical
interests. This reading does much to explain the selective and piecemeal
appropriation of von Neumann and Morgenstern's work in subsequent
decades. It also sets the work broadly in the context of changes taking
place within mathematics during this period, as well as debates within
Austrian economics over the possibility of economic prediction and socialist planning, which are reflected in the economic motivation given
the book in its opening chapters. Von Neumann provided the axiomatic
and set-theoretical method that would help transform the social sciences
into mathematical disciplines in the years ahead, while Morgenstern's

involvement inflected the work with a number of economic concerns and terminology.

Chapter 3, "From Military Worth to Mathematical Programming," explores how von Neumann and Morgenstern's work managed to find its first significant constituency. This was never a given, since, from the perspective of 1944, *Theory of Games and Economic Behavior* was a book without an obvious audience. The theory was too complicated to provide guidance for the casual card player, the most likely reader of a book on "games" in 1940s America, and, as scholarship in the history of economics has noted, members of the American economic mainstream immediately expressed skepticism about the immense, formula-rich tome. Less often noted is that fact that mathematicians, the community with the most obvious technical capacity for cultivating the theory, were also initially hesitant to steer their graduate students toward an area of applied mathematics so far from the cutting edge of pure research. Game theory's survival depended instead on the emergence and persistence of new research activities that bucked traditional academic hierarchies, such as operations research (OR), which brought together interdisciplinary teams of scientists to advise the military on the most effective use of new technologies, from gunsights to airplanes. Out of war-forged relationships between academics and the military developed several postwar research programs, supported by the military, in which game theory would play a prominent role.

The aim of this chapter is to sketch these networks of patronage that supported game theory in the immediate postwar period, and to examine the intellectual place of the theory in the research programs of mathematicians and their supporters. Some patrons, such as the RAND Corporation, have received a great deal of historiographical attention; others, like the Office of Naval Research's logistics program, are less well known, but were nevertheless highly influential in establishing game theory's place in academia through research contracts with a number of major universities. In the military context, game theory proved attractive less for its ability to "solve" particular games to prescribe rational behavior or promote a particular world outlook, and more for the way it might provide a structure for tackling several essentially managerial problems, most notably, how to bring coherence to the kinds of unruly technology assessment projects that first emerged in the context of wartime OR, or how to discipline military budgets during the national security consolidation of the later 1940s. However, the theory's greatest boosters were the mathematicians left stranded between the military and academia when the war ended. For them, *Theory of Games and Economic Behavior*—especially the theory of the two-person zero-sum game—proved capable of pulling together the diverse activities they had performed for the military and putting them under a common mathematical framework that had a certain amount of intellectual coherence and scholarly credibility. The result was a number of military-sponsored research programs that focused primarily on developing the mathematics behind the theory of the two-person zero-sum game, and the emergence of a tradition of research and pedagogy that focused centrally on games related to various optimization problems. In tandem, optimization and consistency of preferences became the hallmarks of game-theoretic rationality.

Chapter 4, "Game Theory and Practice in the Postwar Human Sciences," examines the first efforts to extend game theory's reach beyond applied mathematics to a few tentative beachheads in the postwar social and behavioral sciences. Von Neumann and Morgenstern had originally envisioned game theory's place in such fields as a positive theory of social interaction built up from unobjectionable axioms of rational individual behavior and duly tested against observation. Mathematicians at RAND and at Princeton were among the first to take up this challenge, aiming to test the theory against experiment. The results proved inconclusive, doing little to help the mathematicians decide between the various "solution concepts" for non-zero-sum games then circulating. Moreover, the experiments called into question the issue of what might constitute a true "test" of the theory, given the somewhat artificial scenarios needed to engineer results in conformity with particular solution concepts, and the ability of human subjects to remember, learn, and conform, which problematized the nature of the empirical phenomena against which the theory was to be tested.

However, during the late 1940s and early 1950s, game theory began to appear in conjunction with several research programs in the human sciences that were specifically interested in understanding phenomena of learning, information-processing, and the mapping of social and individual attitudes and values. The chapter examines some of these, most notably the tradition of "group dynamics" and the study of behavior in small groups that grew from the work of social psychologists Kurt Lewin, Leon Festinger, and Alex Bavelas at MIT; it also touches on the history of several explicit attempts to bring game theory into the social and behavioral sciences, most notably an interdisciplinary project aimed at the "measurement of values" that emerged from the University of Michigan, and that developed connections with projects ongoing at the RAND Corporation in 1952. In 1957, the results of such research programs would be synthesized by two young mathematicians, R. Duncan Luce and Howard Raiffa, in their landmark textbook, *Games and Decisions*, which would serve as an overview of research in game theory to that date and would later function as a key conduit for bringing game theory to practitioners in the wider social and behavioral sciences. The game theory of Luce and Raiffa proved substantially different from that of von Neumann and Morgenstern, billing itself less as a grand positive theory of social interaction and more as a heterogeneous notational idiom tying together a cross-disciplinary investigation of "decision-making" or "decision processes" by individuals and groups. The "theory" aspect of game theory thus proved ambiguous in the postwar human sciences, being at once a possible description of human behavior, a prescription for rational conduct on the part of human subjects, and something like a methodology for engineering interesting psychological phenomena.

Chapter 5, "The Brain and the Bomb," explores how, in the late 1950s and 1960s, game theory became caught up in a contentious set of debates over how to keep the world safe in an age overshadowed by the destructive power of nuclear weapons. Some aspects of these debates are well known to historians of the postwar United States, especially the intricate arguments (both in and out of the corridors of power) over the prospects for "atomic diplomacy," the doctrine of "massive retaliation," studies of weapons basing and targeting in the 1950s, the politics of a nuclear test ban and civil defense measures, and the emergence of Mutual Assured Destruction as the foundation of Cold War geopolitical stability. Game theory proper became a high-profile aspect of these debates especially after the publication of Thomas Schelling's *The Strategy of Conflict* (1960), which applied a loosely game-theoretic analysis of bargaining to problems of nuclear strategy. Thereafter, games appeared as a central feature of discussions of nuclear deterrence and arms control, whether in technical reports of the Arms Control and Disarmament Agency, or in a number of popular "exposés" of nuclear strategy and military strategists, such as Norman Moss's sensational *Men Who Play God* (1968), which would connect game theory with the ominous calculations of American nuclear strategy.

In this chapter we see how game theory was pulled into discussions of arms races and superpower relations during this period, to the point that the Cold War came to be seen by many as *the* ultimate game that game theory was meant to analyze. It does this by focusing on one specific community of social and behavioral scientists connected with the Mental Health Research Institute (MHRI) and the Center for Research on Conflict Resolution (CRCR) at the University of Michigan. For these intellectuals, the application of game theory to Cold War geopolitics via the new field of "conflict resolution" represented a natural extension of the kind of interdisciplinary behavioral science promoted by the Ford Foundation and institutionalized at places like the MHRI. Much as it had for the behavioral scientists groping their way toward a general science of decision-making in the early 1950s, game theory provided a common theoretical idiom that allowed them to link laboratory investigations of teamwork, sociological studies of conflict, and studies of international arms races. As a result, games like Prisoner's Dilemma or Chicken became part of a common conversation between researchers from a variety of disciplinary backgrounds as well as members of the general public. Yet while these individuals saw themselves as discussing a common set of problems about rationality, conflict, and the possibilities for superpower cooperation and arms control, their very different disciplinary and political allegiances were paralleled by disagreement about how one might "solve" such games, what the role of scientific knowledge might be in approaching great problems of war and peace, and the sources and character of rationality itself.

Chapter 6, "Game Theory without Rationality," marks a turning point in the postwar history of game theory. From the moment of its creation, game theory was envisioned as a theory of rational decision-making, capturing something of the uniquely human ability to anticipate, value, reason, calculate, and choose. Yet in the later 1960s and 1970s, elements of game theory emerged in evolutionary biology to form an exciting new avenue of theoretical development connected with Robert Trivers's notion of "reciprocal altruism" and John Maynard Smith's concept of an "evolutionarily stable strategy" or ESS. For Trivers, the evolution of a behavioral strategy of reciprocated altruism solved the problem of why cooperation might emerge in repeated plays of the Prisoner's Dilemma game. And in developing his ESS concept, Maynard Smith discovered that populations of organisms effectively "choose" evolutionary strategies—governing any characteristic from the sex ratio of their offspring to the fighting strategies they adopt—to form a particular kind of

Nash equilibrium, a solution concept for multiplayer games that assumes that players cannot communicate and form binding agreements.

This chapter places game theory's emergence in evolutionary biology at the confluence of several developments in both of these fields during the 1960s and 1970s. The debates over arms control and conflict resolution had placed game theory at the center of discussions on the problem of how to rationalize cooperative behavior in a world of rational egoists, a problem that figured less centrally in von Neumann and Morgenstern's original presentation of game theory, or in the work of the RAND mathematicians in the 1940s. Evolutionary biologists like Hamilton, Maynard Smith, and Price encountered game theory, at least in part, through the writings of conflict resolution gurus like Anatol Rapoport, and, fittingly, problems like the evolution of animal combat tactics would prove to be among the earliest applications of game theory in biology. The problems game theory was used to investigate in the context of arms control also proved relevant to arguments unfolding in evolutionary biology over the possibility of "group selection" that had emerged in the wake of the ornithologist V. C. Wynne-Edwards's 1962 book, *Animal Dispersion in Relation to Social Behaviour*. Drawing primarily on his ornithological fieldwork, Wynne-Edwards had suggested that a wide range of adaptations, from reproductive rates to limited aggression, could be explained by reference to sacrifices of individuals to maintain "homeostasis" and balance in a population. Game theory first appeared in the work of evolutionary biologists who found the theory's parsimonious explanations of social life in terms of the actions of self-interested individuals more convincing than the collectivism they associated with Wynne-Edwards's theory of group selection, and who favored a set of working practices involving mathematical models and computer simulations. Yet in a striking irony, these biologists reached for game theory precisely to deny attribution of rationality (in the sense of foresight, memory, and conscious choice) from evolutionary actors undergoing natural selection.

As chapter 7, "Dreams of a Final Theory," makes clear, the foray of game theory into evolutionary biology was associated with a significant shift in game theory's character in the 1970s and 1980s. Although many parts of the social sciences had been relatively slow to internalize game theory, by the mid-1980s game-theoretic concepts had become established at the core of several fields, most notably economics and political science. In particular, the noncooperative Nash equilibrium solution concept for games, grounded as it was in an austere methodological individualism, emerged as a pivotal concept in the social sciences much as it had in evolutionary biology nearly thirty years after Nash's seminal papers on the subject. The reasons for this explosion of interest in game theory in the social sciences and its simultaneous "Nashification" are not entirely clear, although several authors have tried to offer explanations. The publications of Nobel laureates like Reinhard Selten and John Harsanyi during the 1960s did much to build on Nash's work, extending his analysis to encompass subtle issues of information and learning, and improving the applicability of noncooperative game theory in the social sciences. Meanwhile, as historians and practitioners of economics have shown, a number of factors unique to economics appear to have promoted the adoption of game theory in several areas of this field: technical problems with Walrasian general equilibrium theory, the premier theoretical framework for economics until the 1970s; emerging trends in American antitrust jurisprudence and policy; or even macro-level changes in the identity of the discipline—toward the generalized analysis of social institutions—and correspondingly imperial and ambitious aims. In this context, noncooperative game theory held out hope for a theory of social interaction that would reduce the study of all forms of social organization, from markets to governments, to the finding of equilibrium solutions to the relevant games.

Even if economics can perhaps lay claim to being game theory's center of gravity in recent decades, the focus of this chapter is broader: on probing what was at stake in the transformation of game theory outlined above and in chapter 6. As with the history of evolutionary game theory, the legacy of the arms control debate looms large in this story, with many of the seminal works cited by social scientists in the 1980s continuing to gnaw away at fundamental problems posed during that debate, especially how to rationalize cooperation in a world of rational egoists. The role of beliefs and information in bargaining and the significance of learning and anticipation in repeated games likewise proved key areas of research. The postwar penchant for physics-like theoretical rigor in the social sciences also played a role in tilting the balance of interest away from cooperative game theory—von Neumann and Morgenstern's original approach to modeling economic behavior. Yet while the program of noncooperative game theory held out hope for a final theory of social interaction based on elegantly minimalist assumptions of individual rationality, the goal remained elusive, much as it had for the mathematicians of the 1950s. The fact that most games of interest to social scientists possessed multiple Nash equilibrium points led to the proposal of an ever-increasing number of "equilibrium refinements," which sought to assign a unique outcome to each game. As these proposed refinements grew in number and complexity, the gap between the information-gathering and calculational capabilities that could reasonably be expected of rational individuals and the demands of equilibrium selection expanded as well.

The book therefore concludes with the waning of the Cold War and, simultaneously, the eclipse of "rationality" as the concept binding together the diverse programs of inquiry surrounding game theory. Practitioners like Kenneth Binmore or Herbert Gintis retained the core concepts of noncooperative game theory, but justified this retention less by reference to the reasoning abilities of humans and more to blind processes of evolution. Many game theorists thus still saw noncooperative game theory as the ultimate theoretical framework capable of uniting the social sciences—but under the banner of biology. At another extreme, some pioneers of game theory in economics like Reinhard Selten began to move away from models grounded in stylized assumptions of rational choice entirely, and to embrace more context-specific models of human decision-making processes and complex computer simulations of social change. The turn within game theory toward the hunt for equilibrium solutions and the dawning recognition of the problems inherent in this research agenda forms the endpoint of this history, as the threads that bind it together—rationality, the "theoretical tools" of game theory, and the Cold War—at last separate. In the wake of this development, game theory today is a very different thing than it was during its early years. Once its invocation of "rationality," its shimmer of modernity, and its world-building ambitions allowed it to claim the length and breadth of the human sciences and human experience for its demesne; but it is now more a creature of particular disciplines: a modeling technique like any other, to be integrated with solid empirical research and qualified with all kinds of special conditions and caveats. This history continues to unfold, but it is distinct from the one that is the focus of this book.

Acts of Mathematical Creation

In the fall of 1944, *Theory of Games and Economic Behavior* began to appear in Princeton University Press advertisements in the *New York Times*, which introduced it blandly as "a mathematical investigation of economic theory, social organization, and of games."[1] The description is apt in its ambiguity. Works of political economy in those days were typically wordy affairs, full of grave commentary on currencies and investment, trade and unemployment. Von Neumann and Morgenstern's book, by contrast, explicitly eschewed discussion of such sensitive topics as "monopolies" and "combinations," among others. And surely the book was not one of the hundreds of available pamphlets that hawked some new "system" for winning at bridge, poker, and other pasttimes of the weekend card-shark, who was the typical audience for a work on "games" in 1940s America: its 600+ pages of dense mathematical theory made it too formidable for them. The object of the book's inquiry was loftier than mere card games, and indeed its ambitions extended beyond economics as it existed at the time. According to the breathtakingly bold description presented by its authors in the work's opening pages, "we wish to find the mathematically complete principles which define 'rational behavior' for the participants in a social economy, and to derive from them the general characteristics of that behavior."[2] Scholarly reviews of the book began to trickle out over the next few years, but their authors, who brought to bear far more sophisticated academic perspectives than those of the press's anonymous blurb-writer, praised the work for its impressive technical virtuosity, but were less than specific when it came to identifying a conclusive overlap with the state of the art in their fields of specialization.

Today, *Theory of Games* is widely credited as game theory's founding publication. Some have contested the priority of von Neumann and Morgenstern, calling attention to the work of French mathematician Émile Borel, who formulated a "théorie des jeux" along lines similar to von Neumann's in the 1920s.[3] Regardless of the striking similarities and differences between von Neumann's theory of games and Borel's, for decades after 1944, reference to von Neumann and Morgenstern's book was practically the *sine qua non* of publications on "game theory" or the "theory of games." At the same time, game theory as it is presented and practiced today stands at some remove from this seminal text. As a basis for pedagogy, the book was being displaced in many disciplines as early as the 1950s, as a new generation of textbooks began to emerge that brought the theory to audiences beyond the relatively elite group of mathematicians and mathematical economists who were the theory's earliest practitioners. Today, by all appearances, it is rarely cited and rarely read. Especially since the ascendancy of the tradition of "noncooperative game theory" in the 1970s, a recent student of game theory must find the original text antique and unfamiliar in its notations, terminology, style of presentation, and choice and order of topics. An appreciation of these unfamiliar elements of *Theory of Games* is nevertheless important for understanding game theory's appropriation and adaptation in the postwar era. This chapter's focus is therefore on explicating *Theory of Games* itself by grounding aspects of the text in the history of the work's creation.

The essential timeline of the book's history has long been known, both from the long trail of publications that von Neumann and Morgenstern left behind in the prewar years, and from the post facto reminiscences of the authors and their colleagues. In December of 1926, the precocious young Hungarian mathematician von Neumann presented a paper on games before the Göttingen mathematical society. This paper was later published in *Mathematische Annalen* under the unassuming title "Zur Theorie der Gesellschaftsspiele" (Toward a theory of parlor games). In it, von Neumann proposed a mathematical formulation of the general problem of games, developed a fairly complete theory of games involving two players with diametrically opposed interests, and hinted at the outlines of a general theory of multiplayer games.[4] Throughout the 1930s von Neumann apparently did little with games per se, although he occasionally lectured on the topic after he moved to Princeton as a visiting professor in 1930. However, in 1939 at Princeton University, von

Neumann met Oskar Morgenstern, an expatriate Viennese economist, who saw in von Neumann's game theory a chance to introduce rigorous mathematical theory to economics. What had begun as a mathematical theory of parlor games could also describe the behavior of actors in a social economy. The two embarked on the collaboration that would create *Theory of Games*, working hard through the fall of 1941 and the spring of 1942 to produce it. Morgenstern brought the manuscript to Princeton University Press in April of 1943, and the work finally appeared in late fall of 1944.[5]

In recent decades an outpouring of scholarship has fleshed out this story, in far greater detail. Among other things, historians of economics have begun to evaluate Morgenstern's role in the production of *Theory of Games*, drawing especially on his meticulously kept diaries. These have revealed much about Morgenstern's early economic inspirations as well as further details about his collaboration with von Neumann, allowing not simply a finer allocation of credit for their joint publication, but also rich assessments of the place of game theory in Morgenstern's intellectual development, and of the relationship between game theory and the "Austrian" school of economics, of which he is considered a member.[6] As for many of his fellow Austrians, Morgenstern's work was shaped by a struggle with the great ideologies that dominated interwar Europe, especially socialism and nationalism. And, like his colleagues, he reacted to these ideologies by arguing for a depiction of society that was not historicist but rational and scientific, a depiction that took some minimalist assumptions about the behavior of individuals as its starting point, rather than appealing to some irreducible zeitgeist, and sought to derive from them the behavior of society as a whole. Yet Morgenstern's talent was primarily critical in intent and literary in methodology: to actually carry out the work of formulating this new economics, he increasingly turned to mathematicians who would direct him toward von Neumann's work and, ultimately, toward the marriage of formalist mathematics and methodological individualism that would mark so much postwar economic theory.[7]

Meanwhile, a number of works have begun to place game theory among von Neumann's many and varied interests during the 1920s and 1930s. For example, why did von Neumann's 1928 paper appear when it did, only to have him wait until the 1940s to start working on *Theory of Games*? The scholarship has pointed to a range of influences, for example, von Neumann's disenchantment with Hilbertian projects of axiomatization and subsequent embrace of "applied" problems, and his increasing preoccupation during the later 1930s with the deteriorating political situation in Europe, among others.[8] More broadly, what were the game-theoretic ambitions and inspirations of this prolific yet inscrutable genius? Encompassing quantum mechanics and theories of automata, practical problems of computing and military logistics, von Neumann's exceptionally wide range of interests has presented a puzzle to historians trying to make coherent sense of this remarkable polymath. In considering the sources, published and unpublished, that the two men left behind, it is hard not to be struck by the contrast between von Neumann and Morgenstern: the latter scarcely having a thought that he did not commit to paper, and the former's archive and *oeuvre* coming across as comparatively formal, impersonal, and telegraphic.[9]

The aim of this chapter, however, is not to recover further insight into the ambitions and mindset of *Theory of Games'* authors by drawing on their programmatic statements regarding applied mathematics, philosophy of science, or economic methodology, for example. Rather, its aim is to develop a reading of the text itself—which was, after all, most people's entry point into the theory of games—as a diverse collection of terminology, notations, axioms, and results with their own unique genealogies and historical trajectories. Certainly, it is possible to read *Theory of Games* in this way from the vantage of the present, having witnessed the subsequent selective appropriation and adaptation of the various components of the book. Consider the assessment of one of Morgenstern's students, Martin Shubik, writing in the early 1990s:

> Although I did not appreciate it at the time, the book of von Neumann and Morgenstern could be regarded as four important separate pieces of work. They were (1) the theory of measurable utility; (2) the language and description of decision-making encompassing the extensive form and game tree with information sets, and then the reduction of the game tree to the strategic form of the game; (3) the theory of the two person, zero-sum game; and (4) the coalitional (or characteristic function) form of a game and the stable-set solution.[10]

Of these "pieces," the multiplayer game dominated the book, constituting nearly 400 out of the book's 600-odd pages; a further 135 pages explored the two-person zero-sum game, while the theory of utility and the extensive form game made up the balance. Granted, Shubik's reading has the benefit of hindsight; yet such a mapping of *Theory of Games* also reflects fault lines that existed prior to 1944. The mathematical structures, results, and terminology underlying these components of the book were forged at different times; they possessed connections with a variety of different research programs and agendas in both mathematics and economics; they contained somewhat different assumptions about the behavior of "rational" individuals; and they were absorbed into the game theoretic corpus at different moments in its development. Such a reading is therefore not simply the product of hindsight, but reflects features of *Theory of Games'* creation that would prove significant in shaping game theory's subsequent history.

This chapter therefore explores the genesis of the various components of *Theory of Games* to highlight the diversity of mathematical forms and intellectual traditions that lay beneath the seemingly solid edifice of von Neumann and Morgenstern's seminal presentation of game theory. It reads the text primarily as a product of mathematical imagination, albeit one that borrowed liberally from a wide swath of the mathematics of its time: an exercise in world-building that, through notations and utterances, conjured into being a new reality whose fundamental building blocks were objects like sets, orderings, graphs, functions, operators, and matrices. While such a reading suggests that the driving inspiration for much of the content of *Theory of Games* was von Neumann's mathematical concerns, it also notes the influence of Morgenstern in shaping the vocabulary and especially the presentation of the work. Read in this light, the book comes across less as providing a coherent, well-defined program for game theory and more as a concatenation of notational idioms and results that had been on the mind of its authors in the preceding decades. Predictably, under its cover also lay technical, disciplinary, and philosophical fault lines that predated publication, and that are crucial for understanding the subsequent reception and reworking of game theory in the years after 1944. Given this peculiar intellectual heritage, it is hardly a wonder that the work's reviewers were as puzzled as the book's blurb-writer was bland. Was this a work of economics or of mathematics? And given the varied mathematical traditions represented on its pages, what conception of "rational behavior" would emerge in the process and who might express interest? This chapter therefore sheds light on the underlying disunity of *Theory of Games*, highlighting the dissimilar communities of people and divergent disciplinary associations with interests in the work, which was at once a treatise on the mathematics of parlor games, an extended disquisition on the proper methodology for economic inquiry, and a proposal for a new kind of social theory.

Out of Set Theory: Extensive Forms and the Dynamics of Games

What is a game? How can it be described mathematically? This was the question addressed in the second chapter of *Theory of Games*, immediately after the overall introduction and plan of work laid out in chapter 1. Here, von Neumann and Morgenstern proposed a general definition that they felt would cover all classes of game. This definition was causal-dynamic, employing concepts such as "anterior" and "posterior," and considering the changing state of information available to the players. It was also plainly inflected with the vocabulary and concerns of chess:

> The game consists of a sequence of moves. At each move the "umpire" announces to both players whether the preceding move was a "possible" one. If it was not, the next move is a personal move of the same player as the preceding one; if it was, then the next move is the other player's personal move. At each move the player is informed about all of his own anterior choices, about the entire sequence of "possibility" or "impossibility" of all anterior choices of both players, and about all anterior instances where either player threatened to check or took anything. But he knows the identity of his own losses only. . . . Otherwise the game is played like Chess, with a stop rule.[11]

While their initial verbal description of a "game" explicitly referenced chess, von Neumann and Morgenstern's subsequent characterization employed a notational idiom that left the chessboard far behind. Diagrammatically, von Neumann and Morgenstern represented such a game as a branching tree, what they would dub the "extensive form" of a game (see fig. 2.1).

Here, Ω, a_1, a_2, . . . label sequential "moves." The compartments around the different nodes or branchpoints represent the players' "information sets." If a node is not inside the information set, it is not a possible state of the game. If, on the contrary, there are several nodes in an information set, then the player making the move does not have complete information about what is happening and the different nodes represent different possibilities. In a game like chess, in which both players can see

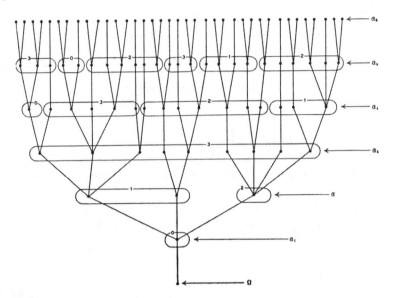

FIGURE 2.1. The extensive form of a game. From *Theory of Games and Economic Behavior*, 78.

the board and obtain perfect information about the moves of their opponent, information sets always contain only one node. Behind the image of a branching tree lay an even more abstract and general set-theoretic formulation of games. From the start of play, the set Ω (down at the root of the tree) contains all possible sequences of moves that the players can make during the game. As the players make moves $a_1, a_2, \ldots,$ the sets $A_1, A_2, \ldots$ containing possible future sequences of moves "shrink." In the language of sets, $\Omega \supseteq A_1 \supseteq A_2 \ldots,$ (where $\supseteq$ means "contains") and so forth. As the players gain information about the sequence of moves, their information sets correspondingly shrink, albeit not as quickly as they would in a game where the players' information captured the narrowing of future possible sequences of moves with perfect accuracy.

This particular description of a game has often been downplayed by latter-day commenters in comparison to von Neumann's treatment of games in "normal form" and the celebrated "minimax theorem," with their rich and varied applications in economics and the social sciences.[12] Certainly, read from the perspective of today, many post-1944 publications related to extensive form games come across principally as exercises in graph theory, constituting a self-referential research program pursued principally by mathematicians at some distance from the kinds of problems that have become the focus of most game theorists since the 1960s.[13] Yet the extensive form of games is worth putting first for a number of reasons. Most obviously, dynamical descriptions of games (albeit rather different ones) come first in both von Neumann's 1928 paper and von Neumann and Morgenstern's 1944 treatise. In the latter work, the extensive form of the game potentially served a useful expository and pedagogical function—as a way of introducing the nonmathematical reader to the basics of set theory that would play such an essential role in the chapters to follow. Certainly, no exposition of elementary set theory was present in von Neumann's earlier paper on game theory, and the numerous figures scattered throughout this part of the book, illustrating graphical approaches to the topic, appear to date from the collaboration with Morgenstern. Thus the images illustrating possible relations between sets on pages 62–65 of the 1944 book include the famous picture of an "elephant" disguised within a set diagram—an inside joke prompted by a remark of Mrs. von Neumann in the course of von Neumann and Morgenstern's intense writing sessions.[14] The graphical representations of sets and trees likewise may indicate a concession to visualizability and intuition on the part of von Neumann, whose personal mathematical style exhibited especially "a feeling for the set-theoretical, formally algebraic basis of mathematical thought," in the assessment of his longtime colleague and friend Stanislaw Ulam.[15]

But the history of this description of "games" also helps illustrate the way in which von Neumann and Morgenstern wrote their book into a number of research programs ongoing within the rarefied community of set theorists, logicians, and topologists of which von Neumann was a part during the 1920s, thereby creating a constituency for the work in some quarters of mathematics and philosophy after 1944. Thus, for example, the RAND Corporation publications on extensive form games and set theory by logician and philosopher W. V. O. Quine in the later 1940s can be read, at least in part, as a manifestation of long-standing connections between set theoretic formalizations of "games" and the prewar interests of set theorists and logicians.[16] More broadly, as von Neumann would note, "The mathematical character of the theory is entirely different from that of mathematical physics—and consequently also from mathematical economics of the conventional kind. In those fields of analysis, in particular the calculus of differential and integral equations etc.,

dominate. The technique of the theory of games is of a more combina-
torial and set-theoretical character."[17] Arguably, it was the intellectual
and disciplinary concerns of this constituency of set theorists and "com-
binatorial" mathematicians, more than an attraction to "application," a
focus on the development of economic theory, or the needs of professional chess players, that animated the initial mathematics behind extensive form games.

The history of these concerns is well known. In the closing decades
of the nineteenth century a number of new and startling mathematical developments—especially connected with the growth of interest in
transfinite numbers stemming from the work of Georg Cantor and Richard Dedekind, among others—led to the discovery of a number of mathematical paradoxes. The classic example of these was uncovered by the
English philosopher Bertrand Russell in 1901, in the midst of ongoing
efforts to put mathematics on a foundation of logic and the analysis of
sets, unordered collections of objects. Consider the set of all sets that
are not elements of themselves: is it an element of itself, or not? Either
possible answer leads to contradiction. Such troublesome examples hint
at a set of problems inherent in the logic of self-referential statements,
and in the extension to infinite situations of naive thinking about apparently commonsense objects like "sets." The paradoxes of set theory
and the "foundational" crises that surrounded them provoked a diverse
set of responses from mathematicians. Some, most notably the Göttingen mathematician David Hilbert, championed a program of formalization: mathematical concepts should be clearly defined by sets of axioms,
from which all relevant results can be proved by applying rules of logic.
These axioms had to be logically consistent, and complete in the sense
that all possible theorems could be derived from them. Thus, in the
1890s, Hilbert had already produced an axiomatic formulation of geometry. In his famous address to the 1900 International Congress of Mathematicians in Paris, he called for the axiomatization of arithmetic. In subsequent years, his colleagues (most notably Ernst Zermelo, who would
join Hilbert at Göttingen in 1905) began providing the axiomatic framework for the theory of sets, the most fundamental mathematical objects
that could serve as the basis for axiomatizing arithmetic and many other
branches of mathematics.[18] Another reaction, often associated with the
work of Dutch mathematician L. E. J. Brouwer, suggested that the paradoxes of set theory were precisely a result of the abstraction that charac-

terized formal approaches to mathematics. Thus Brouwer drew upon the ideas of France's Henri Poincaré in proposing an "intuitionist" philosophy of mathematics, in which ideas like the continuum of real numbers were not logical constructions but were themselves fundamental concepts of mathematical intuition. But even absent such a sweeping alternative research agenda, many mathematicians were nevertheless skeptical of the trademark Hilbertian "nonconstructive" proofs, that is, proofs that showed the logical necessity of the existence of some mathematical object, but did not show how to construct or identify the object. Such proofs often involved the use of the logical "law of the excluded middle" in transfinite situations, or application of Zermelo's controversial "axiom of choice."[19]

Von Neumann appeared on this mathematical scene in the 1920s. Born in 1903, the son of a subsequently ennobled Jewish-Hungarian banker, he had been a mathematical prodigy: stories abound of his extraordinary capacity for mental calculation or feats of memorization.[20] But from the tales of technical virtuosity also emerge a set of philosophical instincts that fit with an age in which "foundations" and absolute definitions of meaning or value appeared increasingly shaky: a delight in paradox, word games, and linguistic jokes, and a fascination with the borders of what was logically possible and impossible.[21] Thus his brother would recall a topic of conversation around the family dinner-table: would an all-powerful deity be able to create a boulder that He could not move?[22] The generalization of "power" from the finite to the infinite, and fascination with the resulting mysteries of logical possibility and hypothetical "creation," all foreshadowed an inclination toward the new mathematics of logic and set theory. Subsequently von Neumann studied mathematics at the University of Budapest, the University of Berlin, and at Göttingen, as an assistant to Hilbert, in the process becoming a leading exponent of Hilbert's axiomatic approach to mathematics and the sciences more generally. Naturally, his first major publications concerned the logical foundations of set theory. In a 1923 paper he introduced a new set-theoretic definition of the ordinal numbers—numbers related by the ordering relations of "greater than" ($>$) "less than" ($<$) or "equal to" ($=$). And two years later, von Neumann published his own axiomatization of set theory, improving on Zermelo's, and incorporating his new definition of the ordinal numbers.[23]

Such set-theoretic concerns provided von Neumann with a possible entry point into the analysis of "games." Less than a decade before von Neumann made his reputation as a young and energetic axiomatizer of set theory, in 1913 Zermelo had published a paper, "Über eine Anwendung der Mengenlehre auf die Theorie des Schachspiels" (On an application of set theory to the theory of chess).[24] Here, Zermelo introduced a general set-theoretic formalization of board games (of which chess was but one example) very similar to von Neumann and Morgenstern's later formulation of extensive form games. Chess, in Zermelo's characterization, consists of a possibly infinite sequence of "positions" $p_1, p_2, \ldots$, each of which is chosen from a (finite) set P of all possible configurations of the pieces on the board. Given this formulation, one interesting question was to identify the "value for one of the players of any position possible while playing . . . as well as the best possible move," as Zermelo put it. But like the nineteenth-century logicians who sought to bound their discipline against the incursions of psychologists, Zermelo sought to answer this question while drawing a bright line between his "mathematical-objective" investigation of such games on the one hand, and analyses that drew upon "more subjectively-psychological notions such as the 'perfect player' and like concepts."[25] Hence Zermelo actually focused his analysis less on uncovering concrete advice for the players on the "best possible move" (the task of various practical manuals or systems for chess written by the masters of the day) and more on understanding how to characterize "winning positions" that assured one player a win in a finite number of moves, regardless of the play of his opponent, simply by the logical nature of the game. Having defined such a "winning position" precisely, Zermelo then sought to determine an upper bound on the number of moves that a player needed to force a win, given that he finds himself in a winning position in the first place. His stated reasons for pursuing this decidedly impractical application of mathematics were especially meaningful in light of the critical assessment given Zermelo's attempts to formalize set theory in the preceding years.[26] In bringing the "theory of sets" and the "logical calculus" to bear on such a concrete problem, Zermelo hoped to demonstrate the "fruitfulness of these mathematical disciplines in a case, where almost exclusively *finite* aggregates are concerned," as he put it.[27]

As economists Paul Walker and Ulrich Schwalbe have shown, Zermelo was only one of a community of set theorists to identify "games" as a possible "application" of such exceedingly abstract mathematical work. In 1927 Dénes König, a pioneer of the emerging mathematical subdiscipline of "graph theory" and faculty member at the University of Budapest, published a paper, "Über eine Schlußweise aus dem Endlichen ins Unendliche [On an inference from the finite to the infinite]." The paper took as its starting point a set-theoretic existence principle especially applicable to games like chess, because it captured the fact that any given "state of the board" arises from a previous state (or possibly several):

> Let E_1, E_2, E_3, . . . be a countably infinite sequence of finite nonempty sets and let R be a binary relation with the property that for each element x_{n+1} of E_{n+1} there exists at least one element x_n in E_n, which stands to x_{n+1} in relation R which is expressed as $x_n R x_{n+1}$. Then we can determine in each set E_n an element such that $a_n R a_{n+1}$ for the infinite sequence a_1, a_2, a_3, . . . always holds.[28]

If the number of sets E is finite, the result is straightforward, since you start with an element of S_n and proceed a finite number of steps by backwards induction; however, transfinite set theory suggests that even when the sequences of "moves" are potentially infinite (as in board games) it is always possible to identify such a sequence. König then proceeded to examine the relevance of the result to a number of concrete, if somewhat artificial, problems: the number of colors needed to paint maps, the relatedness of family members (any person must have at least one previous relative, so that "generations" are analogous to "rounds" or "plays" of a game), and so forth.[29] However, the hint that König apply his theorem to games apparently came from von Neumann, who had suggested that if it is possible for one of the players to win the game in question, then he can win in a finite number of moves irrespective of what his opponent might do. In his paper, König proved the result, and also included in a postscript a set of related results sent to him by Zermelo.[30] Ultimately, König's paper would make a cameo appearance in chapter 2 of *Theory of Games and Economic Behavior*, in an extended footnote on the existence of "stop rules" in board games.[31]

The opening of von Neumann's 1928 "Zur Theorie der Gesellschaftsspiele" bears rather less resemblance to Zermelo and König's work than does the characterization of extensive form games in *Theory of Games*, yet it too reads to some extent as the achievement of a mathematician fresh from his youthful work in the remote abstractions of set theory. Von Neumann's formalization of games mirrored the causal-dynamic depictions of board games like chess found in Zermelo and König's papers:

> A game of strategy consists of a certain series of events each of which may have a finite number of distinct results. In some cases, the outcome depends on chance, i.e. the probabilities with which each of the possible results will occur are known, but nobody can influence them. All other events depend on the free decision of the players $S_1, S_2, \ldots, S_n$. In others words, for each of these events it is known which player, S_m, determines its outcome and what is his state of information with respect to the results of other ("earlier") events at the time when he makes his decision. Eventually, after the outcome of all events is known, one can calculate according to a fixed rule what payments the players $S_1, S_2, \ldots, S_n$ must make to each other.[32]

Specifically, the players would make "moves" subject to requirements of temporal consistency such that the ordering of moves is transitive, like the relationship between ordinal numbers: there are no intransitive "cycles" of moves F_i such that $F_1, \ldots F_n = F_1$. Of course, von Neumann's exposition was more evocative of a game like poker than chess, since it contained references to "draws," in which chance determined the outcome, and "steps," in which one of the players was in charge. Such a game could be completely specified by listing z "draws" that choose one of a set of outcomes according to a probability distribution; s "steps," for which the player controlling the step is known; and n functions that determine the amount of money won by player n in a complete sequence of draws and steps. The sum of all these functions must be zero, since within the game, money is conserved.

Von Neumann's 1928 paper thus also marked a leap beyond the highly stylized "applications" of set theory appearing in the work of Zermelo and König. "Gesellschaftsspielen" (literally "society-games" or "parlor games") per se, rather than particular lemmas from set theory or graph theory, form the paper's subject of study. The exposition and approach reads virtually as an advertisement for the Hilbertian program of formalization and axiomatization. As von Neumann would argue, not only did his mathematical framework capture the essence of games "from roulette to chess, from baccarat to bridge," but any human situation: for, "after all, any event—given the external conditions and the participants in the situation (providing the latter are acting of their own free will)—may be regarded as a game of strategy if one looks at the effect it has on the participants." The rubric of "games" also encompassed economic behavior: as von Neumann would hint in a footnote, "This is the principal problem of classical economics: how is the absolutely selfish 'homo economicus' going to act under given external conditions?" The great virtue of abstraction and formalization was that it asked the question, "What element do all these things have in common?" and then proceeded to establish generally applicable results.[33]

More importantly, von Neumann's introduction of additional mathematical objects—especially functions specifying the "payoffs" received by players at each step of the game, and the probabilities associated with them—beyond the austere calculus of sets, orderings, and relations found in the work of Zermelo and König marked a key turning point in the mathematical analysis of games of strategy. This was a first step toward answering the essential question posed by Zermelo and von Neumann alike: "n players $S_1, S_2, \ldots, S_n$ are playing a given game, G. How must one of the participants, S_m, play in order to achieve a most advantageous result?" In his analysis of chess, Zermelo had not come close to providing an answer. Von Neumann, by contrast, took inspiration from the theory of probability, which had analyzed games in which the money won by the players depends on chance alone, not the decisions of other players. In such games, he noted, the "best" method of play for a player is to maximize the expected value of the game, that is, the sum of the probabilities of possible outcomes multiplied by the monetary value of each outcome. However, the meaning of "best" remained poorly defined when several players make decisions that affect one another; this would be the subject of rest of von Neumann's paper.

The Concept of "Strategy" and the Normal Form of the Two-Person Zero-Sum Game

The exposition of the "extensive form" of games aside, in both von Neumann's 1928 paper and in von Neumann and Morgenstern's 1944 book, the authors gave relatively short shrift to this representation of games. Even before the close of the second chapter of *Theory of Games,* von Neumann and Morgenstern's analysis was beginning to move on toward much simpler situations in which "each player has one move, and one move only."[34] As a further simplification, the subsequent two chapters focused entirely on developing the theory of games with two players, in which the winnings of one were the losses of the other, and vice versa. This theory of the "two-person zero-sum game in normal form" was the centerpiece of von Neumann's 1928 article, and its celebrated "minimax theorem" has been taken by many latter-day commentators to be the heart of von Neumann and Morgenstern's achievement.[35] The theory of such games was the starting point for von Neumann and Morgenstern's efforts in 1944 to "build up a positive theory" by considering games "according to their increasing degree of complication." It also provided an intellectual foundation for the theory of multiplayer and non-zero sum games.[36]

The emblematic hieroglyph for this component of the book was not the ramifying tree of the extensive form game, but rather a "table" or matrix representing the possible combinations of winnings each player could receive—for example, figure 2.2. The rows and columns simply list the "strategies" ("1" and "2") available to the game's two players; the matrix entries list the winnings of one of the players, in this case the player whose strategies are listed in the rows. While the matrix seems at first glance an unremarkable notational device, its influence as a technique for accounting for human interaction has been extraordinary. Many if not most readers of *Theory of Games* outside the mathematical community were ill-equipped to grasp all the subtleties of von Neumann's proofs and derivations but, as we shall see, in a number of cases the matrix alone proved capable of spurring the development of game theory–inspired research programs in disciplines outside mathematics. Game matrices and tables as such were an innovation of *Theory of Games*: von Neumann's 1928 article instead expressed the players' winnings in the two-person zero-sum game principally as *functions* of the

	1	2
1	100	0
2	−50	100

FIGURE 2.2. The game matrix. After *Theory of Games and Economic Behavior*, 177.

strategic choices of players 1 and 2, associating with those ordered pairs of strategic choices a unique number.

This particular game matrix had special meaning for Morgenstern because it succinctly notated a situation described in Arthur Conan Doyle's short story, "The Final Problem":

> Sherlock Holmes desires to proceed from London to Dover and hence to the Continent in order to escape from Professor Moriarty who pursues him. Having boarded the train he observes, as the train pulls out, the appearance of Professor Moriarty on the platform. Sherlock Holmes takes it for granted—and in this he is assumed to be fully justified—that his adversary, who has seen him, might secure a special train and overtake him. Sherlock Holmes is faced with the alternative of going to Dover or leaving the train at Canterbury, the only intermediate station. His adversary—whose intelligence is assumed to be fully adequate to visualize these possibilities—has the same choice. Both opponents must choose the place of their detrainment in ignorance of the other's corresponding decision. If, as a result of these measures, they should find themselves, *in fine*, on the same platform, Sherlock Holmes may with certainty expect to be killed by Moriarty. If Sherlock Holmes reaches Dover unharmed he can make good his escape.[37]

In the above matrix, strategy 1 corresponds to "go to Dover" and strategy 2, "go to Canterbury." Moriarty's strategic options correspond to the rows of the matrix, and positive numbers refer to winnings while negative numbers refer to losses, while Holmes must choose a column and receives the negative of whatever Moriarty gets; the numbers in the boxes are arbitrary units chosen to make rough intuitive sense. The punchline of von Neumann and Morgenstern's analysis in the subsequent text was that in such situations, even (or perhaps especially) if the two protagonists were as hyperlogical and near-omniscient as Holmes and Moriarty, the outcome of their interaction was ultimately a matter of chance. In the midst of perfect rationality and full knowledge about the state of the world, uncertainty would spring up irrepressibly by the very nature of human interaction.

The story of Holmes and Moriarty was one that Morgenstern was fond of repeating, long before he and von Neumann cast it in matrix form. It appeared first in his 1928 book, *Wirtschaftsprognose: Eine Untersuchung ihrer Voraussetzungen und Möglichkeiten* (Economic prediction: An investigation of its assumptions and possibilities), and subsequently in a 1935 article, "Perfect Foresight and Economic Equilibrium," in which he again called attention to the problems associated with predictions in economic theory. Hence the story forms something of a key for interpreting his inclinations and influences as an economist.[38] Born in 1902, Morgenstern grew up in Vienna and studied political science and economics at the University of Vienna in the early 1920s. As a result, his early intellectual development occurred against the backdrop of several key intellectual fault lines that marked Viennese economic thought in the interwar period. One was the lingering influence of the rivalry that developed in the late nineteenth century between German historicists and followers of the "Austrian School." German romantic philosophy and historicist economics tended to view society as a functionally integrated organic whole that evolved through historical processes; the economist's role was to understand the course of the development of societies in their historical and cultural specificity. Such a view was often associated with a fundamental Germanic conservatism: a preference for corporatist state-society relations that reflected the character of the age, and a corresponding distrust of liberalism or the search for ahistorically "natural" or "rational" social institutions.[39] The Austrian School emerged in part from a critique of historicism, namely, the "Methodenstreit" touched off by Viennese economist Carl Menger's *Investigations into the Method of the Social Sciences* (1883). This work championed a vision for economics as a science that looked for "exact laws" of society, independent of a particular place and time, and then attempted to understand the evolution of social institutions as a result of these laws. Through his clout as a professor, Menger cultivated a new generation of economists that included Eugen von Böhm-Bawerk and Friedrich von Wieser, both of whom would prove influential on Morgenstern.[40]

In the interwar period, with the fall of monarchic rule in Austro-Hungary, the Austrian-German rivalry merged into a tempestuous series of battles between conservative nationalists, liberals, and the left-of-center parties who briefly held power in much of postwar German-speaking Europe. The subsequent reaction to the economic writings of the newly empowered socialists was spearheaded by the liberal Menger student Ludwig von Mises, whose work helped launch the "socialist calculation debate." One aspect of this debate would focus on the ability of state planners to calculate outputs, prices, and wages for the economy in question—in other words, on the solvability of the equations of general economic equilibrium that sought to relate these quantities.[41] However, Mises argued that even if such planning calculations could be done in principle, the planned economy would not accurately mimic the operations of a market economy. Planners simply could not act "as if" they were capitalists because they lacked realistic incentives to uncover information and new productive capacities, so that only the uncoordinated activities of individuals could create the efficiencies of the market. For Mises, liberalism and antisocialism aligned with the traditional methodological stance of the Austrian School. Mises' writings, especially his 1922 *Die Gemeinwirtschaft* (Socialism), had a huge impact among the new generation of economists coming to study in Vienna. As Morgenstern's contemporary Friedrich Hayek would later recall, "When *Socialism* first appeared in 1922, its impact was profound. It gradually but fundamentally altered the outlook of many of the young idealists returning to their university studies after World War I. I know, for I was one of them."[42] Both Mises and Hayek would go on to make their careers by championing a rational and individualistic approach in social science, a perspective they saw as a counter to the twin specters of romantic nationalism and socialism.

Drawing upon Morgenstern's meticulously kept diary, Robert Leonard has painted a vibrant picture of Morgenstern's idiosyncratic journey through these debates.[43] At first, as a university student, Morgenstern gravitated toward the teaching and ideas of Othmar Spann, whose social philosophy rejected the Vienna School's traditional methodological individualism and "causal-genetic" explanations in favor of an organic and holistic conception of society that he hoped would overcome the social alienation and ethnic pluralism that were straining the politics of interwar Vienna.[44] While aspects of Spann's message appealed to the young Morgenstern, he nevertheless soon moved toward the Menger protégés and Vienna School standard-bearers Böhm-Bawerk and Wieser, the prime targets of Spann's critiques. He also encountered Mises, who would serve as his adviser at Vienna. Although Morgenstern would remain far from a pure *laissez-faire* liberal, continuing to harbor some of the more conservative instincts of his middle-class German-Viennese upbringing, the methodological perspectives of Böhm-Bawerk, Wieser, and Mises would carry through into his subsequent research.[45] After graduating in 1925, Morgenstern also spent time abroad in Britain and the United States on a fellowship from the Laura Spelman Rockefeller Memorial Foundation. Guiding his travels and study at this time was his growing interest in the nature and causes of business cycles. A number of

American economists—most notably Wesley Mitchell at Columbia University, Henry Ludwell Moore, and Warren Persons at Harvard—were developing an institutionalist perspective on business cycles, seeking statistical indicators that might assist in the prediction of economic indicators and the control of business cycles.[46] To liberals like Mises and Hayek, American institutionalism represented another offshoot of German historicism and an intellectual rationalization of government intervention in the economy.[47]

Morgenstern's first book, *Wirtschaftsprognose* (1928), represented a characteristically "Austrian" response to the work of Mitchell, Moore, and Persons. Much of the book was in effect a critical survey of these economists' work. Yet its underlying thesis was that economic prediction was both impossible and useless, and that the entire apparatus of econometric data-gathering and statistical analysis could never comprehend the complexity of an actual economy. Morgenstern's argument naturally drew strength from his own theory of the behavior and decision-making processes of businessmen, economists, and government officials. Each makes decisions on the basis of information that—crucially—includes not only conventional economic indicators and statistics, but also his or her expectations and "anticipations" of the future behavior of other participants in the economy. In this context, when one economic actor (for example, the government or a monopolist) makes a prediction or introduces an economic plan, others will modify their expectations and act upon them, leading to a chain reaction of predictions and counterpredictions—literally, "der sich zu Tode jagenden Prognosen"—that eventually makes the original prediction inaccurate, the plan untenable, and the future inherently ineffable. It was here that Morgenstern presented the story of Holmes and Moriarty as an "amusing analogy" to the interdependence of predictions in a society, a kind of parable about the limitations of economic prediction. And in a comment that appears pregnant in light of his subsequent work, Morgenstern concluded his passage on Holmes and Moriarty by noting that "Examples of such kind can be drawn from everywhere. Chess, strategy, etc.," all of which captured the dynamic nature of interactions between economic actors.[48]

Nevertheless, for one so focused on the limitations of economic planning and prediction, Morgenstern's subsequent career moved in a remarkably technocratic direction. Upon returning from his fellowship abroad, he quickly rose through the establishment of Viennese economics. In 1929, he assumed the editorship of *Zeitschrift für Natio-nalökonomie*; in 1931, he succeeded his friend Hayek as director of the Austrian Institute for Business Cycle Research (Österreiches Institut für Konjunkturforschung), which was funded primarily by the Rockefeller Foundation. From this perch, he produced a number of papers and a book that reflected his theoretical interest in economic problems of prediction and time.[49] He also became known as an expert on statistical economics and economic forecasting, serving from 1932 as an advisor to the Austrian National Bank, and from 1936 as an advisor to the Austrian Ministry of Commerce and a member of the Committee of Statistical Experts of the League of Nations.[50]

Von Neumann, in his 1928 paper, saw the "normal" matrix form of games was first and foremost a product of the mathematician's art: a simplification of the more general and realistic "extensive" form of a game, to which, he argued, the extensive form was essentially equivalent. Such a simplification was made possible by two assumptions, which von Neumann spent a not-insignificant amount of space explaining and defending. The first flowed from the notion of a "strategy," a complete contingency plan covering all possible choices a player would need to make during the game. Assuming the adoption of strategies did not compromise the "free will" of the players in the course of playing out the game, having the players choose between *sequences* of moves would permit von Neumann to collapse the causal-dynamic chain of the game (which captured precisely the kinds of complex interactions and interdependent decisions that fascinated Morgenstern) into one moment. The second assumption von Neumann invoked was practically a mathematical convention since the development of the calculus of probabilities in the eighteenth century: that one could replace "draws," or the outcome of chance events, with the expected values of those events (the values of the possible outcomes weighted by their probabilities). While von Neumann would note the existence of "well known objections to the use of expected value (and the ensuing attempts to replace the latter by the so-called moral expectation or similar concepts)," he nevertheless bypassed these with a footnote only.[51] Thus complete information about any game could be given simply by listing the expected value received by each player under each combination of their strategies. This list (or matrix) constituted the "normal form" of the game. In the very simplest case in which there are two players, each of whom has available two possible strategies, the amount won by player 1 is a function of the possible choices made by the two players, that is, it is $g_1(x, y)$, where x and y represent the strategies chosen by players 1 and 2. Because $g_1 + g_2 = 0, g_1 = -g_2$ (the zero-sum condition).

Von Neumann's subsequent examination of this simplest kind of game walked a fine line between a couple of distinct analytic perspectives. The first took inspiration from the question he posed at the outset (much as had Zermelo in his earlier paper on chess): "How must one of the participants, S_m, play in order to achieve a most advantageous result?" However, in the case of games involving two players with "free will," von Neumann was adamant that this problem could not be reduced to the maximization of expected winnings treated in the traditional mathematical theory of games of chance. In contrast to the approach taken subsequently by Bayesian decision theorists, one could not simply hang a probability distribution on the moves of one's opponent and maximize against it, in effect treating opponents as stochastic forces of nature. But given this fact, there was no mathematically unambiguous answer to von Neumann's question, since neither player could control both x and y. As von Neumann would write, "It is easy to picture the forces struggling with each other in such a two-person game. The value of $g(x, y)$ is being tugged at from two sides, by S_1 who wants to maximize it, and by S_2 who wants to minimize it. S_1 controls the variable x, S_2 the variable y." Hence von Neumann shifted to a slightly different question—"What *will* happen?"—even as he continued to offer his players copious advice on what they should (or could) do.[52] Consider the first player. Irrespective of the actions of his opponent, he *can* choose a strategy x so that

$$g(x, y) \geq \text{Max}_x \, \text{Min}_y \, g(x, y)$$

Similarly, the second player can choose a strategy y so that

$$g(x, y) \leq \text{Min}_y \, \text{Max}_x \, g(x, y)$$

So the value of $g(x,y)$ is sandwiched between the MaxMin (lower bound) and the MinMax (upper bound). If the two are equal, then that number is the determinate value of the game. In a passage reminiscent of Zermelo's eschewal of "psychology" in his paper on chess, von Neumann argued that, in such a situation, "it makes no difference which of the two players is the better psychologist, the game is so insensitive that the result is always the same."[53]

However, in the much more likely case of a gap between the upper and lower bounds, the outcome remained uncertain, since it was unclear who would obtain this surplus value. One player might "find out" his opponent's intended course of action through some connivance, allowing for him to tilt the game in his favor—although this once again would presumably depend on psychological factors that von Neumann was loath to introduce. And it was here that he instead proposed "by use of an artifice, to force the equality of the two above-mentioned expressions." Instead of restricting the players to a choice of "pure" strategies (determinate courses of action), von Neumann allowed them to choose their course of action according to a probability distribution—that is, to play "mixed strategies." "[The player] only has to specify Σ_1 probabilities . . . and then draw the numbers $1, 2, \ldots, \Sigma_1$ from an urn containing these numbers with the probabilities," von Neumann suggested. "This may look like a restriction of his free will: it is not he who determines x," von Neumann noted, anticipating the arguments of those who might find the use of "artifice" to "force" a desired outcome on the game-players a step too far. "But this is not so. Because if he really wants to get a particular x, he can specify $\xi_x = 1, \xi_u = 0$ (for $u \neq x$). On the other hand, he is protected against his adversary "finding him out"; for, if, e.g., $\xi_1 = \xi_2 = 1/2$, nobody (not even himself!) can predict whether he is going to choose 1 or 2!"[54] Thus, von Neumann concluded, "Although in Section 1 chance was eliminated from the games of strategy under consideration . . . it has now made a spontaneous reappearance." Even if the "rules of the game" do not explicitly refer to the spinning of roulette wheels, the random selection of cards, or the drawing of balls from urns, "the dependence on chance (the 'statistical element') is such an intrinsic part of the game itself (if not of the world) that there is no need to introduce it artificially by way of the rules of the game: even if the formal rules contain no trace of it, it still will assert itself."[55]

This remarkable passage arguably constitutes the punchline of von Neumann's analysis of the two-person zero-sum game in normal form, and indeed perhaps of the entire 1928 paper. The initial "artifice" introducing the element of chance into the players' strategies may have been von Neumann's alone; but in this result, he drew a general lesson about the irreducibly stochastic nature of games and indeed "of the world" more broadly. Certainly, it resonated with one emerging from another subject (besides set theory and parlor games) that would occupy von Neumann's mental energies during the 1920s: quantum mechanics. At the time that von Neumann gave his paper on games at Göttingen in December of 1926, he had embarked on a series of publications that were intended to provide an axiomatic formulation of quantum mechanics in terms of the algebraic properties of operators on infinite-dimensional "Hilbert spaces." This work would eventually allow him (among other things) to reconcile the two competing quantum theories of the day: Werner Heisenberg's "matrix mechanics" and Erwin Schrödinger's "wave mechanics."[56] As with the theory of games, in which "what will happen" became uncertain despite perfect certainty about the rules of the game and simply as a logical consequence of human interaction, measurement in a quantum-mechanical world appeared not just error-prone, but irreducibly stochastic. This was at least the conclusion of the crowning achievement of von Neumann's *Mathematische Grundlagen der Quantenmechanik*: a theorem in which, starting from five apparently reasonable axioms, he demonstrated that no "hidden variables" could eliminate the uncertainty inherent in quantum measurements.[57]

Beyond calling attention to these similarities, a number of commentators have suggested a direct connection between von Neumann's work in quantum mechanics and the development of game theory. Thus the mathematician and former von Neumann student Merrill Flood would later assert "that von Neumann was apparently led to his basic result [about the two-person zero-sum game] by considerations arising from his studies in the logical foundations of quantum mechanics."[58] Historian of economics Philip Mirowski has also called attention to the way that quantum mechanics proved inspirational for both von Neumann and Morgenstern during their collaboration as they tried to forge a new theoretical apparatus for economics. Deep in discussions of games with "Johnny," Morgenstern would tell his diary, "It is becoming clearer to me that the whole thing is like quantum mechanics. It will only last if that is absorbed. We both know that."[59] Morgenstern was almost certainly wrong in his assessment: game theory would survive, even thrive, without being hailed as the quantum mechanics of parlor games. But the skeptical epistemological outlook shared by quantum theory and Austrian economics appears frequently in *Theory of Games*. Consider their analysis of the Holmes-Moriarty game: in a footnote they would add that "indeed, while standing on the platform, Holmes is only 48%

alive," since the probability that Holmes and Moriarty end up on the same platform is 3/5 × 2/5 + 2/5 × 3/5 = 48/100.[60] As with the parable of Schrödinger's cat, von Neumann's musings on the immovable boulders produced by an omnipotent God, or perhaps the plot of a film noir, the paradoxes of existence and probability, possibility and impossibility, form the leitmotif of von Neumann and Morgenstern's book.

The Minimax Theorem and Game Theory's Early Connections with Economics

The year 1928 would prove pivotal in establishing a connection between economics and the mathematical analysis of games. In this year Morgenstern suggested that chess, because of its dynamic nature, could serve as a plausible analogy to the economics of anticipation and forecasting. And in a footnote to his 1928 *Mathematische Annalen* paper, von Neumann suggested that the starting question of his game theory, "How must one of the participants . . . play in order to achieve the most advantageous result?," was identical to "the principal problem of classical economics: how is the absolutely selfish 'homo economicus' going to act under given external circumstances?"[61] The game-player may have captured the selfishness of the economic actor, but the theory of the two-person zero-sum game, the centerpiece of von Neumann's 1928 article, remained far from constituting a recognizable framework for economic analysis. Moreover, to a mathematical reader, von Neumann's article must have come off as somewhat *sui generis*. Its central result, the much-celebrated "minimax theorem," was specifically targeted at the analysis of the two-person zero-sum game in normal form: specifically, it showed that if one permits players to play mixed (probabilistic) strategies, there always exist strategies x and y such that $Max_x\, Min_y\, g(x, y) = Min_y\, Max_x\, g(x, y)$. And while von Neumann drew upon a number of existing concepts and results in his proof of the theorem, the result was not obviously connected to any other significant problems or theorems in mathematics. Or at least, if it was, von Neumann did not trumpet the fact: the 1928 paper was glossed first and foremost as an exploration of games per se. Understandably, the survival of game theory, and especially the widespread interest in the mathematics of the two-person zero-sum game after World War II, would have been unlikely had not von Neumann spent the years between 1928 and 1944 building connections between this aspect of the theory and several research programs emerging in mathematics and economics alike.

From a number of facts, we know that von Neumann's engagement with mathematical economics deepened significantly in the later 1920s. Not long after he was appointed Privatdozent at the University of Berlin in 1927, he expressed an interest in the field to Nicholas Kaldor, a fellow Hungarian who would later become an economics professor at Cambridge. Kaldor steered von Neumann to Knut Wicksell's *Value, Capital, and Rent*, which drew upon the theories of French economist Leon Walras. Walras (and his followers of the "Lausanne School") developed models of general economic equilibrium based on systems of simultaneous equations relating quantities of goods produced by various productive processes, quantities demanded by individuals and firms, and the prices at which exchanges took place. Solving these equations would ideally provide a listing of prices and quantities of all goods produced in an economy. Further developed by Gustav Cassel in the 1910s, the equations were often referred to as the Walras-Cassel equations of general equilibrium.[62] According to Kaldor, von Neumann was immediately critical of the Walras-Cassel formulation. After all, the equations could be inconsistent, resulting in no solution, and that they did not exclude the possibility of negative prices.[63] Other scholars have connected von Neumann's growing interest in economics with a mathematician named Robert Remak, who, like von Neumann, was a Privatdozent in mathematics at Berlin. During his time there, Remak wrote a couple of articles on mathematical economics; one proposed to identify prices that solved the equations of general economic equilibrium, in effect solving a problem necessary for the implementation of socialist planning.[64] Finally, at Berlin in the late 1920s, the physicist Leo Szilard created a study group, consisting of a number of mathematicians and physicists, to invite lecturers to speak on the use of mathematics outside physics. At one meeting of the study group, Jacob Marschak—a Russian-born economist who initially made a name for himself as a critic of Mises in the socialist calculation debates—gave a presentation on the Walras-Cassel equations of general equilibrium.[65] According to Marschak's memories of the occasion, "One of the mathematicians became extremely agitated and began a stream of interruptions, arguing that the equilibrium relationships should be described by inequalities instead of equations. That mathematician was John von Neumann."[66]

Subsequently, in 1932, von Neumann gave a talk on the equations of economic equilibrium before the Princeton mathematical colloquium. The venue reflected von Neumann's own calculations of professional advantage: in 1930, seeking a way out of central Europe's congested academic job market, he had accepted the post of visiting associate professor at Princeton; in 1931, the position was made permanent, although he continued to divide his time between Princeton and Berlin until 1933 when he was invited to be an inaugural member of Princeton's newly founded Institute for Advanced Study.[67] The talk was later edited and published in 1937 under the title, "Über ein ökonomisches Gleichungssytem und eine Verallgemeinerung des Brouwerschen Fixpunktsatzes" (On a system of economic equations and a generalization of the Brouwer fixed-point theorem).[68] In the paper, von Neumann considered an economy that produces goods $G_1 \ldots G_n$ using productive processes $P_1 \ldots P_m$, where it is possible that $m > n$ (i.e. there could be more than one distinct way to produce a given good). There are constant returns to scale and no technological change. Each good has a nonnegative price, and each process is pursued with a nonnegative "intensity" that reflects the relative frequency with which the process is employed in the economy. (If the process is not used at all, the "intensity" is zero.) Finally, over a unit period of time, there is an interest factor $\beta = 1 + i$ (where i is the prevailing interest rate in percent), and the economy expands in each period by a factor α. Since there are potentially more productive processes than goods, the systems of linear equations that govern prices, intensities, the interest rate, and the expansion rate would not necessarily yield solutions; instead, von Neumann modeled the situation as a system of inequalities and proposed to investigate the possibility of solving it to determine prices, intensities, interest factor, and expansion rate.

Von Neumann's analysis of this system of inequalities drew upon methods identical to those in his treatment of the two-person zero-sum game. One can interpret the "game" in this case as arising from the conflict between the interest rate and the rate of expansion of production, that is, between saving and spending in any given period, with the former controlling the prices associated with different goods and the latter controlling the intensities of production. The players' "pure strategies" in this instance are simply the different goods produced by the economy and the technically feasible distinct processes of production, respectively; the probabilities of playing these pure strategies selected by the two players correspond to the prices of the goods and intensities of the productive processes since, like probabilities, these numbers are nonnegative and can be made to sum to one by dividing through by the gross income and the gross output, respectively. The "value" of the game for the interest-rate "player" is the interest factor, while the value of the game for the expansion-rate player is the expansion factor. Finally, as an analogue of the function $g(x, y)$ (the payoff to the players from choosing strategies x and y) von Neumann identified a function $\varphi(x, y)$ that captured the relationships between the factors of production in the economy for a given level of prices and intensities. With saving and spending, interest rate and expansion rate locked in their zero-sum struggle, the solution to the system of inequalities occurs when $\varphi = \alpha = \beta$; in other words, in equilibrium, the rate of expansion is equal to the prevailing rate of interest. The challenge was to prove that prices and intensities of production existed for which this equality held—a problem analogous to proving the existence of mixed strategies that solved the two-person zero-sum game. Thus as von Neumann would note, "The question of whether our problem has a solution is oddly connected with that of a problem occurring in the Theory of Games dealt with elsewhere." In the case of the expanding economy, like the theory of the two-person zero-sum game, "[the equation for φ] does not lead to a simple maximum or minimum problem, the possibility of a solution of which would be evident, but to a problem of the saddle point or minimum-maximum type, where the question of a possible solution is far more profound."[69]

The solutions may have been analogous, but in his 1937 paper von Neumann emphasized that his results represented a significant generalization and conceptual deepening of those presented in his earlier work on games. Referring to the 1928 paper, he argued that "the problem there is a special case . . . and is solved here in a new way."[70] First, von Neumann's new proof made the minimax theorem appear not as a specialized result about the commutativity of "max" and "min" operators acting on the payoff function of a game, but as a corollary of a fundamental result of topology, the study of the continuous transformation and deformation of multidimensional objects. The result in question was a "fixed-point theorem," first introduced by Brouwer in 1911, which specified the conditions under which a function f would possess a point such that $x = f(x)$.[71] As historian of mathematics Tinne Kjeldsen has shown, this fact is remarkable given that many retrospective accounts claim incorrectly that Brouwer's theorem was already fundamental to von Neumann's 1928 proof of the minimax theorem.[72] Yet a reader of von Neumann's 1937 paper could be forgiven for believing this, given the way von Neumann advertised the centrality of Brouwer's theorem to his results. "The way in which our questions [about the economy] are put leads of necessity to a system of inequalities . . . the possibility of a solution of which is not evident, i.e., *it cannot be proved by any qualitative argument*," von Neumann asserted. "The mathematical proof is possible only by means of a generalisation of Brouwer's Fix-Point Theorem, i.e. by the use of very fundamental *topological* facts." The connection between the economic problem at hand and topology might be "surprising," but was in fact "natural" given "the occurrence of a certain 'minimum-maximum' problem" that had not only cropped up in the theory of games, but that also was "familiar from the calculus of variations," a longstanding area of mathematical study that identified the extremals of functionals—for example, the maxima or minima of integrals along curves in space.[73]

Thus, as with the extensive form of games, which von Neumann and Morgenstern would weave together with a particular research tradition emanating from the theory of sets and graphs, the theory of the two-person zero-sum game in normal form now possessed connections with an economic problem as well as increasingly well-established programs of study in mathematics. Princeton's topologists, who counted among their number giants in the field like Solomon Lefschetz, James Waddell Alexander, and Oswald Veblen, were by the 1930s a force in this rapidly emerging mathematical subdiscipline, with Veblen's 1922 *Analysis Situs* and Lefschetz's 1930 *Topology* becoming canonical texts. As one joint reviewer of the two books would note, "As late as 1925 it was possible for the writer of a widely used text-book to refer to topology as the 'Sorglingkind der Mathematik,'" a disorderly hodgepodge of results to be drawn on at will when proving results in better-established adjacent fields. But this had changed in recent years: among other things, Lefschetz's own generalization of Brouwer's theorem in the later 1920s had drawn upon a number of key existing results to provide "an instrument of great power for resolving problems in a variety of branches of mathematics," especially "Algebraic Geometry and the Calculus of Variations."[74] Among topologists, fixed-point theorems and related results were thus key objects of study in their own right, in their quest to classify objects by the features they possess that remain invariant after continuous deformations.[75] Albert Tucker, who would emerge as a standard-bearer of game theory on the Princeton mathematics faculty in the later 1940s, first caught the eye of Lefschetz in a seminar by providing an improved proof of Lefschetz's own famous fixed-point results.[76] Or consider the work of Shizuo Kakutani, a Japanese analyst who visited Princeton in 1940–42. In a 1941 paper on "A Generalization of Brouwer's Fixed Point Theorem," Kakutani extended the Brouwer theorem to deal with functions associating points with *sets*, and proceeded to give conditions under which there existed a point x that was contained in $f(x)$. Kakutani's theorem had the useful property that von Neumann's two theorems— the minimax theorem of 1928 and the more general result of 1932—were simple corollaries, derivable from Kakutani's theorem in proofs of a couple lines each.[77]

Von Neumann's foray into the equations of economic equilibrium thus linked a new set of potential audiences for the theory of games: topologists interested in applications of the core results of their fields, and economists wrestling with models of the macroeconomy. After 1944, these groups would again meet under the rubric of "programming" and the patronage of the American military. But even in the prewar period, the inequalities of economic equilibrium began to accumulate a social world of their own. A few years after Marschak's talk in Berlin, the Viennese economist Karl Schlesinger would turn to his colleague, the mathematician Karl Menger, for advice in connection with the Walras-Cassel equations. Menger, the son of the founder of the Viennese economic tradition Carl Menger, put Schlesinger in touch with one of his students, Abraham Wald, who could serve as a mathematical "tutor." Rewriting the equations as systems of inequalities, Wald was able to prove that the systems did in fact possess solutions. The result was later published in the proceedings of Menger's mathematical colloquium at the University of Vienna.[78] And within a few years of his Princeton talk, von Neumann would publish his model of an expanding economy in the *Ergebnisse* of Menger's mathematical colloquium. What had begun as the mathematics of parlor games had emerged as a nexus linking new areas of mathematics and economic theory.

Multiplayer Games and the Problem of "Solutions"

Despite the connection between the theory of the two-person zero-sum game and the model of an expanding economy that von Neumann developed in the 1930s, by the end of chapter 4 of *Theory of Games* the book's authors nevertheless remained far from their goal as laid out in the opening pages of the book. The proper subject of theories of "economic behavior" was indeed the kind of phenomena explored in von Neumann's model: "the very complicated mechanism of prices and production, and of the gaining and spending of incomes." Yet as von Neumann and Morgenstern noted in the next breath, "it is now well-nigh universally agreed, that an approach to this vast problem is gained by the analysis of the behavior of the individuals which constitute the economic community," as opposed to aggregate interactions between interest and growth, money and goods.[79] By contrast, *Theory of Games* was to provide the foundation for a theory of society built up from postulates of rational individual behavior that were universal, as valid for poker players as for businesses and consumers. "For economic and social problems the games fulfill—or should fulfill—the same function which various geometrico-mathematical models have successfully performed in the physical sciences."[80] Specifically, von Neumann and Morgenstern sought to equate the task of building economic theory with the problem of "solving" games involving many players in the same way that one might solve a mathematical equation.

The idiosyncratic "solutions" that von Neumann and Morgenstern proposed to such games would prove one of the most puzzling features of their book. In their opening chapter, von Neumann and Morgenstern suggested that a solution "is plausibly a set of rules for each participant which tell him how to behave in every situation which may conceivably arise," that is to say, a specification of what constituted rational conduct in each possible situation.[81] But the nature of these rules changed dramatically as the focus shifted from the isolated decision maker to the two-person zero-sum game, and finally to the three-person zero-sum game, the obvious next category in von Neumann and Morgenstern's taxonomy of games. Such games, as the authors noted at the start of chapter 5, "bring entirely new viewpoints into play," since "the passage from the zero-sum two-person game to the zero-sum three-person game obliterates the pure opposition of interest" between the players in question.[82] Specifically, "as soon as there is a possibility of choosing with whom to establish parallel interests, this becomes a case of choosing an ally. When alliances are formed, it is to be expected that some kind of a mutual understanding between the two players involved will be necessary," whereas "an opposition of interests . . . requires presumably no more than that a player who has elected this alternative act independently in his own interest."[83] Von Neumann and Morgenstern were aware that the assumption of "alliances," "agreements," "understandings," or "coalitions" could be objectionable: what, for example, might enforce "the 'sanctity' of such agreements?" in a single play of the game, they wondered.[84] And they hinted that in future work they would start to "investigate what theoretical structures are required in order to eliminate these concepts," specifically, a generalization of the utility concept so that it would no longer be transferable and freely divisible like money.[85] But for the moment at least they turned for inspiration to their initial conception of "solutions" as complete listings of rules for rational conduct, arguing almost by default that "if we do not allow for [coalitions], then it is hard to see what, if anything, will govern the conduct of a player" in multiplayer games.[86] Having assumed that players will form coalitions, it followed that the "rules" that constituted a solution needed to tell the players whether to join an "alliance," "understanding," or "coalition" with another player or group of players, and under what terms—that is, what combination of inducements should be required to join, and hence, winnings and losses should be apportioned among the players.

These apportionments of winnings among the players von Neumann and Morgenstern called "imputations." The term was drawn from the theory of value that was a trademark of the Viennese economic tradition as expressed in the work of Menger, Böhm-Bawerk, and Weiser, Morgenstern's intellectual forebears. While classical economists sought the value of a good in the values of the inputs needed to produce it, thereby tracing the root of all value back to labor or land, marginalists inverted this reasoning, arguing that the value of a good was determined by the expected price of the goods it could be used to produce. The imputation problem (*zurechnungsproblem*) thus concerned how to divide up the value of the product between the different factors of production— much as the players in a game sought to divide the fruits of their collaboration or collusion among themselves. Von Neumann and Morgenstern saw their work as addressing one of the fundamental problems posed by the Viennese theory of value and prices, a problem that, "in its general form, has neither been properly formulated nor solved in the economic literature."[87] And indeed, earlier approaches had proven problematic. For example, Weiser (in his 1889 *Der natürliche Wert*) had tackled the problem by solving a system of n equations for the unknown values of n inputs, where the equations represented n distinct production functions for particular commodities. This approach faced difficulties similar to those faced by the Walras-Cassel approach to economic equilibrium, namely, that in more realistic economic models the solution to the equations may be overdetermined or underdetermined, resulting either in no solutions or in too many.

In the case of the two-person zero-sum game, the imputation appeared to be well defined, if not purely deterministic: one player received the unique expected value of the game, while the other received the negative of that value. However, in games with more than two players the distribution of winnings generally depended on which coalitions the players joined. Consider the three-player "simple majority game," the initial example by which von Neumann and Morgenstern probed the features of the general three-person zero-sum game, starting in the fifth chapter of *Theory of Games*.[88] In this game, by forming a coalition, any two players can win at the expense of the third player. Hence the winnings of coalitions are generally greater than the winnings of the individuals in that coalition playing independently.[89] To describe this possibility, von Neumann and Morgenstern proposed to represent multiplayer games by a "characteristic function" $v(S)$, that associated a numerical value with every possible *coalition*, that is, every possible subset S of the set of players, rather than simply with individual players as they had in the two-person zero-sum game. This "characteristic function form" of the game (as opposed to the earlier extensive and normal forms) captured the possibilities for coalition-formation within the game (see fig. 2.3).

But here, despite suggesting a concise mathematical representation of the problem, the assumption of individual rationality did not provide anything like the "complete set of rules" needed to tell the players what to do. Players concerned about losing could potentially bribe away (or in later game theoretical parlance, offer "side payments" to) possible members of a winning coalition, thereby leading to a new system of coalitions and a new distribution of payoffs. In this unstable process, it was possible that no single imputation would "dominate" the others; each would give way to another in a never-ending intransitive cycle of "betrayal, turncoatism, and deception," as von Neumann's colleague Norbert Wie-

$$v(S) = \begin{cases} 0 \\ -1 \\ 1 \\ 0 \end{cases} \text{ when } S \text{ has } \begin{cases} 0 \\ 1 \\ 2 \\ 3 \end{cases} \text{ elements}$$

FIGURE 2.3. The three-player simple majority game in characteristic function form: v is the value or winnings of coalition S, which can contain one, two, or three of the players. After *Theory of Games and Economic Behavior*, 260.

ner would later put it.[90] Hence there was in general no unique "best" imputation as there had been with the two-person zero-sum game, and von Neumann and Morgenstern proposed to define a "solution" as a *set* of imputations S with two properties:

No y contained in S is dominated by an x contained in S.

Every y not contained in S is dominated by some x contained in S.

Or more compactly, "The elements of S are precisely those elements undominated by elements of S."[91]

Even in the case of the three-player simple majority game, this seemingly neat solution allowed for an extraordinary wealth of possible game outcomes. Not only were the "solutions" not unique, but they often consisted of infinite sets of nondominated imputations. In these, von Neumann and Morgenstern could discern a wide variety of social arrangements, both objective and "inobjective," equitable and inequitable. One solution, for example, constituted a system of "discrimination": "One of the players . . . is being discriminated against by the two others. . . . They *assign* to him the amount which he gets, c," and then somehow divide up the remaining winnings among themselves.[92] A particular "solution" thus was a "system of imputations" that corresponded with what von Neumann and Morgenstern repeatedly referred to as an "established order of society" or "accepted standard of behavior."[93] And for yet higher numbers of players they conjectured that these might include "a much greater wealth of possibilities for all sorts of schemes of discrimination, prejudices, privileges, etc."[94] In this way, von Neumann and Morgenstern once again appeared to have discovered a major aspect of uncertainty in social relationships, one that they felt could not be eliminated except by reference to broader "cultural norms," which in turn may have emerged via the dynamic evolution of a society. But this possibility posed a problem for the future, since they admitted that their current theory was "thoroughly static." Although "a dynamic theory would unquestionably be more complete and therefore preferable," they were careful to say that this was a topic beyond the scope of *Theory of Games*.[95]

The mathematical apparatus of multiplayer games (characteristic functions and solutions as nondominated sets of imputations) that von Neumann and Morgenstern presented in *Theory of Games* greatly expanded upon the ideas in von Neumann's 1928 paper. There, in sketching the development of a theory for more than two players, von Neumann assumed that two players could form a coalition and "rob the third without any ado."[96] Mathematically, this had the convenient effect of reducing the three-person game to three two-person zero-sum games, where one player in each of the games was a coalition of two players. Von Neumann noted that the possibility of side payments made it difficult to tell which of these subgames would finally wind up playing out, so that a more encompassing definition of game "solutions" would need to encompass this fact. Yet there were plenty of other very interesting mathematical questions that he could ask about multiplayer games, and in the long final section of that paper, following the proof of the minimax theorem, he turned to some of these. For example, could one identify any major categories of games involving three, four, or five players that were equivalent in some sense, that is, that possessed similar strategic properties? Finally, in indicating directions for the future study of n-person games, he explored the properties of the characteristic function and conjectured that the "characteristic function form" of a game would prove equivalent to the "normal form" of a game.

Similar questions guided the treatment of multiplayer games in *Theory of Games*. Here, there were some signs that von Neumann and Morgenstern hoped to wring deep and generally applicable insights out of the mathematical form of the characteristic function. For zero-sum games, the characteristic function had an axiomatic structure familiar to those acquainted with "the mathematical theory of measure," the study of how to specify a consistent "length" for sets of real numbers that proved critical for developing modern theories of integration and probability. Specifically, v(all players) = 0; v(all players not in S) = $-v(S)$; and for two nonoverlapping coalitions S and T, $v(S) + v(T) \leq v(S$ and $T)$.[97] The parallel is of course inexact, since, as the last condition states, the value obtained by a coalition is greater than or equal to the sum of the values obtained by its members individually, where in measure theory the measure of a set is in general the sum of the measures of its disjunct subsets. Von Neumann and Morgenstern therefore concluded that "the general characteristic functions $v(S)$ are a new generalization of the concept of measure."[98] Each such function corresponded to a game, and each game possessed such a function, making their study a natural first step toward a general theory of multiplayer games. Thus Lloyd Shapley, one of the first generation of postwar game theorists at the RAND Corporation and namesake of the "Shapley value" solution concept for cooperative games, would complete his 1953 Princeton dissertation on precisely such set functions.[99]

Yet in *Theory of Games*, von Neumann and Morgenstern's approach seemed to be rapidly overwhelmed by the complexities that emerged as the number of players grew larger. After solving completely the three-player game in chapters 5–6, and developing an "inexhaustive and chiefly casuistic treatment" of four- and five-player games in chapter 7,[100] by chapter 9, they had to admit that despite being "voluminous," their results were only "fragmentary," a fact that "very seriously limits their usefulness in informing us about the general possibilities of the theory." They therefore proposed to "find some technique for the attack on games with higher *n*" by jettisoning the search for "anything systematic or exhaustive" and instead identifying "some special classes of games involving many participants that can be decisively dealt with," since "it is a general experience in many parts of the exact and natural sciences that a complete understanding of suitable special cases—which are technically manageable, but which embody the essential principles—has a good chance to be the pacemaker for the advance of the systematic and exhaustive theory."[101] Hence, in chapter 9, they explored the possibility of decomposing games into smaller constituents and probed the relationship between the imputations of the constituents and those of the larger game. And in chapter 10, they mapped the solutions of a major class of games called "simple games," in which coalitions could be neatly separated into winners or losers. But in the end, once again, it seemed that a "systematic and exhaustive theory" of multiplayer games (or at least, one founded on the characteristic function form of the game) remained elusive.

While the mathematics tilted at a comprehensive theory of games, especially for Morgenstern, the interpretation of game theoretic "solutions" as "orders of society" or "standards of behavior" was crucially important, and it reflected ideas that he had been exploring since the early 1930s. One key set of influences on Morgenstern's thinking during this period can be traced to his friendship with the University of Vienna mathematician Karl Menger. As he had with the economist Karl Schlesinger, Menger referred Morgenstern to his perennially unemployed student Abraham Wald for mathematical coaching, and to serve as a statistical consultant to Morgenstern's Institute for Business Cycle Research.[102] Menger also provided Morgenstern with a conduit to a new philosophical world, especially the famous "Vienna Circle," of which he was accounted a member along with philosophers like Moritz Schlick and Rudolf Carnap, mathematicians like Menger's teacher Hans Hahn, and even social scientists like Otto Neurath.[103] And finally, Morgenstern also began to attend Menger's regular mathematical colloquium at the university, which numbered Wald and the logician Kurt Gödel among its participants, where he could observe this group's forays into the application of mathematics to the social sciences during the early 1930s.[104]

Among these forays was a small 1934 book by Menger himself, *Moral, Wille, und Weltgestaltung* (later published under the English title of *Morality, Decision, and Social Organization*). Menger's book reflected in part the spirit of the Vienna Circle: its dedication to the synthesis of logic and mathematics with positivism and empiricism, and to the ideal of a "unified science" spanning both the natural and social sciences. Here the starting point for extending the scientific approach to ethics and the study of social behavior was Menger's home discipline of geometry. Could one identify logically consistent systems of ethical norms, whether those norms were biblical commandments or the Kantian categorical imperative? As a partial answer, Menger introduced a set-theoretic calculus for interrogating the compatibility of different norms. For instance, suppose that members of a society can be "opposed to," "in favor of," or "indifferent to" a given norm. If there are n norms, then there are $3n$ distinct possible groups of "consentience." This estimate represented an upper bound, since for any group the norms might be logically incompatible, thereby eliminating themselves from consideration, or individuals with different attitudes toward norms could be compatible with each other precisely because of their differences. As a result, Menger was led to consider the possibilities of "compatibility groups," groups in which adherents to different norms might nonetheless prove compatible.[105] Even with this restriction, from a logical perspective it was impossible to determine which particular system of norms would arise. Menger's ultimate suggestion was that just as geometry possessed Euclidean and non-Euclidean variants, there might be many possible systems of norms that were internally consistent but mutually inconsistent. Individuals simply had to choose between these systems: "An individual's profession of a particular system of norms is an expression of feeling. The actual modeling of his behavior according to some morality is based on a decision."[106] The patchwork of these decisions was in turn responsible for the diversity of the social order, a diversity that, when coupled with freedom of association, made it possible for individuals to find the social organizations and systems of norms that best suited their particular needs.

As Menger asserts in his autobiographical works, his interest in extending scientific approaches to ethics and the social sciences reflected the tumultuous Viennese political environment of 1933–34 street fighting between government forces and Social Democrats, and an attempted Nazi *putsch* as Hitler came to power in neighboring Germany.[107] Despite the difficult political situation of the time, Menger's mathematical colloquium did meet to discuss several applications of mathematics to the social sciences and moral philosophy. On March 19, Karl Schlesinger presented the first part of his work "On the Production Equations of the Economic Theory of Value" with Wald providing a proof of the existence of the solution to these equations during the same meeting. Shortly thereafter, in the colloquium of May 15, Menger presented "A Theorem on Finite Sets with Applications to Formal Ethics," which would appear as a key result in his "Five Logico-Mathematical Notes on Voluntary Associations." Finally, on June 19, Menger gave a brief talk on "Bernoullian Value Theory and the Petersburg Game," an investigation of value under conditions of risk that he would later publish, at Morgenstern's invitation, in *Zeitschrift für Nationalökonomie*.[108] Also during that difficult year, Menger shared his manuscript of *Moral, Wille und Weltgestaltung* with Morgenstern, who had expressed interest in the project.[109]

The ideas circulating within Menger's mathematical colloquium resonated strongly with Morgenstern's economic thinking during these years. Morgenstern's sweeping evaluation of the state of economics as applied to policy, *Die Grenzen der Wirtschaftspolitik* (1934), effectively turned Menger's logical razor on the major systems of economic organization of the day: liberalism and socialism. Both systems, he argued, were hopelessly mired in contradictions, both in terms of their inner logic and their economic effects. Morgenstern therefore devoted much of the book to exploring the possibility of developing consistent systems of economic policy. While the traditional foundation of such policy had been the pursuit of an "optimum of social welfare," Morgenstern argued that such an optimum would be impossible to identify in an ever-changing social and economic situation. "Even if, in spite of all the difficulties, a theoretical scale of values could be drawn up," he suggested, "empirically-factually there is no unique determinacy of the 'optimum of social value.'" The most that the economist could do without imposing his own definition of social value was to ensure that economic policy was logically consistent, even if many logically consistent systems of policy existed.[110] Or consider a 1934 article in *Zeitschrift für Nationalöko-*

nomie, titled "Perfect Foresight and Economic Equilibrium," in which Morgenstern praised *Moral, Wille und Weltgestaltung* as a plausible step toward developing the mathematics necessary for this task.[111] Again, almost word for word, he reprised the standoff between Holmes and Moriarty and the "infinite chain of mutually supposed reactions and counterreactions" that had marked Morgenstern's earlier critical assessment of the possibilities for economic prediction. But this time Morgenstern suggested a distinctly Mengerian possibility for escaping from this situation. "This chain can never be broken off by an act of knowledge, but always only *by an arbitrary act, by a decision*. Unrestricted foresight and economic equilibrium are thus incompatible." However, he was quick to add, equilibrium might emerge if "defective, dissimilar, arbitrarily distributed foresight" was scattered about.[112] To show this required developing a new mathematics by which the interdependencies and differences in perspective between individuals could be captured in a realistic manner. Such a mathematics did not yet exist in the economic literature, but Menger's *Moral, Wille, und Weltgestaltung,* though "limited to the case of independent individuals," would "hopefully be recognized in time by economists and sociologists for its fundamental significance."[113]

Thus, in his articles from the mid-1930s, Morgenstern took on the role of prophet for and instigator of a "Mengerian" turn in the social sciences: one that privileged logic as the economist's tool for identifying coherent systems of policy or economic systems, but that left open the possibility of multiple social orders separated only by "an arbitrary act," "decision," revolution, or social upheaval. Addressing the economists, he would call for an axiomatization of economic theory in his 1936 "Logistik und Sozialwissenschaften," introducing readers to the basics of set theory and mathematical logic (including reference to Gödel's recent "incompleteness theorem," which had so undermined the Hilbertian dream of a complete and consistent axiomatization of mathematics).[114] All the while, his involvement with the mathematicians deepened. After publication of "Perfect Foresight and Economic Equilibrium," Moritz Schlick invited Morgenstern to present the article at a meeting of the Vienna Circle. He also presented his work at Menger's mathematical colloquium. According to Morgenstern, "after the meeting broke up, a mathematician named Eduard Čech came up to me and said that the questions I had raised were identical with those dealt with by John von Neumann in a paper on the Theory of Games published in 1928." After hearing more about the paper, Morgenstern was eager to meet von Neumann, but in spite of efforts by Menger to find a common occasion for them to meet, this would not happen until 1939, after Morgenstern had moved to Princeton.[115]

Between Mathematics and Economics: Utility

Von Neumann and Morgenstern at last got acquainted in Princeton in February of 1939. Morgenstern then gave "an after luncheon talk . . . on business cycles at the Nassau Club"; in the audience were von Neumann, Oswald Veblen, and the physicist Niels Bohr, among others.[116] The setting reflected not only the genteel world of private dining societies ringing America's Ivy League campuses, but also the great "intellectual migration" from Europe to the United States that had been underway throughout the interwar period, picking up speed amid the deteriorating political situation of the later 1930s. Von Neumann had been a relatively early arrival, and Bohr would remain in Denmark until his dramatic wartime escape from the Nazi occupation in 1943 to join a nuclear research program in England; Weyl had left Göttingen for Princeton in 1933 as the Nazis came to power in Germany, and Morgenstern patched together funding from the Rockefeller Foundation and a temporary Princeton lectureship and moved to the States, even before he was formally dismissed from the University of Vienna and his Institute by the Nazis in 1938. Similar stories accompanied many of the other individuals already appearing in this history. Menger also left, like many of his fellow members of the Vienna Circle. He took up a teaching appointment at Notre Dame in 1937, finally resigning his appointment at Vienna in March of 1938 shortly after the *Anschluss*. Other members of the mathematical colloquium also left at this time, most notably Gödel, who took a position at the Institute for Advanced Study in Princeton, and Wald, who ultimately ended up at Columbia. Jacob Marschak, who had given the talk on equations of economic equilibrium that so agitated the precocious von Neumann, left Germany for England not long after the Nazi takeover in 1933. In 1938, sensing the imminence of war in Europe, he accepted a Rockefeller Foundation traveling fellowship and left for the United States with his family. Initially, he found a post in the graduate program of the New School for Social Research, an organization that had been set up largely to accommodate the flood of eminent scholars fleeing Europe.[117]

These recent arrivals were absorbed into very particular locations in the late 1930s' academic ecosystem. Among other things, their areas of expertise—mathematical economics and statistics—were relatively poorly institutionalized. As one student of Marschak's, the future Nobel laureate in economics Kenneth Arrow, would reflect on the environment less than a decade later, "A small, but increasing, band of young economists was finding mathematics and formal thinking useful. Statistical theory, moreover, was just beginning to attract significant numbers of students. Still, neither of these groups was in great demand . . . mathematical economics was little regarded; and theoretical statistics had yet to find a suitable home."[118] This situation highlights the significance of one key institution, the Cowles Commission for Research in Economics, associated with the University of Chicago, that was devoted to the development of economic theory and the creation of new statistical techniques for econometric measurement. Founded in 1932 by Alfred Cowles, a wealthy investment advisor, the Cowles Commission and associated Econometric Society (with its flagship journal *Econometrica*) lay at the heart of mathematical economics in the United States at this time, with its conferences and the meetings of the Econometric Society attracting theoretical economists from all over the world. Thus Wald attended the fourth annual summer conference in 1937, traveling to Colorado Springs from his temporary post as an economic researcher at the Geneva Research Center in Switzerland. The next year, in 1938–39, he would serve as a research fellow at Cowles, although he soon left for a fellowship at Columbia. Marschak also came from England to attend the 1937 conference, and would later became Cowles's research director in 1942.[119] Especially under Marschak's leadership, Cowles's integration of mathematical theory and economic measurement stood out in the American social sciences, just as its staff and students—many of them Jewish, European in extraction, and theoretical in orientation—were typically outsiders in American academia more generally. Moreover, the firmest institutional bridgeheads of the Cowles brand of economic theory and measurement were in the small number of mathematics departments where statisticians had managed to find appointments. During the late 1930s there were relatively few of these, with Harold Hotelling at Columbia, Jerzy Neyman at Berkeley, and Samuel S. Wilks at Princeton being the most notable.

The collaboration between von Neumann and Morgenstern—which emerged as their initial meeting led to exchanges of articles and drafts, joint writing sessions, and ultimately a book proposal—reflected in microcosm some of the dilemmas facing this broader community of transplanted academics. Working between the mathematics and economics of their day, their collaboration always represented to some extent a negotiation between the demands of their disciplines. Much of *Theory of Games*, even up to the end of chapter 10, had largely followed the rough blueprint for the mathematical investigation of parlor games laid out in von Neumann's 1928 article. This was true even if the von Neumann–Morgenstern collaboration was evidently responsible for a number of innovations in exposition, such as the new proof of the minimax theorem based on considerations of convexity, which (as with the many other figures included in *Theory of Games*) made the result more "pictorial and simple to grasp."[120] On the other hand, the extension of the theory to games with many players and the interpretation of solutions as "norms of behavior" had been a major step toward Morgenstern's dream of capturing the complex interdependencies of economic actors in a new mathematical idiom. But it was two other closely related innovations, found in the closing chapters of *Theory of Games*, that most directly brought von Neumann's mathematics into contact with existing economic theory: the dropping of the zero-sum condition on games and the development of the mathematical theory of utility, the economically meaningful unit for measuring the winnings and losses experienced by the game players. The beginning of chapter 11 thus marked the transition between "zero-sum and non-zero-sum games" that corresponded roughly to the "distinction between purely social and social-economic questions," a move from parlor games to economics, and away from the zero-sum game's emphasis on "the problem of apportionment to the detriment of problems of 'productivity' proper."[121] Subsequently, in chapter 12, the authors would begin to reconsider their assumption—so essential to the existing theory—that players had access to a transferable and divisible medium of exchange for facilitating side payments between coalition members.

Von Neumann and Morgenstern developed the theory of the non-zero-sum game by means of "a mathematical device" that managed to salvage the existing machinery of the multiplayer game as they had developed it in the preceding pages. Their trick was, of course, to replace the general n-player non-zero-sum game with a $n + 1$ player game by adding a final "fictitious player" who would "make the sum of the amounts obtained by the players equal to zero." Because the player was purely "a formal device for a formal purpose," it was "therefore absolutely essential that he should have no influence whatever on the course of the game."[122] Such a player clearly could not make any "moves," but he could certainly be *paid* compensations through other players' "noncooperation with others," so that "the fictitious player gets into the game in spite of his inability to influence its course directly by moves of his own."[123] Having the "fictitious player" receive payments (rather than be exploited to the hilt) amounted to "a self-denial in exploiting a possible collective advantage" that could not be excluded *a priori*, "and we have had several instances showing that a stable standard of behavior can require such conduct." Indeed, von Neumann and Morgenstern suggested that "the reader who is familiar with the existing sociological literature" would know that the debate over whether members of a group will achieve "the maximum collective benefit" "is far from concluded." Nevertheless, they would go on to argue that their existing scheme suggested that the fictitious player's winnings "would be restricted to its minimum value," or, as they added in a footnote, "the self-denial in question does not occur and the maximum social benefit is always obtained." Such a result was striking, but "not as sweeping as it may seem, since we are assuming a numerical and unrestrictedly transferable utility, as well as complete information."[124] Nevertheless, from this perspective, von Neumann and Morgenstern could go on to suggest some game-theoretic models for a two-person market, and a three-person market involving a seller and two buyers. As these models drew on the previously discussed solutions to the three- and four-player game, the "solutions" did not provide determinate prices at which transactions would be settled but rather intervals in which the prices might be found.[125]

However, such assumptions about "utility" proved to be something of a barrier to bridging mathematics and economics both during the von Neumann–Morgenstern collaboration and during the theory's ensuing reception. Prior to meeting Morgenstern, von Neumann had been relatively informal about what exactly his players were receiving when they played games, although his 1928 paper did tacitly assume the existence of some divisible and transferable unit of value in order to perform expected value calculations and implement side-payments. The issue of what this unit was apparently arose early in the collaboration, as Morgenstern recalled:

> We were in need of a number for the pay-off matrices. We had the choice of merely putting in a number, calling it money, and making money equal for

both participants and unrestrictedly transferable. I was not very happy about this, knowing the importance of the utility concept, and I insisted that we do more. At first, we were intending merely to postulate a numerical utility, but then I said that, as I knew my fellow economists, they would find this impossible to accept and old-fashioned, in view of the predominance of indifference curve analysis, which neither of us liked.[126]

Morgenstern was keenly aware of the state of utility theory in economics, being just then in the process of penning a scathing review of John Hicks's *Value and Capital* (1939). "The student of the literature of marginal-utility theory," Morgenstern wrote, "must be fully aware that there has been unanimity for probably thirty years that the assumption of measurability does not conform to the facts, is not needed in the subsequent exposition, and, indeed, has not only not been used but has definitely been eliminated from the theory of value."[127] Yet Hicks apparently continued to feel that the "immeasurability" of utility remained a stumbling block to developing a theory of "consumer's choice," and promoted precisely his own "indifference curve" analysis that proceeded instead from assumptions about the rates at which consumers are willing to substitute one bundle of goods for another. This approach, too, Morgenstern found not entirely original when compared to the traditional assumption of "diminishing marginal utility," stating that it "adds not one iota to the theory or the facts."[128]

In the end, however, von Neumann and Morgenstern would defy conventional economic wisdom to develop a theory of measureable utility grounded in a set of assumptions about individual preferences. As with von Neumann's solution of the two-person zero-sum game, which depended on broadening the space of available strategies by the inclusion of chance, von Neumann and Morgenstern solved the problem of utility by permitting game players to have preferences between "events" or "anticipations," not simply over "goods," "products," or any of the other traditional objects of economic analysis. Events could be perfectly determinate ("receive $5") or probabilistic, like lottery tickets ("get $5 if a coin toss is heads, $2 if it is tails"). Adding the dimension of risk suggested a way to measure strength of preference. Assuming a person were indifferent between an event B and an event that gave him an even chance at events A and C, this would suggest that the difference in utility between A and B was identical to the difference in utility between B and C, thereby permitting the construction of a numerical utility function from knowledge of a player's preferences between events. This function had the useful property that the expected value of the utilities for a set of events was equal to the utility of the expected values of the events, thereby preserving large chunks of von Neumann's original theory of games.

For von Neumann, this result was apparently a simple matter. Upon writing down the axioms for their utility system, Morgenstern remembers him pushing back his chair and exclaiming, "Ja hat denn das niemand gesehen?"—"But didn't anyone see that?" Doubtless the axioms were fairly elementary from a mathematical perspective: some reasonable assumptions about the properties of preference ordering (e.g., transitivity) and the continuity of the derived utility function sufficed to demonstrate the existence of a function mapping "events" to numerical utilities that was unique up to a linear transformation. So preliminary was the result that he and Morgenstern did not publish fully rigorous treatment of utility in the 1944 edition of *Theory of Games*, only including it as an appendix to the second edition of 1947 after an outpouring of interest from economists.[129] Instead, their original presentation—a mere sixteen pages in chapter 1 of *Theory of Games*—stressed the plausibility of their results from the perspective of the history and philosophy of science. "It is clear that every measurement—or rather every claim of measurability—must ultimately be based on some immediate sensation," they wrote. "In the case of utility the immediate sensation of preference . . . provides this basis."[130] Given this fact, the situation in economics "is strongly reminiscent of the conditions existent at the beginning of the theory of heat: that too was based on the intuitively clear concept of one body feeling warmer than another, yet there was no immediate way to express significantly how much, or how many times, or in what sense." Thus, "even if utilities look very unnumerical today," "the historical development of the theory of heat indicates that one must be extremely careful in making negative assertions about any concept with the claim to finality."[131]

Simple though the mathematics may have appeared to von Neumann, the Neumann–Morgenstern theory of utility ultimately received a share of attention from the economists who reviewed *Theory of Games* vastly out of proportion to the space devoted to it in the book itself. On the one hand, utility cemented the relationship between the conception of the individual player prevalent in the new theory of games, and the traditional conception of economic rationality as individual maximization of expected utility. Jacob Marschak, in an early and laudatory review, actually went so far as to equate the maximization of "expected profits or utilities" by individuals and firms with "*the* principle of rational behavior" in economic theory, and praised von Neumann and Morgenstern's aim of deriving economic theory from assumptions of individual utility maximization, even if they had not quite yet achieved it.[132]

Harvard economist Carl Kaysen, in his more skeptical response, "A Revolution in Economic Theory?," likewise credited von Neumann and Morgenstern with attempting to bridge the two major problems of economic theory: "maximization problems" involving the utility of lone individuals or firms, and "market problems" involving the prices that arise from the interaction of many individuals. Yet Kaysen found the Neumann–Morgenstern theory of utility the Achilles' heel of the entire work. Among other things, the utility postulates "involve an assumption about economic behavior which is contrary to experience: that there is no specific utility or disutility attached to risk itself." He therefore suggested that, while measurable and transferable utility is conceivable for firms and entrepreneurs whose focus is single-mindedly on money, it broke down for households and individuals, who instead "seek to maximise a personal, non-measurable utility."[133] Morgenstern would later admit that some of Kaysen's criticisms were technically correct, if irrelevant in his opinion to the general functioning of the theory.[134]

By the early 1950s a vigorous debate about the possibility (and desirability) of restoring measurable utility to economic theory had emerged, drawing in economists from Morgenstern's Princeton colleague William Baumol to Milton Friedman.[135] As Kaysen's student, the young Daniel Ellsberg, would note in his critical examination of utility theory in 1953, von Neumann and Morgenstern's book led to a new split along old lines between those who supported a "cardinal" theory of utility and those who felt that ordinal differences only were important, a divide made more bitter by the implications of a "transferable utility" for utilitarian wealth redistribution programs.[136] In sum, whether embraced or panned for its descriptive or predictive inadequacy, the theory of utility would constitute a key point of entry for economists engaging with *Theory of Games.*

Intriguingly, the closing chapter of von Neumann and Morgenstern's book also focused critical attention on their utility concept. At this point, nearly 600 pages into their text, the authors appeared to express a certain frustration with the plodding pace at which they were moving toward developing a unified, general theory of games. At every step, despite the significant differences in the mathematical structures of the extensive, normal, and characteristic function forms of games, the theory's central concepts were especially those of "imputation, domination, and solution." Yet as they moved from zero-sum to constant-sum games, and again from constant- to variable-sum games, they had needed to generalize these concepts in ways that "represented a real conceptual widening of the theory and not a mere technical convenience."[137] Similar generalizations might also be expected of utility, which had previously been treated in "a rather narrow and dogmatic way," given that "our theory of games divides clearly into two distinct phases": the solution of the two-person zero-sum game in normal form to obtain the "value" for such games, and solution of the zero-sum n-person game as described by the characteristic function. The latter phase—in particular, the construction of characteristic functions—was built on the former, and each phase "makes use of specific properties of the utility concept": measurability (for computing expected utilities) and transferability (for exploring side-payments).[138] This division of the existing theory into "two stages" may have permitted von Neumann and Morgenstern initially to "divide the difficulties in order to overcome them," but now these differences were emerging as a "weakness," a stumbling block on the road to a totally general theory of all games. Perhaps, von Neumann and Morgenstern tentatively suggested, "a unified treatment for the entire theory of the n-person game" would become possible by relaxing some of the assumptions made of utility, and thereby developing "a more adequate mathematical approach to the phenomenon of bargaining." But this remained a project for the future.[139]

Conclusion: The Disunity of *Theory of Games*

In the end, even in von Neumann and Morgenstern's eyes, a complete and general "theory of games" remained elusive. The book's final chapter, filled with pregnant hints about bargaining and modifications of the utility concept, was peppered with references to things that they might discuss in time to come, indications of some possible future grand unification. But the book as it stood had the feel of a collection of special cases and incompletely welded-together mathematical structures. The extensive form, with its roots in von Neumann's early work on logic and set theory, developed a subtle treatment of "information" and the role of knowledge in games and economic behavior, but remained relatively cut off from the balance of the work and largely a preliminary to the study of the game in normal form. The theory of the two-person zero-sum game in normal form, meanwhile, developed a set of connections linking the emerging mathematical subdiscipline of topology, the study of certain systems of inequalities, and highly simplified poker games, while the characteristic function form of the game drew together the study of measure and set functions and the general mathematics of "domination" and ordering of imputations. Morgenstern's economics only further added to the medley of themes and terminology running through the book: the Holmes-Moriarty story and the way the two-person zero-sum game helped address the paradoxes of time and anticipation that he detected in existing economic theory; the "imputation problem" and the underdetermined nature of the social order; the insistence on addressing head-on problems connected with the measurability and transferability of utility.

Hence while the response to the book was generally positive, even spilling over into laudatory coverage in the *New York Times* and a series of popular articles by *Fortune* writer John McDonald later in the decade, the praise was often more for the overall spirit of the work than for the specific contribution that it made to any particular discipline.[140] Von Neumann's towering reputation in his field and his deployment of concepts from set theory, measure theory, or topology might stir some interest among mathematicians, even if the specific results he invoked were starting to recede somewhat behind the cutting edge of "pure" mathematical research. Statisticians did express some enthusiasm for the work: "games" had after all been the probabilists' muse since the emergence of a mathematical theory of probability in the seventeenth and eighteenth centuries, and "in both economics and statistics the problem of rational behavior is a fundamental one," as Leonid Hurwicz would assert. And by the mid-1940s, having contributed to the American war effort as a statistician, Abraham Wald had begun to connect the theory of the two-person zero-sum game with statistics, by interpreting decision-making on the basis of statistical evidence as something like a "game against nature."[141] Likewise, among economists, the theory of utility and the two-person zero-sum game's characterization of the rational economic actor as an expected utility maximizer attracted some interest by virtue of its familiarity. And some reviewers in the economics literature were indeed fervent in their expressions of admiration for the book's technical virtuosity.[142] Yet *Theory of Games* was really only a good start toward von Neumann and Morgenstern's aim of revolutionizing economic theory, as even the most friendly reviewers of the book in the economics community would admit: it could not arrive at unique "solutions" (and thus testable predictions) to even relatively simple problems of monopoly or duopoly, much less market games involving more than three participants. This was of course the aim of Walrasian or Marshallian macroeconomics, which assumed the negligibility of individual influence ("perfect competition") to arrive at a unique economic equilibrium.[143] Kaysen likewise criticized this point, stating that "a complete theory would specify in detail these forces [determining the final solution] and their weights, thus leading to a single price (or perhaps a probability distribution of prices) as the solution."[144]

For reviewers less committed to the Walrasian or Marshallian brand of theorizing in the social sciences, the long-run significance of *Theory of Games* seemed still more doubtful. The theory's presentation, which had become somewhat more graphical, intuitive, and pedagogical in the course of von Neumann's collaboration with Morgenstern, would still put off many less mathematically trained readers.[145] And, as one reader suggested, von Neumann and Morgenstern "limit their science of economics to that body of economic doctrine which is amenable to their theory of games," thereby bypassing many of the most interesting problems of economic inquiry: the problems of production, the economics of noncapitalist societies, "the instability of economic systems," and "crises and unemployment."[146]

Theory of Games in 1944 thus was less a secure achievement than a promissory note. Its survival depended on acceptance of its broader agenda and style of analysis, if not particular results—"the promises for the morrow," as one reviewer put it.[147] One could perhaps go even further, to argue that *Theory of Games*'s appeal lay precisely *not* in providing a complete system or definitive piece of work, but in its heterogeneous collection of notations, metaphors, terminology, and results that could be appropriated, reinterpreted, and put to work in a variety of contexts. As Marschak pointed out in the *Journal of Political Economy*, "the main achievement of the book lies, more than in its concrete results, in its having introduced into economics the tools of modern logic and in using them with an astounding power of generalization."[148] Herbert Simon, a rising star in political science whose research would in time span the breadth of the social and behavioral sciences, likewise noted that the promise of the theory lay in the future: in its ability to provide "a fundamental tool of analysis for the social sciences" by suggesting terminology that would in time serve as an alternative to sociologists' parsing of "ends" and "means" in social action, or by providing a finer understanding of terms like "cooperation" and "competition" that "have become such important categories of sociological, political, and economic theory."[149]

Given this situation, it also makes sense that game theory would appeal most to the younger generation in many disciplines: young men returning from the war as it was ending and entering the academy, economists and mathematicians coming of age in a world where older disciplinary imperatives were also being transformed by new inflows of ideas and patronage. Thus Martin Shubik, who would emerge as one of the first wave of postwar game theory practitioners, initially encountered *Theory of Games* in the library at the University of Toronto as an undergraduate sometime in 1948. Reading through the first pages, he understood almost none of it. But he was convinced that somewhere, buried deep in the pages of murky symbols and equations, lay the future theories of "economics, political science, and sociology."[150]

CHAPTER THREE

From "Military Worth" to Mathematical Programming

G ame theory's postwar career was launched by a book, but also by a bomber: the B-29 bomber, to be precise. This was the plane that would form the core strength of the 20th US Air Force, commanded from the Mariana Islands during the last year of World War II by cigar-chomping future Strategic Air Command chief Curtis LeMay. Under LeMay's command, B-29s spearheaded the firebombing of Tokyo in the spring of 1945; they also dropped the first atomic bombs on Hiroshima and Nagasaki that August. Beyond its association with some of the key actions of the war, the B-29 represented one of American air power's great technological breakthroughs. Developed in great secrecy and at great cost, the plane with its extraordinary capabilities proved the perfect vehicle for demonstrating the potential of strategic bombing—bombing that bypassed the front lines in favor of direct attacks on an enemy's industrial base and civilan populations—on which America's top air commander, General Henry Harley "Hap" Arnold, had staked much of his reputation. It was also a harbinger of things to come as world war gave way to Cold War, and as the combination of strategic air power and nuclear weapons emerged as a cornerstone of American national security policy.

However, the B-29 was from the start plagued with problems related to production, maintenance, training, and use. Early trials of the plane were nothing short of disastrous, with one test plane in early 1943 catching fire in the air and crashing, killing Boeing's flight crew and a number of people on the ground. More broadly, the plane was pushing the boundaries of what was technologically and humanly possible, flying at high speeds and altitudes at ranges of up to 3,700 miles, and employing state-of-the art remote-controlled guns and bombsights that required extraordinary efforts on the part of its crews. General Haywood "Possum" Hansell, who served as the first commander of the B-29 force in the Mariana Islands, described the situation thus: "Our people had no gunnery experience, and the gunnery equipment for the B-29 was extremely sophisticated. It was so complex that the average individual simply couldn't operate it. . . . [The] technology really did outrun itself."[1] Despite these challenges, President Roosevelt had promised Chiang Kai-Shek that he would supply the planes to bases in China by April of 1944; with so little time to spare between production and use in the Pacific Theater, an extended period of trial-and-error learning was impossible. Against this backdrop, in the spring of 1944, the Army Air Forces initiated a project, coordinated by the Princeton-based mathematician Merrill Flood, to muster scientific and engineering expertise in determining "the most effective formations and flight procedures for the B-29 airplane."[2] By the fall of that year, reference to von Neumann and Morgenstern's book would appear in the Project's internal memoranda—a portent of things to come, much like the B-29 itself.

It is difficult to imagine that game theory would have survived and thrived as it did in the postwar era without the surge in national security funding that was a legacy of World War II and of the subsequent Cold War, as Philip Mirowski has rightly emphasized.[3] Yet the appeal of von Neumann and Morgenstern's "theory of games" in this context was ambiguous to say the least. Did it provide a set of rules for how one ought to behave in a range of situations, like a practical "system" for winning at poker or bridge? If so, for whom was this "system" intended? Alternatively, was it a positive theory of human social interaction, as Morgenstern's economic gloss on the book maintained—or at least, a somewhat skeletal and incomplete first cut? Game theory could indeed be used to shed some light on how to "solve" some highly stylized and simplified problems that were potentially of interest to military patrons: how best to defend a bomber against an enemy fighter plane, or how to hunt for a submarine that is itself intent on evading detection. But it is not obvious why generals and admirals might reach for game theory's arcane mathematical axioms over the innumerable regulations, equipment manuals, and the like that they promulgated to govern the practical complexity of day-to-day military operations. Perhaps instead, as Robert Leonard has provocatively suggested, game theory—like war gaming and the rigorous methodologies of systems analysis—spoke to the psychological needs of the military's scientists. Their ritualistic activities of theorizing and experimenting were, to some extent, "collective meditations, so to speak, at a time of anxiety and strain."[4]

This chapter builds on these extant accounts by seeking a sense of the subtler interplay between the aspirations of the individuals who were quickest to embrace von Neumann and Morgenstern's work, and the rapidly evolving institutional environment linking the military and academic experts between World War II and the Cold War that lay beyond. The standard-bearers of game theory in the postwar period were mostly trained as mathematicians and statisticians, although some had backgrounds in fields like economics. And while the precise nature of their work varied depending on context, it was often associated with the rubric of "operations research," originating in a set of activities aimed at assessing the operational efficacy of new tactics and technologies on the battlefield. This group's embrace of game theory makes the most sense in the context of several interlinked developments unfolding in the waning years of World War II and the later 1940s: the managerial problems facing operations researchers as they were called upon to assess the effectiveness of increasingly complex technologies; the shift from hot war to Cold War, from the real-time analysis of bountiful combat-generated data, to the anticipation of (and budgeting for) wars that would never actually be fought and technologies that could not be directly tested on the battlefield; and, finally, the struggles of academics involved in the war effort to find a place for their esoteric newfound skills in the postwar disciplinary and institutional order. In this context, more than "solving" any particular problem to provide prescriptions for rational conduct, game theory and related mathematical techniques proved critical in satisfying the professional aspirations of "applied" mathematicians and operations researchers, putting them at the center of a number of previously unrelated activities that were now held together by a common mathematical framework.

At the same time, the understanding of "game theory" that emerged in this context was the result of a highly selective appropriation of von Neumann and Morgenstern's work. It has long been clear that the bulk of the publications on game theory to emerge in the immediate postwar era focused principally on developing the theory of the two-person zero-sum game and on solving related problems of mathematical optimization.[5] This chapter further evaluates this development in terms of the interaction between the interests of the military and those of its mathematicians outlined in the preceding paragraph. Agendas of research into "military worth" or "programming" indeed steered mathematicians toward developing theories of optimization and metrics of military performance in the service of a broad range of activities related to logistics and to technology selection, design, and training. Moreover, the mathematics involved in solving the two-person zero-sum game—topology, the geometry of the simplex, the implications of convexity, fixed-point theorems, and numerical methods of solution—not only lent itself readily to such programs of research, but provided relatively fertile ground for the kind of theorem-proving toward which this particular group of mathematicians were most inclined. By contrast, the fate of another key component of von Neumann and Morgenstern's work—the theory of the non-zero-sum and multiplayer game—demonstrates the limits of game theory's relevance to postwar mathematicians and their military patrons. The various alternative "solution concepts" for non-zero-sum games that emerged during this period out of critiques of the von Neumann–Morgenstern solution to such games plausibly reflects not only the influence of the military context in reorienting the trajectory of game theory's subsequent development, but also, to some extent, game theorists' inclination toward the mathematical machinery of the two-person zero-sum game.

The War, Operations Research, and American Mathematics

In contrast to mainstream economists, who were relatively slow to warm to game theory, many mathematicians appear to have welcomed *Theory of Games*. Certainly, the most visible and technically capable enthusiasts for game theory after the war came from backgrounds in mathematics and statistics. (In the 1930s and 1940s these fields were related through a small but prominent group of mathematical statisticians, but in general they were not identical.) Yet their enthusiasm deserves some qualification, since game theory's foothold in mainstream academic mathematics after the war was in fact fairly precarious. John Nash's early publications on game theory were generally hailed by fellow members of the Princeton mathematical community, yet Nash quickly felt pressure to publish something in an area of pure mathematics, like number theory or abstract algebra, in order to obtain a professorship.[6] In this regard, Nash was not unique: self-identified "game theorists" apparently had difficulty getting funding from the National Science Foundation's mathematics program or landing positions in academic math departments. Thus John Isbell, a promising young mathematician who, like Nash, had studied game theory with Albert Tucker at Princeton in the early 1950s, later remembered his struggles to find a job. Immediately after receiving his PhD in 1954, he had applied for a postdoctoral fellowship to do research in game theory, but was turned down. According to a friend, the NSF committee that considered his application was indignant at the prospect of funding a game theorist, and it was only after Isbell reframed his work as "topology"—a recognized specialty within pure mathematics—that he received support.[7]

Thus it would prove fortuitous that von Neumann and Morgenstern's book appeared just at the end of World War II, a conflict that had a dramatic impact on the American mathematical community on a number of interconnected levels: institutions, professional cadres, and ideas. Traditionally, the *raison d'etre* of advanced training in mathematics was generally to train teachers for university or collegiate setting. The exceptions to this were "industrial" or "applied" mathematics, and mathematical statistics. The former were relatively small and certainly not well organized until after the war.[8] The latter, while expanding rapidly during the 1930s, was still a relatively new field with no stable disciplinary base, hovering as it did on the border of the social sciences, business, biology (especially agricultural science), and mathematics. Nevertheless, during the 1930s, some statisticians received appointments in departments of mathematics.[9] Corresponding to this overall institutional situation, mainstream American mathematics tended to be characterized by its commitment to "purity," its distance from practical problems of any kind.[10] Even "applied mathematics," with the exception of mathematical statistics, typically meant "mathematics applied to theoretical physics," especially the theory of the partial differential equations that were the standard language of the physical sciences in the nineteenth and early twentieth centuries.[11]

Military-related funding during the war changed this situation significantly. From 1940 onward, mathematicians found themselves doing novel kinds of work in a number of new institutional locations. Operations research (OR)—a heterogeneous set of activities connected with assessing the effectiveness of new technologies on the battlefield—appeared in the United States in the early 1940s. As OR was institutionalized in units attached to various military divisions, mathematicians and physicists played central roles in its development.[12] Other mathematicians found work through various divisions of Vannevar Bush's National Defense Research Committee (NDRC)—after 1941 a unit of the Office of Scientific Research and Development (OSRD). However, in 1942, Bush established an Applied Mathematics Panel (AMP) to serve as a central agency for fielding mathematical questions posed by the services and other OSRD units, and to coordinate the work of many of the pre-existing mathematical research groups and avoid duplication of effort.

The AMP in particular helped to cultivate a new generation of practical mathematicians, men and women who were talented managers and salespeople, and who had fewer qualms about violating disciplinary norms. This was entirely intentional: for, as Larry Owens has shown in his examination of the problems facing the AMP's administration, established research mathematicians often proved difficult to manage. Thus Warren Weaver, who assumed leadership of the AMP after directing the Rockefeller Foundation's Division of Natural Sciences during the 1930s, complained early in the war of the "dreamy moonchildren, the prima donnas, the a-social geniuses" that he felt were all too common among the academic elite.[13] The generation that filled the ranks of the operations research units or staffed the various AMP projects was different: its members were younger; they were less likely to have been employed in their field during the depression years; and many of them had been relatively marginal—for a wide variety of reasons—within the universities. Working with the military was, by contrast, a much more rewarding option. Mina Rees, who earlier had had difficulty advancing as an academic on account of her gender, served as a technical assistant to the AMP (and to Warren Weaver in particular) from 1943 until the official end of the panel's work in 1946, in the process winning the President's Certificate of Merit and the (British) King's Medal for Service in the Cause of Freedom for her work. After the war, as the head of the mathematics division of the Office of Naval Research, Rees generally remained the only woman among the power brokers of science funding; years later, thinking of one meeting with top Navy brass, she recalled that "the only discrimination she felt was from a civilian."[14]

Likewise, Merrill Flood, an AMP consultant for much of the war, had had trouble finding a position as a mathematician after earning his PhD from Princeton during the lean years of the 1930s. Having started a family at an early age, he applied his considerable ingenuity to finding a way to make a living in mathematics outside academia, serving as a statistical consultant to the governor of West Virginia, founding a mathematical consulting company, and proposing the creation of a "mathematical clinic" at Princeton. Princeton was not nearly as enthusiastic about the endeavor as he was, although this kind of intellectual entrepreneurship would make him an omnipresent administrator and researcher working for various defense agencies after the war.[15] John D. Williams, another AMP technical assistant, had never even completed his PhD at Princeton "due to some personal reason." However, his facility with mathematical statistics and his "good feel for the practical aspects" of problems and made him an ideal wartime mathematician/ administrator.[16] After the war, he would emerge as the RAND Corporation's top mathematician, in which capacity he would oversee a large percentage of game theory research. Other relatively young mathematicians who flocked to leadership positions in the AMP were Edwin S. Paxson, B. O. Koopman, and John W. Tukey, all of whom would be deeply involved with game theory's cultivation in the postwar world. Von Neumann was also a consultant for the AMP, although his specialty was typically seen as the solution of complex differential equations.

Just as the AMP was staffed by a group of nontraditional mathematicians, it also embraced an intellectual hierarchy substantially different from that of academic mathematics. By the end of the war, AMP leadership could identify four major areas of research in which the panel was involved. Naturally, the first area listed involved "mathematical studies based upon certain classical fields of applied mathematics," by which they meant "classical dynamics and the dynamics of rigid bodies, the theory of elasticity and plasticity, fluid dynamics, electrodynamics, and thermodynamics." However, "probability and statistical studies" were now given a heading of their own for their insights into "the effectiveness of bombing; various aspects of naval warfare, including fire effect analysis and the performance of torpedoes; the design of experiments; sampling inspection; and analyses of many types of data collected by the Armed Services." Computational services—the production of numerical tables, the supervision of human computers, and the de-

velopment of fully automatic computing equipment—were another pronounced area of growth, even if a number of programs in these areas
had existed before the war.[17] But one area was almost completely new,
reflecting the influence of practical problems of military operations on
the mathematicians' work: "Analytical studies in aerial warfare, including assessment of the performance of sights and antiaircraft fire control
equipment; studies relating to the vulnerability of aircraft to plane-to-
plane and to anti-aircraft fire and the optimal defense of the airplane
against these; and analyses of problems arising from the use of rockets
in air warfare."[18]

The distance between the AMP's areas of research and the emphasis of prewar academic mathematics thus appeared to leave little hope
for the operations researchers and the AMP staff in the context of postwar military reconversion. As the war ended, the AMP and similar units
faced the prospect of being permanently discontinued, and their leaders
turned to the task of winding down their contracts and writing their final
reports. As a result, the latter exercise looked to the future as well to the
past: in synthesizing their experience of the preceding years, they were
also trying to assure a place for the mathematical innovations of the war
years in the postwar environment, possibly in academia, but more likely
in some new hybrid program of military-sponsored research. Against
this backdrop, several powerful wartime administrators looked to game
theory to serve as an example of the promise that mathematical investigation held for the military. Game theory appeared in two postwar publications associated with the NDRC: a revised and declassified version
of the summary technical report of NDRC Division 6 (undersea warfare), written by physicists Philip Morse and George Kimball, "Methods
of Operations Research"; and one chapter (with no indicated author) in
a volume from the summary technical report of the AMP, titled "Analytical Studies in Aerial Warfare." The volume by Morse and Kimball
would emerge as a highly influential textbook of operations research—a
significant contribution to the professionalization of OR via the establishment of a standard pedagogy focused on the mastery of arcane mathematical techniques.[19] Yet although the chapter from the AMP report
cut a lower public profile, it was probably more immediately influential
on account of its powerful (though unacknowledged) author. That author was Warren Weaver—the AMP chief himself—and his chapter laid
out a vision for the future of game theory that would prove prophetic.

The Problem of "Military Worth"

The ideas in Weaver's chapter had roots in the challenges that the AMP mathematicians had begun to encounter almost a year and a half earlier, during that fateful spring of 1944. With the ever-buggy B-29 bomber in the process of being rushed from test flights into actual combat missions in the Pacific, in late May the Army Air Forces (AAF) Board convened a conference in Orlando, Florida, that called attention to a number of aspect of the new bomber's performance that needed immediate study, especially "the preparation of a single manual on the use and maintenance of the fire control equipment of the B-29 airplane," "the specifications of the necessary gunnery training," and more broadly, "the consideration of the tactical use of the B-29 airplane." With respect to this last point, rather than simply analyzing data on the entire plane's performance after action in the field and systematically investigating the impact of various changes in technology and tactics (an impossibility given the tight deadlines involved), the AAF had already begun a collaboration between a Dr. E. J. Workman, of the University of New Mexico's Physics Department, and the 2nd Air Force to study a number of problems related to the use of the plane's guns in simulated combat.[20] In the wake of the conference, the AAF proposed a sweeping contract with the NDRC to "organize and outline scientifically controlled and evaluated investigation of the most effective formations to provide maximum utilization of the B-29 airplane's equipment" and more generally to "determine the most advantageous manner in which B-29 aircraft can be flown."[21] This contract was ultimately handed to the Applied Mathematics Panel and given number AC-92.[22]

The bulletins of project AC-92's steering committee—which met for the first time in July 1944—describe a project of extraordinary scope and complexity. This scope was reflected, among other things, in the composition of the steering committee: its core members were from the AMP, including Warren Weaver (who served as chair) and Merrill Flood (the committee's executive secretary, based in the AMP's Princeton office), but it also included the president of the AAF board, a representative from the fire control division of the NDRC, various military liaisons, and a representative from the OR unit attached to the 20th Air Force which would be the prime recipient of the new B-29.[23] As the project went along, it also rapidly drew in technical personnel from a number of AMP divisions and research programs scattered throughout the country. Full-scale experimentation on the problem of optimal bomber formations continued under Workman's direction at the University of New Mexico; another group based at the Mt. Wilson Observatory worked with a reduced-scale model to study formation fire power; proving ground experiments on the guns were run at Eglin Field; and a major parallel study, conducted by the Applied Psychology Panel, sought to evaluate (and possibly improve) the airplane's fire control and communication systems via experimentation on human subjects, and to develop training and crew selection techniques.[24] In addition, AMP groups at Brown and Columbia worked on analyzing data from both Europe and Japan that could give insight into the risk to various bomber formations from flak, or the effectiveness of airborne rockets. A number of AMP staff (most notably, Samuel S. Wilks and John D. Williams) would also be assigned to work with experimental efforts, such as Workman's, to assist with the planning of experiments, data collection, and data analysis. Finally, at the center of all this activity was Merrill Flood and his staff at Princeton, who were tasked with extracting regular progress reports from the various research groups and ensuring a steady circulation of information within the project.[25]

AC-92 thus was a qualitatively new kind of project for the AMP. Most of the hundred-odd studies performed by this organization were done in response to fairly specific requests: produce tables of an obscure mathematical function, solve this set of equations, analyze this set of data sent back by an OR unit working closer to the front lines. Among these, AC-92 stands out due to the extraordinary scope of its mandate—to "determine the most advantageous manner in which B-29 aircraft can be flown," a problem that potentially touched upon every aspect of the aircraft's engineering, staffing, and deployment. It is also remarkable how central the AMP mathematicians were to this process, dominating the project's steering committee, coordinating the schedule and circulating results, and sitting in on almost all of the key discussions about how the study would be designed. Moreover, given the ambiguities inherent in their mandate, it is small wonder that the project rapidly became an administrative headache for Weaver and Flood, who stood at the center of this sprawling, heterogeneous enterprise, nor is it surprising that members of the scattered scientific working groups found cause to resist the encroachments of the mathematicians at the core of the whole process. While it is difficult to uncover specific technical disagreements in the AMP memoranda, egos were clearly bruised; at one point, a meeting with Workman in New Mexico almost ended in blows as the mathematicians, experimentalists, and military officials struggled for control of the project.[26]

This, then, was the situation facing the executive secretary of AC-92's steering committee, Merrill Flood, in late 1944, when he tried to clarify the project's guiding philosophy in a memorandum issued under the aegis of the AMP's Fire Control Research Office in Princeton.[27] In this memo Flood first invoked the "theory of games"—which he had learned "by attending one of [von Neumann's] pre-war Princeton lectures on game theory"—to solve a highly simplified model of aerial combat involving an offensive bomber and a fighter plane, both flying at the same altitude, with the latter defending "two targets separated so widely that the fighter can be used to defend only one of the targets at any particular time." In line with von Neumann's results on the two-person zero-sum game, Flood identified an optimal strategy that made strategic use of uncertainty, thereby concluding that "the best way to carry out an operation will often involve the use of deliberate randomness."[28] However, Flood's memo made it clear that the use of game theory was merely illustrative of a new kind of analysis that he hoped would strengthen the mathematicians' hand in the ongoing struggles over the B-29 study. As he noted, "The broad purpose of project AC-92 is the determination of optimum tactics for use in the employment of very heavy bombers in operations against Japan," but at the current time "it is not possible either to state this problem much more precisely or to state a program for its solution in anything approaching precise terms." What was needed was a mathematical formulation of the problem that could "provide a broad working framework for discussions, more careful planning, theoretical and experimental work, and decisions on early recommendations made under the project."[29] To this end, Flood proposed the development of a standard measure of "military worth" that would capture in one index the gains and losses associated with a particular operation. This measure would permit a taxonomy of factors determining gains or losses in military worth as "dominant," "subdominant," and "negligible," but also "controllable" and "uncontrollable." Critically, Flood noted that this analysis could rule out the need for further development of military tactics and technologies, especially if "the effect of a subdominant but controllable factor is outweighed by the effect of an uncontrollable dominant factor." Armed with a suitable calculus of military worth, the mathematicians could potentially shoot down research proposals from the scientists and engineers that did not contribute to the completion of the overall project at hand.[30]

This notion of "military worth" would form the centerpiece of Warren Weaver's "Analytical Studies in Aerial Warfare," which appeared over a year later in the AMP's summary technical report. For Weaver, the point of formulating a mathematical framework for the study of aerial warfare was precisely to help manage the complex and multidisciplinary process of equipment testing, design, and training of the kind that had characterized AC-92 (see figure 3.1). Consider, he suggested, the example of a colonel at Wright Field who is trying to decide which of several gunsights he should order to equip his planes. To make this decision, he must take into account an extraordinary array of knowledge: statistical estimates of bombing accuracies, "the logistics of the theatres in which these sights are to be used," the targets under consideration, experimental data on the "psychology and physiology of operation of bombsights," and much more. What was needed was a rigorous way of bringing together knowledge about such performance factors, and exploring the in-

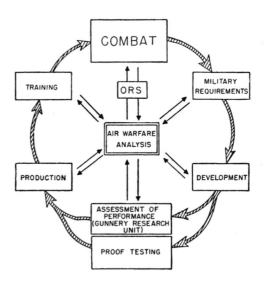

FIGURE 3.1. The centrality of "Air Warfare Analysis" to the process of military technology assessment. From Office of Scientific Research and Development, "Aircraft Fire Control Developments in the AAF," in *Summary Technical Report of the Applied Mathematics Panel NDRC*, vol. 2, *Analytical Studies in Aerial Warfare*, ed. Mina Rees (Washington, DC, 1946), 223.

terconnections between them, and assessing the relative contribution of each of them to a wing's ultimate success on the battlefield. Weaver's abstraction of the situation was as follows: suppose we want to perform an operation O according to plan P that is completely specified by of a number of decisions. These decisions depend on a number of variables, some of which we control (decision variables D), but many of which are "basic variables" V controlled by "nature" or even by an enemy. Now, the "central feature of a general theory of air warfare" can be stated as follows: "the plan P should, in fact, be characterized by the set of decisions D which is *worth most to us*." While Weaver acknowledged that the term "worth" is usually associated with economic contexts, in this case he referred to a concept of "military worth" in which "destruction and harm to the enemy will be counted as the more valuable."[31]

How could one find the best combination of decision variables? Both practical and conceptual problems presented themselves. To actually carry out calculations of military worth, Weaver envisioned "a great *Tactical-Strategic Computer* [TSC]" with input dials for the decision variables, constant registers for the "basic variables," and an output dial showing "military worth" for each combination. Given breakthroughs in digital computing and "multistage servo systems" stimulated by the war, even if such a device sounded a bit like a "Rube Goldberg machine," nevertheless "the present author would like to risk the prophecy that just as World War I saw the birth, and World War II the high development and effective use of fire control predictors which, so to speak, solve the immediate tactical problem for a single gun, so the next war (should there be one) may result in the development and use of general tactical computers for use in the field."[32]

His comments on computers and servomechanisms aside, Weaver felt that the physical construction of such a "Rube Goldberg Machine" would be less challenging than the development of the mathematics that underlay it, in particular, the formulation of a suitable calculus of military worth that could capture the functional relationships between the various decision variables involved in a problem. Here, it was important to develop nothing less than a new kind of mathematics, a mathematics designed to suit the managerial realities of operations research. Game theory was the perfect example:

Military worth, as the phrase is here used, is closely related to the general concept of *utility* in economic theory. And the reader is warmly encouraged to read the discussion of a numerical theory of utility given (on pages 15–29 and elsewhere) in *Theory of Games in Economic Behavior* by John von Neumann and Oskar Morgenstern. This pioneering and brilliant book is, it should be pointed out, connected in a most important way with the viewpoint here being presented, for it develops a large part of the mathematics necessary for theories of competitive processes.[33]

Later in the chapter Weaver again praised what he read as "the powerful analysis of competitive processes carried out by von Neumann and Morgenstern" and suggested also that similar mathematical frameworks were being developed by "quantitative ecologists" such as Vito Volterra, Alfred Lotka, Georges Teissier, and G. F. Gause, who were interested in "competitive biological systems" and related concepts like the ecological "niche." It seemed likely that the cultivation of entirely new types of mathematics, associated with new applications, would be necessary in meeting military needs in the postwar era.[34]

Given the strains that had emerged over the B-29 project, Weaver suspected how controversial were these dreams of military-worth computers that poured forth from his fevered mind. He therefore took pains to circulate the draft among his closest associates from the AMP, among them Flood, Thornton Fry, Koopman, Paxson, Rees, Tukey, Oswald Veblen, W. Allen Wallis, Samuel S. Wilks, and Williams. "You may think that this draft is, in places, too informal in character," he wrote, but its chances of being read by people "in a position to effect the future policies" are better "if a few raisins are put in with the oatmeal."[35] Yet on the whole the respondents liked the raisins and agreed that the controversial points that they represented were important ones. Veblen, writing from his office near von Neumann at Princeton's Institute for Advanced Study, noted the necessity for a general theory of warfare, but suggested that

> what is really lacking is some person of sufficient ability and interest, who has the time and energy to develop the kind of a theory which you adumbrate. Perhaps it might better be worked out in connection with some type of competitive activity other than warfare. There must be other situations in which a theory involving basic variables, decision variables, and ____ worth, would have significant application.[36]

But Paxson, a member of the younger generation of applied mathematicians coming up through the AMP, was particularly enthusiastic for continuing the war-forged relationship with the military, noting that, in the current military planning environment, "the need for a TSC at top planning levels becomes very obvious" and urging the peacetime development of such machines, even if "the human elements" in the military "make one pessimistic about any such development."[37]

Instituting the Study of Military Worth at RAND

Veblen and Paxson's concern about finding an institutional home for the analysis of military worth turned out to be overdone given Warren Weaver's contacts within the military bureaucracy. For example, in corresponding with Pentagon officials interested in the development of research programs on fire-control effectiveness and other problems in operations research, Weaver apparently shared ideas very similar to those that would appear in the AMP final report with Edward Bowles, a technical advisor to the secretary of war.[38] Subsequently Bowles would prove instrumental in creating "Project RAND" by contract with the Santa Monica–based Douglas Aircraft Company. Much like the AMP's guidelines for AC-92, the terms of the original contract with Douglas were exceptionally broad, calling for "study and research on the broad subject of intercontinental warfare, other than surface, with the object of recommending to the Army Air Forces preferred techniques and instrumentalities for the purpose." The initial budget was set at $500,000 for three years, although that was quickly upped to $10 million, a sum that would last until the conclusion of the original contract in the spring of 1950. Thereafter, successive Air Force contracts (and consulting contracts with other agencies such as the Atomic Energy Commission) would provide RAND with an ever-growing river of funds.[39]

Weaver and his understudy, John Williams, were also well acquainted with Frank Collbohm, the newly appointed director of Project RAND, from the course of their time with the AMP. After writing up the final AMP reports, Williams was moving west to take a job as chief statistician at the Naval Ordnance Test Station at Inyokern, California, when he received a telephone call from Collbohm, who told him about the RAND contract and urged him to join the Project. Collbohm said that he greatly admired the AMP's work, and told Williams "that he would like to have RAND permeated by the kind of analyses that the Applied Mathematics Panel had done during the war."[40] Inspired by the open-ended nature of RAND's mandate, Williams agreed to head to Santa Monica. He also dispatched a note to Olaf Helmer, an AMP colleague of his who was just then moving from New York to join Williams at Inyokern. Helmer later remembered that he received Williams's missive when he was about halfway across the country. The message: never mind the Navy, I meant *Air Force*.[41] Helmer, Williams, and Paxson would be among the first members of the project—beside the cadre of engineers from Douglas Aircraft—and in the spring of 1946 they formed the nucleus of a new section of the Project devoted to the "Evaluation of Military Worth" (later shortened to "Military Worth" or "Evaluation"). By 1948 still more mathematicians would join. The list of new arrivals— George Brown, Melvin Dresher, Henri Bohnenblust, T. E. Harris, J. C. C. McKinsey, and Lloyd S. Shapley—would constitute a veritable "who's who" of postwar game theory.

Weaver never joined the RAND staff, but as Martin Collins has shown in his definitive study of RAND's early history, his brand of "military worth" analysis proved a strong intellectual influence on the new project. Among other things, during those first, formative years of the RAND contract, Williams and Weaver were central figures in a network of consultants and staffers—including other old AMP standbys like Wilks, Wallis, and Helmer, as well as statistician Frederick Mosteller— whose advice would play a significant role in developing the project's program of research into "military worth." Such a program would naturally involve further development of the mathematics of "competitive processes" like game theory, but it would also require the recruitment of social scientists willing to contribute their expertise on the economic, political, psychological, and social impacts of changes in military technology.[42] Subsequently, Weaver presided over a RAND-sponsored Conference of Social Scientists, held at the Hotel New Yorker in New York City, from September 14–19, 1947. The invitees were all by now luminaries in their respective fields: economists Ansley Coale, Charles Hitch, and Jacob Viner; psychologists Donald Marquis and Samuel Stouffer, political scientists Franz Neumann and Bernard Brodie, and anthropologist Margaret Mead. Also present were the members of the study group and the fledgling program at RAND, most notably Helmer, Mosteller, Wallis, Weaver, Wilks, and Williams.[43]

Weaver's opening address to the conference concisely articulates the methodological perspective of military worth analysis and thus sheds light on the distinctive modality of game theory's use at RAND. Wartime experience, Weaver argued, had led to the recognition that interdisciplinary research teams were essential to the solution of what he called "problems of organized complexity." A good example was the "convoy problem"—"How do we get stuff across the Atlantic?" Operations research in Britain provided an excellent model of how to address such a problem: throw together a team of researchers from a variety of disciplines—psychology, economics, physics, and beyond—and see what they come up with. Such patterns of collaboration showed great promise for the practical improvement of warfare, not to mention the broader study of complex social and biological problems.[44] While such cooperation between specialists had in the past "resulted only from the pressure and necessity of war," Project RAND sought to continue it in peacetime by providing a free-wheeling institutional environment coupled with the discipline of a "quantitative logical analytical theory" for "analyzing general theories of air warfare." Such theories would further develop Weaver's concept of military worth by developing "quantitative indices for a gadget, a tactic, or a strategy, so that one can compare it with available alternatives and guide decisions by analysis, to the extent that analysis can be shown to be relevant and possible."[45]

But, as Weaver was quick to add, such quantitative theories were unlikely to produce definitive advice on how to act in a particular situation; rather they acted as a kind of logical brake on the analysis of complex problems. The analysis of military worth thus might constitute a first step toward living what Weaver called "the rational life"—not exactly the mindless application of decision rules and quantification to problems of decision-making, but rather the judicious blending of "some experience, some insight, and some analysis of problems, as compared with living in a state of ignorance, superstition, and drifting-into-whatever-may-come."[46] Such analysis might, for example, determine the sensitivity of a decision to its myriad input parameters; the military worth analyst might be able to determine "the sensitivity, so to speak, of the boundaries of knowledge with respect to different items of information." Speaking of the relationship between the mathematical analyst and the disciplinary specialist, Weaver continued: "We can say to you, 'Please don't strain yourself indefinitely to try to find out *this* particular fact because that is not going to move the boundary of ignorance very far.'"[47] Thus Weaver stressed that any general "theory" of air warfare did not represent a technocratic substitute for specialized expertise but rather a way of coordinating and policing an interdisciplinary research effort. In consequence, inasmuch as the RAND mathematicians leaned on game theory to develop the analysis of military worth, game theory's prescriptions were intended to apply less to generals or fighter pilots, and more to the military's scientists themselves.

Apart from introducing the leading lights of American social science to the kind of work that was being done for RAND, the conference helped guarantee a supply of experts to staff the Project and to serve as consultants. And indeed, many of the conference participants would reappear throughout RAND's subsequent history as staff researchers, consultants, or both. Also about this time, as RAND's collection of staff and consultants began a period of major expansion, von Neumann and Morgenstern began to be aware of the scale of the research agenda that their work had inspired. On December 16, 1947, Williams wrote to his old professor von Neumann, proposing the terms of a consultancy. "The members of the Project with problems in your line (i.e. the wide world) could discuss them with you, by mail and in person," he wrote. "We would send you all working papers and reports of RAND which we think would interest you, expecting you to react (with frown, hint, or suggestion) when you had a reaction." In an oft-quoted line, Williams reassured von Neumann that "the only part of your thinking time we'd like to bid for systematically is that which you spend shaving."[48] Von Neumann's response is revealing, suggesting his relative unfamiliarity with the Project even at that late date: "The very sketchy information I have about [your work] makes it sound interesting and attractive," he wrote.[49] While von Neumann would visit Santa Monica in the later 1940s, he soon became more of a distant eminence.[50]

As the research staff affiliated with RAND began to swell after 1947, the organizational structure of the Project began to shift as well. In 1948 RAND was substantially reorganized and expanded, and the initial Air Force contract was transferred from Douglas Aircraft to a new non-profit corporation—the RAND Corporation—that had been specially chartered for the purpose in the summer of that year.[51] By the spring of 1949 the corporation was divided into eight divisions; half were devoted to "hardware" R&D (aircraft, missiles, nuclear energy, and electronics), while four others (economics, systems analysis, social science, and mathematics) represented the legacy of the original "military worth" program and the Conference of Social Scientists. Two conference attendees were installed as chiefs of new divisions: Hitch in economics, and Hans Speier, a sociologist from the New School of Social Research, in social science. Meanwhile, two old AMP hands also became division chiefs, with Paxson in systems analysis and Williams and in a new (and slightly pruned) mathematics division. While the corporation thus increasingly seemed to organize expertise along more traditional academic lines, R. D. Specht, a RAND mathematician, would reminisce in the later 1950s, "These compartments into which RAND is organized are not to be taken too literally," with "the occasional political scientist, astronomer, physiologist, or psychologist" cropping up in the Engineering Division, while "Mathematicians, in turn, have infiltrated most Divisions." Moreover, projects that began in one division—"a project on Strategic Air Command operations" or a study of "limited war"—had a way of spreading out to envelop others.[52] Mathematicians certainly remained a key element of RAND's research force: by the spring of 1948 the corporation employed 23 mathematicians full-time, out of a total technical staff of 145. Throughout the 1940s and 1950s, mathematicians and logicians would comprise a roughly constant percentage of the RAND workforce, and a significantly larger proportion of the technical staff, despite this being a period of tremendous growth and diversification of the overall RAND budget.[53]

With mathematicians such a significant presence inside Project RAND, game theory per se became institutionalized as an active area of research, so that even though many of the resulting papers bore the mark of Weaver's "military worth" agenda, the internal logic of the mathematicians' discipline also began to assert itself. Upon his arrival in the spring of 1946, Williams pushed the mathematicians under his direction toward research in game theory that might be relevant to the development of mathematical theories of air warfare. One manifestation of this agenda was an extensive series of publications laying out highly simplified models of air duels, or reconnaissance missions in which a commander might learn some information about his opponent at a cost to himself.[54] This former tradition would prove remarkably long-lived: the so-called problem of noisy duels, in which each plane knows when the other has commenced fire, would only be considered solved in the later 1960s.[55] The models were highly simplified—certainly far too simple to be of direct applicability to actual combat situations—but the mathematics proved exceptionally complex. In such duels, the planes might choose from a continuum of pure strategies: for instance, by selecting a time at which to begin firing that will maximize damage to the opponent while minimizing damage to themselves. However, such games presented a number of technical difficulties not present in von Neumann and Morgenstern's original conception of game theory. As a result, many of the dozens of research memoranda that the Project produced from 1946 onward began to look like the typical traces left behind by an academic enterprise: sober and workmanlike, consisting of long strings of theorems, lemmas, and proofs, they plowed through many of the standard kinds of questions mathematicians asked about the identification of various classes of games, and the existence of solutions.

By early 1948, RAND's mathematicians issued several progress reports synthesizing their work and indicating profitable new directions for the theory, including a five-part series on "Contributions to the Theory of Games" authored by the mathematicians as a group.[56] Their programmatic statements in these documents reflected the confluence of mathematical impulses and "military worth analysis" found at RAND. Consider, for example, the remarks of Olaf Helmer at a meeting of the Institute for Mathematical Statistics in June 1948. While von Neumann and Morgenstern had focused on proving the existence of solutions for the two-person zero-sum game with discrete strategies, research was urgently needed on several other topics that had become salient since 1944. These included simpler methods for actually calculating mixed-strategy solutions in games with discrete strategies, as well as solutions for games with *continuous* payoff functions, possibly drawing upon principles from the calculus of variations—a reflection of RAND's interest in problems like "games of timing." Solutions also were sought for discrete games where the matrix entries were no longer simple expected values but were instead probability distributions. And finally, there was demand for improved results on games with more than two players. In the military context the formation of coalitions—even where coalitions were mutually beneficial and the third player was "nature"— seemed "utopian." It therefore seemed likely that useful and relevant results could be found by developing a theory of multiplayer games in which coalitions were banned or restricted.[57] Already, in the context of Weaver's aims for an interdisciplinary analysis of "military worth," game theory's center of gravity was beginning to pivot away from that found in *Theory of Games*.

From "Military Worth" to Dollars and Cents: Programming

It was at this point, in late 1947 and early 1948, that Project RAND's military worth analysis suddenly ran together with another research initiative originating on the other side of the country, in the Air Comptroller's office at the Pentagon, and associated with the rubric of "programming." Weaver's "military worth" analysis reflected wartime pressures surrounding the development of the B-29: the need to coordinate research activities in an environment where time was the limiting factor. Programming, by contrast, was from the start colored by the postwar military establishment's signature worries about the availability of money and manpower. With President Harry S. Truman presiding over rapid demobilization and military reconversion following the end of the war, and with further budgetary consolidation just around the corner with the passage of the National Security Act of 1947, the military found itself operating in an unaccustomed atmosphere of competition for resources.[58] As E. W. Rawlings, the Air Comptroller, would lament in a memorandum to top Army Air Forces general Carl Spaatz from March 1947, while wartime planning was governed by purely military considerations with "no limit on our budget," now "stringent limitations have been placed on our resources through budgetary and manpower ceilings." Existing strategic plans set by top-level staff officers implicitly required a peacetime budget of $8 billion, according to the comptroller's office, which was charged with translating general "plans of action" into "coordinated *programs* for implementation" by "subordinate commands." Yet with budget projections for the upcoming year estimated at under $3 billion and further cuts contemplated, Rawlings complained that he was increasingly forced to make policy decisions "each time I am asked to recompute a budget, since the deletion or reduction of items in the budget later controls the operating program." He therefore proposed a significant overhaul of the process by which plans were translated into programs, by the development of "general yardsticks, factors, and standards" for budgeting that would help the Air Staff prioritize the distribution of available resources.[59]

Air Force budgeting in the 1940s was indeed a daunting bureaucratic process. Once plans were handed off by the air staff, the generation of a final "program" specifying payments and requisitions of equipment or munitions, manpower and training requirements, and movements of materiel and personnel took roughly seven months, as different planning organs labored to assemble the requisite information and calculations (see fig. 3.2). From this environment of frenetic budgeting activity, "programming" emerged as a field of formal mathematical inquiry. A central figure in this development was George B. Dantzig, a former statistics student of Jerzy Neyman at Berkeley, who had served as an advisor to the US Army Air Forces statistical control office as wartime liaison between the Army Air Forces and the AMP. Provoked by his Pentagon colleagues, Dantzig set himself the problem of automating the production of "a time-staged deployment, training, and logistical supply program," using available computing technologies like "analogue devices or punch card equipment" to speed up the process. Dantzig's first mathematical model of the programming process captured only the functional relationships between the various resources the Air Force needed and the

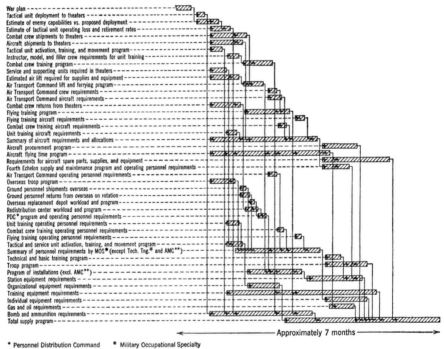

* Personnel Distribution Command   * Military Occupational Specialty
★ Technical Training   ** Air Materiel Command

FIGURE 3.2. The Air Force programming process before the advent of linear programming. From Tjalling C. Koopmans, *Activity Analysis of Production and Allocation* (New York: Wiley, 1951), 191.

activities that produced (or consumed) them. To choose the most suitable combination of activities and resources from the vast set of technologically feasible combinations, a number of additional "ad hoc ground rules" or "edicts" that the military leadership and budget authorities imposed were also necessary. But by the summer of 1947, Dantzig had begun to formulate the programming problem in terms of the explicit maximization of a function of the decision variables that captured Air Force objectives, subject to constraints on the various combinations of activities and resources at the Air Force's disposal. With the problem thus formulated, starting in the fall of 1947, he began actually to calculate a program, albeit a highly simplified one: a diet, consisting of proportions of available foodstuffs, that met the basic nutritional requirements for a human being at minimum cost. The calculations were carried out using the most advanced technology then generally available: a group of human computers using mechanical tabulating machines and a pencil-and-paper notational system.[60]

In early October, daunted by the immensity of the computational task involved in even this relatively simple problem, Dantzig visited John von Neumann at Princeton in hope of improving his methods of solution—a decision that made sense in light of von Neumann's expertise in computing techniques. Having written out his formulation of the programming problem on the blackboard in von Neumann's office, Dantzig recalled:

> Von Neumann stood up and said "Oh that!" Then for the next hour and a half, he proceeded to give me a lecture on the mathematical theory of linear programs. At one point seeing me sitting there with my eyes popping and my mouth open (after all I had searched the literature and found nothing), von Neumann said: "I don't want you to think I am pulling all this out of my sleeve on the spur of the moment like a magician. I have just recently completed a book with Oskar Morgenstern on the theory of games. What I am doing is conjecturing that the two problems are equivalent. The theory that I am outlining for your problem is an analogue to the one we have developed for games."[61]

Von Neumann's essential insight was to realize that the problems faced by the players in a two-person zero-sum game and the particular maximization problem faced by the Pentagon programmers were the same. In a game, the two players are trying to solve an interconnected optimization problem: given a choice of the strategies available, the row player wants to maximize the value of the game, while the column player wants to minimize the expected value of the game. Dantzig's "linear program" possessed a similar structure: to every such program, there is a "dual" program that corresponds to the optimization problem of the second player in the two-person zero-sum game. In his conversation with Dantzig, von Neumann was able to outline conditions under which both problems had solutions, much as he had in his 1937 paper on the equations of economic equilibrium.[62]

Von Neumann could develop insights into the nature of the programming problem by drawing analogies to the theory of the two-person zero-sum game, but he was less successful in identifying new numerical techniques for computing particular programs. His proposed algorithm, sent to Dantzig several months after their October meeting, was actually slower than the one Dantzig had devised the previous summer (now famous as the "simplex method").[63] However, beyond von Neumann's somewhat disappointing contribution to the numerical methods used to solve games and programs, the major result of Dantzig's visit to Princeton was to connect two areas of mathematical endeavor that had previously been quite isolated. Dantzig had not found von Neumann's book on "games" when he searched the mathematical literature for results relevant to what he thought was his problem, and the "military worth" analyst first envisioned in the context of the B-29 bomber study occupied a social world distinct from that of the Pentagon bean-counters. Yet the intellectual connection forged at this fortuitous meeting between games, programs, and the computer—while foreseen in Weaver's "Rube Goldberg Machine"—would quickly come to dominate mathematical research into game theory at RAND and elsewhere.

The connection between games and programming, "military worth" and Pentagon budgeting, was cemented in July of 1948, when RAND hosted two month-long colloquia simultaneously, one on game theory and the other on the "Theory of Planning," an umbrella term for the kind of work being done by Dantzig's programming project. The colloquium on planning drew in some thirty-eight mathematicians or statisticians, both from RAND and from academia; top-level operations researchers, both in and out of uniform, working in the Air Force and Navy; and a number of economists. The mathematicians were dominated by RAND's mathematics staff, most of whom were present, but new arrivals included Merrill Flood (then working out of the American Statistical Association's Washington office), Richard Bellman (a graduate student in mathematics at Princeton who would eventually join RAND's mathematics division in 1952), David Blackwell (a mathematical statistician at Howard University) and Tukey (a collaborator of Samuel S. Wilks at Princeton). From the services came a number of top operations researchers and administrators: a colonel and a couple of lieutenant colonels from the Plans and Operations Division of the US General Staff; Dr. William J. Horvath of the Operations Evaluation Group of the Navy Department; and naturally, George Dantzig and his colleague Marshall Wood, both from the Office of the Air Comptroller.

However, the vast majority of new arrivals were economists, most of whom were drawn from the Cowles Commission for Research in Economics at the University of Chicago. This organization, which had been the preeminent center for mathematical economics and econometric research in America during the 1930s and 1940s, had been involved in Dantzig's linear programming project since the previous summer. At that time, Dantzig had visited Cowles to consult with research associate Tjalling C. Koopmans, a Dutch physicist-turned-economist who had developed a mathematical model of transportation systems for the wartime Allied Shipping Board. Koopmans and his colleagues had been unable to give immediate advice on solving Dantzig's problem, but began to steer the researchers at Cowles toward programming problems. As a result, Koopmans and his colleague Kenneth Arrow joined RAND's Hitch as the nucleus of the economics delegation to the conference. By contrast, the colloquium on game theory was a smaller affair. In attendance were B. O. Koopman, Frederick Mosteller, Edwin Paxson, Ivan Sokolnikoff, John Tukey, John von Neumann, Abraham Wald, W. Allen Wallis, and Samuel S. Wilks; Warren Weaver was invited, but could not attend. Interestingly, a number of the participants had also attended the planning colloquium, namely Koopmans, as well as many of the military operations researchers such as Horvath, Wood, and Zimmerman. The reading list for the seminar consisted of a long list of RAND research memoranda, virtually a catalogue of the game theory publications produced by Williams's mathematicians.[64]

The influx of economists was on display at the first major academic symposium devoted to game theory, which took place at the summer session of the Econometric Society, held on September 9, 1948, at the University of Wisconsin–Madison. Present were many of the major players from the RAND seminars the previous summer. Von Neumann presented the theory of the two-person zero-sum game, while Morgenstern discussed its applications to economics via the study of logistics.[65] Girshick and Paxson gave the audience a survey of the ongoing work on games at RAND; Tukey presented an application of game theory to strategic problems; and Dantzig discussed the relationship between games and the problem of "programming in a linear structure." The ensuing roundtable discussion included a number of additional RAND game theorists.[66] During the subsequent year, RAND and Cowles continued to develop the connection between the military worth research agenda at RAND and the economic aspects of planning that had taken on new salience. A subcontract from RAND to Cowles in January of 1949 for a project on the "Theory of Resources Allocation" helped game theory and mathematical programming emerge as major components of the Cowles research portfolio, and Cowles economists (most notably Arrow and Koopmans) became perennial fixtures as RAND consultants. The following March, a conference on applications of game theory to tactical problems was held at Chicago. RAND was represented by the usual members of its mathematics department (Helmer, Paxson, and Shapley), as well as Arrow, a consultant from Cowles; the other attendees consisted of the operations research representatives of the various services.[67]

Logistics and Mathematical Programming at the Office of Naval Research

The years 1948 and 1949 thus saw the convergence of several lines of work that had formerly been distinct, which had the effect of placing the mathematics of game theory and its practitioners at the heart of Air Force innovation on budgeting procedures and technology assessment alike. Game theory provided a common mathematical structure for games and programs; yet as the lopsided attendance at the colloquia held at RAND in the summer of 1948 suggests, "programming" was set to eclipse "game theory" per se as an object of military funding by the later 1940s. To see further how this took place, we now turn to the program of mathematical research promoted by another major funding source for game theory in this period, the Office of Naval Research (ONR).

    While the historiography of game theory and operations research has paid much attention to RAND and the Air Force, it has paid relatively little to the Navy and to the ONR in particular.[68] RAND certainly would become the public face of game theory as the corporation's experts became visibly involved in debates over nuclear strategy in the 1950s, but, given its size and patterns of patronage, the ONR was at least as significant in terms of its ability to reshape the disciplinary landscape surrounding the theory. In contrast with RAND, which funded applied research in a corporate context, the ONR also focused on supporting university-based scholarship via relatively open-ended contracts with particular academics and teams of collaborators. With an annual expenditure specifically for mathematics in the low millions of dollars, its influence among the community of American mathematicians was extraordinary. By November of 1950, the director of the ONR's Mathematical Sciences Division could estimate that roughly a sixth of "productive research mathematicians" in the United States were currently receiving support from the ONR, and that about a sixth of papers presented to the American Mathematical Society between 1948 and 1950 grew from work sponsored by the ONR.[69] In addition, as a central contract-based funding agency with a very broad mandate for supporting scientific research, the ONR proved itself capable of transforming university-based research programs on a grand scale. As a result, in the later 1940s, the office developed long-term funding relationships with a number of prominent game theorists, most notably, Albert Tucker and Oskar Morgenstern.

In the 1940s and early 1950s, the director of the ONR's Mathematical Sciences Division (before 1949, the Mathematics Branch) was none other than Mina Rees, formerly principal technical aide to the AMP, assistant to Warren Weaver, and editor of Weaver's chapter on "A General Theory of Air Warfare." Her work presiding over the postwar termination of the AMP's contracts with the Navy proved a transition to her being offered the job with ONR when it was established in 1946. Shortly thereafter, according to her reminiscences, she quickly saw the significance of Dantzig's work on programming and made a presentation (accompanied by Dantzig) to the top Navy brass that urged the cultivation of similar methods under the aegis of the ONR.[70]

By the spring of 1948, the Navy General Board, with Rees's assistance, assembled a committee to comment on a document proposing systematic logistical-tactical research to be run through the ONR's mathematics division. The committee consisted overwhelmingly of mathematicians—most notably von Neumann, C. B. Tompkins, logician and computer scientist Barkley Rosser, Tukey, and statistician Wilks—but also included Oskar Morgenstern and the director of Princeton's Institute for Advanced Study and hero of America's wartime nuclear weapons program, J. Robert Oppenheimer. It is possible to infer the substance of the board's proposal from the responses of committee members.[71] In a joint reply dated April 26, 1948, the committee recognized that the Navy's central interest in contacting them was in "drawing numerical conclusions concerning the best use and distribution of manpower, ships, food, bases, etc."[72] While acknowledging the quantitative approaches to such problems that the military was already devising, the committee argued for the necessity of theoretical research, arguing that "the general pattern into which these data are to be fitted is at least as important and as demanding of investigation as the data themselves." The committee thus called for the involvement of scientists (and mathematicians in particular) in investigations of truly basic problems in warfare and logistics.

Their best example of such a general theory of warfare was von Neumann and Morgenstern's game theory, and their description of the theory mirrored that in Weaver's AMP report chapter. As the committee noted, *Theory of Games* "described structures with many properties similar to some exhibited by war," and indeed, the book's broader rationale for developing mathematical economics was much like that of the Navy in searching out a new science of war. Among other things, its opening two chapters demonstrated the power and promise of quantitative methods through its discussion of utilities and mathematization in the sciences, while the theory of the two-person zero-sum game explored in chapters 3 and 4 demonstrated that "the most advantageous prosecution of a war is not necessarily achieved by the most vigorous prosecution of each factor at every time," that is, that tradeoffs between interrelated "factors" always have to be made. Already, as the committee members pointed out to the presumably all-too-well-aware naval officers, these theoretical techniques were gaining a following in the office of the Air Comptroller, as well as in Project Rand. But while the Air Comptroller was mostly interested in problems of a "logistic character" and RAND's perspective "has to be chiefly that of the Air Forces," ultimately, "all of these approaches will have to be brought on a common denominator," a common set of mathematical frameworks and results provided by the *Theory of Games*.[73] Finally, much like Weaver earlier, the committee called for the creation of a working group to study the issue, applying their theoretical acumen to clarifying the nature of "the most advantageous action in warfare" and calculational techniques for identifying it.[74]

Perhaps even during the committee meeting itself, Morgenstern sketched the outlines of a grant proposal to the ONR on index cards and scraps of paper. It included $6,000 for the project leader, $6,000 for a research associate, $10,000 for three research assistants, and $2,500 for travel, communications, and materials. Together with university overhead, the contract would be worth $35,500. This meeting would thus mark the beginning of Morgenstern's long and treasured relationship with the ONR, a relationship that would support many of Morgenstern's graduate students, most notably Martin Shubik.[75] Partly on the strength of his ONR contract, Morgenstern began to cultivate connections with RAND throughout 1948. In the summer, he was invited to the joint colloquia on games and the theory of planning held at RAND, but was unable to attend.[76] Later that fall, he wrote to Charles Hitch, RAND's top economist, to introduce himself and his research and to ask for information about RAND's work since he knew little about what was happening there. Hitch replied that he had discussed Morgenstern's work with Rees, who had visited RAND in the preceding week, and that "while we are not planning any major research program within RAND on linear economic models, we are letting a sub-contract to the Cowles Commission on the theory of resources allocation, and our statisticians have been doing some work . . . on computational problems. I hope that you and Koopmans will be able to keep each other informed of progress as your projects develop."[77] By June of 1949, Morgenstern had become a RAND consultant and embarked on a letter-writing campaign to Olaf Helmer in hopes of receiving RAND's mathematical publications relating to game theory and linear programming.[78] Over the next few years, as a consultant, Morgenstern would produce several RAND research memoranda outlining a vision for a general "theory of organization" that he hoped would provide the theoretical foundation for the study of logistics. Typically wordy and discursive, they stand out stylistically among RAND's more aridly technical publications from this period.[79]

During the later 1940s, the ONR also developed funding relationships with a number of mathematicians who would prove important in sustaining academic research into game theory. In the spring of 1946, Rees arranged an ONR contract with Princeton topologist Solomon Lefschetz to study methods of solution for the nonlinear differential equations of hydrodynamics, a topic relevant to naval weapons development. Initially, Lefschetz used the funding (some $15,000 annually) to support himself and a graduate student, Richard Bellman, as a project assistant.[80] However, two years later, Lefschetz's protégé Albert Tucker—then newly hired by the Princeton mathematics department—proposed a significant "extension and broadening" of the project on nonlinear differential equations. It is not entirely clear how Tucker initially described this extension to Rees, but on May 21, 1948, Rees wrote to Tucker:

> On the basis of our telephone conversation yesterday and of the information contained in your letter which just reached me, it is clear that the Mathematics Branch should proceed, as promptly as possible, to recommend a widening of the scope of the ONR work at Princeton to include more general phases of mathematical analysis, and certain investigations in combinatorial analysis related to problems of logistics.[81]

Given that Rees had overseen the work of the Princeton panel that helped set up the ONR's research program in logistics only one month prior to this letter, her enthusiastic response to Tucker's proposal is not surprising. In negotiating the expansion of Lefschetz's project, she would note: "In the case of Professor Tucker, since the work which he will undertake is of pressing interest to the Defense Establishment, we will make every effort to find the necessary funds" quickly.[82]

While Tucker's original project proposal is not extant in the ONR archives, his correspondence with Rees and Fred Rigby, the new head of the "Logistics Branch" within the ONR's Mathematical Science Division, reveals the direction his work took after the spring of 1948. In a letter to Rigby from the spring of 1949, he stated that his ONR work could best be described as "investigations into linear-convex problems related to the theory of games, econometric planning, and logistics."[83] Thus Tucker, like von Neumann in his meeting with Dantzig a year earlier, saw himself exploring less game theory per se and more the common mathematical structures that underlay both the two-person zero-sum game and mathematical programming. During the summer of 1948 Tucker and two of his graduate students, David Gale and Harold Kuhn, conducted considerable work in game theory. One result was a paper on solutions of the two-person zero-sum game, "A Geometric Approach to the Theory of Games," which Tucker presented at a meeting of the Philadelphia Section of the Mathematical Association of America in November of 1948.[84] At the February 1949 meeting of the American Mathematical Society, Tucker presented a second paper explicitly connecting the mathematics of games and programming, titled "Some Linear-Convex Problems Equivalent to a Game Problem."[85] Also during this period, Tucker inaugurated a regular seminar on game theory (held in Princeton's Fine Hall) which would serve as the institutional focal point of the up-and-coming generation of game theorists: David Gale, Harold Kuhn, Lloyd Shapley, John Isbell, Martin Shubik, and John Nash. Indeed, in the spring of 1949, Tucker wrote to Rees to ask if David Gale could submit some of his work for the project as a thesis on game theory, the first of many that Tucker would help supervise.[86]

Much as it had with Morgenstern, the ONR provided Tucker with connections to the community of people doing research on game theory and logistical problems at RAND, the Cowles Commission, and the Pentagon. In January of 1949, Tucker traveled to Washington, DC, to meet with George Dantzig at the Pentagon, and by later that spring, the two were on a first-name basis, trading mathematical results by mail. In a letter of March 16, 1949, Tucker referred Dantzig to some recent work by von Neumann on computing solutions to symmetric games and invited Dantzig to give a talk at the Fine Hall games seminar. Carbon copies of the letter were sent to George Brown, Tjalling Koopmans, Mina Rees, and John von Neumann.[87] During the spring of 1949 Tucker also pursued connections with the group at RAND in hopes of spending some time there when he went on sabbatical at Stanford during the 1949–50 academic year. In this respect, Tucker was only one of the first of many Princeton mathematicians and economists who would join RAND for the summer as consultants. In addition, while Princeton was certainly the most important locus for the academic cultivation of game theory during the 1940s and 1950s, it was only one node in a vast network of logistics research funded by the ONR during this period. The 1948 committee meeting also marked the first step toward the creation of a Logistics Branch within the Mathematical Sciences Division of ONR, headed by mathematician Fred Rigby. By the early 1950s ONR was funding "logistics research projects" at a number of universities including Brown, Princeton, and George Washington University. The last project, under C. B. Tompkins, involved building a "logistics computer" for the automatic computation of programs.[88] By 1954 the volume of work on these logistical/tactical questions reached the point that a new journal, the *Naval Research Logistics Quarterly*, was established (with Morgenstern on the editorial board) as a prime outlet for publications on game theory.

Game Theory and the Military: An Intellectual Settlement

Having outlined the networks of patronage at RAND and the ONR that supported postwar game theory, it is worth stepping back and assessing how this influx of funding reshaped the theory itself. Von Neumann and Morgenstern's book had put forward a number of distinct mathematical formalisms for the analysis of "games": the theory of utility and preference orderings; the extensive form of games; a collection of results concerning the existence, uniqueness, and properties of solutions to two-person zero-sum games, related systems of linear inequalities, and maximization problems; and a formulation and solution concept for addressing the general problem of multiplayer and non-zero-sum games. In the immediate postwar period, the game theoretic literature focused on certain portions of the original *Theory of Games* rather than others, in the process reinterpreting the theory and developing those portions in directions not anticipated in game theory's urtext. We can ascertain the impulses behind this selective appropriation and reinterpretation of game theory in a number of ways—most obviously by examining stated rationales for such a choice of topics. But it is still more revealing to look at game theorists' reactions to the portions of *Theory of Games* that were *not* as well received in the military context, especially von Neumann and Morgenstern's treatment of multiplayer games, and their attempts to overcome some of the problems they perceived with the existing theory. Much of this section therefore examines the study of multiplayer games in the later 1940s and early 1950s with an eye to understanding the bounds of the intellectual settlement involving mathematicians, game theory, and the military. The various "solution concepts" that emerged as alternatives to von Neumann and Morgenstern's theory of the *n*-person game during the immediate postwar period were to some extent unique, reflections of the individual experiences and concerns of the mathematicians who proposed them, but they also share a number of common features that reflect the underlying context of their creation.

Probably the most obvious feature of game-theoretic literature from the immediate postwar era is its focus on the theory of the two-person zero-sum game in normal form. Thus, for example, in one of the first formal courses on game theory taught at Princeton in the spring of 1952, Harold Kuhn focused the overwhelming majority of his lectures on such games, whether with a finite or an infinite number of strategies.[89] And in an addendum to the first volume of the *Annals of Mathematics* series on game theory (edited by Kuhn and Tucker in 1950), Kuhn noted that "until now, the base of the theory has been the theory of the finite zero-sum two-person game" with other portions of von Neumann and Morgenstern's book attracting relatively less attention.[90] Certainly, the theory of the two-person zero-sum game exhibited features that made it attractive to the various military-funded research agendas outlined above—as well as to the mathematicians pursuing these agendas. Thus the fundamental connection between the theory of the two-person zero-sum game and the linear programming problem proved critical in shaping the subsequent fortunes of game theory both at RAND and at Princeton as the problem of "programming" and resource allocation swelled in significance at RAND and the ONR. Indeed, by the later 1940s, applications of game theory in the RAND publications were eclipsed by applications of linear programming, a pattern that is reflected in operations research publications more generally. Consider, for example, a joint bibliography of "Unclassified References on Game Theory and Linear Programming" compiled in the spring of 1950 by Robert Dorfman, assistant for operations research at USAF headquarters, and distributed to the various military operations research programs, and the research groups at Princeton and RAND. Of the 119 entries in the bibliography, 73 were classified as linear programming studies; of the balance of game theory studies, nearly all concerned the two-person zero-sum game, and many specifically addressed the commonalities between game problems and programming problems.[91] An example is the so-called Colonel Blotto Games popular at RAND, in which two opponents must allocate their forces between different bases so as to win a campaign.[92] The theory of the two-person zero-sum game, in this context, became a theory of maximization or optimization—a fundamental transformation in the meaning of the theory, since von Neumann and Morgenstern had imagined game theory as an alternative to the naïve optimization theories they saw in the mathematical economics of their day.

Even within the theory of the two-person zero-sum game, the postwar era saw several shifts away from the presentation of this theory laid out in *Theory of Games and Economic Behavior*. Mathematicians like Tucker and Kuhn, who were working on logistical problems, quickly moved beyond the systems of linear inequalities found in von Neumann and Morgenstern's original theory of the two-person zero-sum game to consider general optimization problems, in which the functions to be maximized and the "feasible set" of possible solutions could take on a wider variety of forms. For example, the maximized function might be a quadratic rather than a linear function of the decision variables, or the feasible set might be bounded by something other than linear inequalities. Additionally, as Kuhn would write in 1950, in the theory of games "one problem overshadows all others," namely, "to find a computational technique of general applicability for finite zero-sum two-person games with large numbers of pure strategies."[93] Such a technique, which would ideally identify minimax strategies in a small, finite number of calculations, was the focus of many of the RAND research memoranda into the 1950s, as well as von Neumann's forays into game theory in the later 1940s. As his letters to T. C. Koopmans and George Dantzig attest, von Neumann had been working up a draft paper on methods for the numerical solution of two-person zero-sum games with a large number of strategies since the time of his 1947 meeting with Dantzig. This paper appears to have been later absorbed into publications with Dantzig and George Brown, while von Neumann turned to other similar problems, most notably finding solutions to a "game equivalent to an optimal assignment problem," which finally appeared in the second installment of *Contributions to the Theory of Games* in 1953. Algorithms for approximating an optimal strategy in a finite sequence of steps were essential to putting game theory in the service of logistics or the analysis of military worth.[94]

While the problems posed by the two-person zero-sum game and linear programming were similar, it is worth repeating that the study of the two-person zero-sum game was hardly coextensive with the broader ambitions of military worth analysis. Nor was linear programming a ready-made replacement for the techniques of resource allocation required by actual military budgeting. Early applications, like the "diet problem," focused on much more manageable numbers of variables in order to be solvable using the relatively primitive computing facilities available at the time, nor did they incorporate dynamic, intertemporal considerations. Among other things, such considerations would inspire Richard Bellman's far more complex theory of "dynamic programming" in the 1950s.[95] The RAND mathematicians' crude models of duels and reconnaissance missions were also presumably too simple to advise actual generals and aviators as to specific courses of action. Their papers and research memoranda were essays in the craft, following the natural inclinations of the mathematicians to axiomatize and prove. They were also promises tantalizing indications of insights a theory of warfare might yield, much like Merrill Flood's original memorandum written as part of the B-29 bomber study, or like *Theory of Games* itself. They kept alive dreams of a world in which the seemingly chaotic "edicts" and "ad hoc ground rules" of Pentagon budgeters would cohere in the computation of consistent plans of action, or in which the competing agendas of scientists and engineers working together on problems of "organized complexity" would be subordinated to the deliberative scrutiny of military worth analysis. The theory of the two-person zero-sum game as developed in the later 1940s thus reflected a set of inclinations, stemming both from the military context and the instincts of its mathematical practitioners, as much as it directly answered questions or solved problems set by this context. Theoretical elegance, mathematical tractability, and the ability to calculate from seemingly inarguable assumptions of rationality through to optimal strategies and determinate values for games—these were the features that the postwar game theorists found so appealing.

The force of these inclinations becomes most visible if we examine mathematicians' explicit frustrations concerning the portion of von Neumann and Morgenstern's work that was *not* immediately embraced in the military context: the characteristic function form of the game and the "stable set" solution concept to multiplayer games. In part, criticisms of these aspects of the theory reflected doubts about the empirical adequacy of game theory as a theory of decision-making behavior that would come to the fore in the years after 1952—a story to which we will return in subsequent chapters. Thus, writing in 1959, Albert Tucker and coauthor R. Duncan Luce would note that "the intuitive basis and empirical meaning of the von Neumann–Morgenstern solution notion has remained tantalizingly elusive." Given that their solution could not in general identify a particular distribution of payments among the players, but simply a "stable set" of such distributions, they "seem neither to prescribe rational behavior nor to predict behavior with sufficient precision to be of empirical value." Even worse, mathematicians increasingly found that the von Neumann–Morgenstern theory was neither "mathematically deep" nor "elegant," providing an embarrassment of riches in terms of possible social configurations but enabling none of the stylistically appealing proofs that would retain the interest of those drawn to mathematics for its own sake.[96] Von Neumann, for instance, had suggested that the most important unsolved problem in the *n*-person theory was whether all finite games have stable set solutions—a fairly fundamental question—yet the solution was still not forthcoming fifteen years after *Theory of Games*. In fact, in the late 1960s, mathematicians would exhibit games without such solutions.[97]

Beyond such explicit criticisms of the von Neumann–Morgenstern solution concept, perhaps nothing more clearly reveals the inclinations of the military-funded mathematicians of the later 1940s than the hunt for alternative solution concepts for the multiplayer or non-zero-sum game that began almost immediately after 1944. Consider Lloyd Shapley's introductory remarks in a series of RAND research memoranda on multiplayer and non-zero-sum games, published in the summer of 1951. While Shapley embraced the framework of coalitions proposed by von Neumann and Morgenstern, he focused attention on their theory's inability to calculate the expected value of participating in such game for each of its players—a fitting focus given the military worth agenda guiding RAND's research. "In attempting to apply game theory to (say) economic or military behavior, we of necessity introduce into the class of relevant economic or military situations the prospect of being required to play a game," that is, the expected value of playing a game. In military applications, this was particularly necessary since operations were best conceptualized as *sequences* of games, between which choices had to be made. Thus, "if the theory is unable to assign values to the games which occur most commonly in the field of intended application, then only isolated situations . . . can be eligible for game-theoretic analysis."[98] Shapley's solution was to define a value for multiplayer games that possessed a number of reasonable properties in terms of the way it captured the marginal value individual players brought to their coalitions.[99]

A more radical departure from the von Neumann–Morgenstern solution concept, albeit one inflected with very similar concerns as Shapley's, emerged in the work of John Nash, another of the young *wunderkinder* orbiting Tucker's games seminar at Princeton. Nash's writings on game theory (laid out in several brief papers and a PhD dissertation, all appearing between 1950 and 1953) are impressive for their remarkable consistency of vision in the way that they sought to extend the most attractive features of the theory of the two-person zero-sum game to non-zero-sum and multiplayer games. The argument laid out in Nash's first paper, "Equilibrium Points in *n*-Person Games" (communicated by Solomon Lefschetz to the National Academy of Sciences in the fall of 1949) is framed in precisely this way. Each player maximizes his gain with respect to that of the other players on the assumption that the other players will do the same: it is the minimax concept writ large. Using Kaku-tani's fixed-point theorem, Nash was able to prove that at least one set of strategies that meets this condition exists; these he named "equilibrium points." He concluded by noting that, in the two-person zero-sum case, his proof reduced to von Neumann and Morgenstern's proof of the existence of minimax strategies.[100] But although Nash's proof was mathematically elegant and terse, the first paper by itself hardly seemed the stuff of Nobel prizes. Apparently, von Neumann did not think so when Nash paid him a visit at his office that fall. As the story goes, he rapidly grasped the essentials of Nash's argument and anticipated the punch line: "That's trivial, you know. That's just a fixed point theorem." Likewise, Nash's advisor, Albert Tucker, did not remember being particularly impressed by the initial idea.[101]

During the ensuing year, as he expanded on his brief note to the National Academy, Nash gave increasing attention to the broader implications of his solution concept.[102] Thus, in his PhD dissertation, Nash introduced a broad new category of games, "noncooperative games," in which no communication, coalitions, or side payments are possible between players, and contrasted it with the "cooperative games" that were the focus of von Neumann and Morgenstern's n-person theory. Von Neumann and Morgenstern's theory had assumed that communication and collusion between players could occur, and also assumed that the most interesting games would be ones in which it did happen. Nash's noncooperative games had the advantage of possessing mathematically tractable solutions, as he had demonstrated in his NAS communication; in the dissertation, he presented a more involved proof of his earlier result using Brouwer's fixed-point theorem. However, the most novel feature of the dissertation is Nash's assertion of the broad applicability of his noncooperative theory. Certainly, noncooperative approaches are suitable for games like poker, where collusion is forbidden by the rules of the game, and in fact Nash used his theory to analyze a simplified three-player poker game.[103] But beyond this, Nash suggested that any *cooperative* situation could be reduced to a noncooperative situation by enlarging the game to include the pregame bargaining that is necessary to form coalitions. Von Neumann and Morgenstern's model had assumed that players would jockey for position and form binding coalitions and patterns of side payments prior to the actual play of the game. This process lay outside of the game play itself, a situation that Nash found problematic for several reasons. First, he noted that this model assumed the possibility of communication between the players as well as the existence of an "umpire" to enforce coalitions—an assumption that presumably was far from universal. But, second and more important, von Neumann and Morgenstern had assumed that, during the process of meting out side-payments, the utilities of the players were not only comparable but effectively transferable as well. The process of working out comparisons of utilities, Nash argued, should be subject to negotiation in the course of the game itself, so he proposed to absorb the entire pregame process of coalition formation and side-payment into the larger noncooperative game. As a result, he could suggest that any cooperative game could be transformed into a noncooperative game. This transformation had the added advantage of yielding values for two-player cooperative games in which each player has a finite number of pure strategies.[104]

In subsequent years, as he spent summers in Santa Monica and began climbing the hierarchy of academic mathematics at MIT, Nash took several major steps toward fleshing out this vision of reformulating cooperative game theory in terms of noncooperative game theory. Specifically, Nash published two papers on the "finite two-person cooperative games" alluded to in the conclusion to his dissertation, and these would serve as first steps toward this theory of noncooperative pregame negotiation. The first of these papers, "The Bargaining Problem" (1950), essentially tackled the problem of how side payments would be allocated in any pregame negotiations prior to a play of a two-person non-zero-sum game. Unlike the cooperative model, the two players are in a pregame situation where they "can barter goods but have no money to facilitate an exchange." They can, however, create randomized combinations of the goods that they hold. Starting from some very simple assumptions—in particular, that the players knew one another's preferences, that they had identical bargaining ability, and that they would not permit joint gains to be unrealized—Nash argued that, if they decide to cooperate, they will choose the joint strategy that maximizes the product of the players' utilities (in the sense of von Neumann and Morgenstern).[105] Finally, Nash's last paper on game theory, "Two-Person Cooperative Games" (1953), represented a first stab at the final problem involved in reducing cooperative to noncooperative games: how to clarify the role of the "umpire" who enforces contracts. In his article, Nash laid out a formal "negotiation model," a sequence of steps in which players make "threats" and corresponding "demands" of the other player for joining the coalition in a two-person non-zero-sum game. The two-person cooperative game, involving only a single move, is thereby transformed into a two-step noncooperative game in which the players threaten each other into cooperation. The role of "umpire" here is central, not so much as a way of forcing the players to cooperate, but in forcing them to carry through with threats that would presumably hurt themselves as well as their opponent. Based on this negotiation model, Nash proceeded to identify the values of the game for the participants, as well as their optimal threats and demands.[106]

Nash's work has attracted much attention in recent decades—a Nobel Prize in economics for Nash himself, and a blockbuster biography and film about his life—as his eponymous Nash equilibrium concept has emerged as a standard theoretical fixture in economics and political science. As a result, much scholarship has sought to explain comments like von Neumann's and the broader perception that Nash's work was initially less well received than it would be subsequently.[107] Viewed through the concerns of the RAND and ONR mathematicians of the 1940s, however, Nash's initial impulses seem quite natural—for example, his focus on obtaining *values* for games and at least proving the existence of solutions, or the way that noncooperative game theory answered Helmer's 1948 call to investigate multiplayer games "where coalitions are either prohibited altogether or restricted." But pursuing this agenda ultimately led Nash into intellectual territory that was a more problematic fit with the context of military worth analysis or the study of logistics. In jettisoning the structure of coalitions, the assumption of a common currency to facilitate side payments, and the like, Nash nevertheless had to fall back on further assumptions: about the relative "bargaining ability" of the game players, about their knowledge of each other's preferences, or about the existence of "umpires" who would force players to carry out threats that were painful to execute. Problems about the interrelationships between power, trust, and knowledge in social interactions— seemingly overcome in the theory of the two-person zero-sum game— would emerge persistently in the case of multiplayer games.

A prime example of the epistemological and methodological problems surrounding non-zero-sum games emerges from the early work of Howard Raiffa, who first encountered game theory as a young PhD student in mathematics at the University of Michigan.[108] Through his mentors Robert Thrall and Arthur Copeland, Raiffa took a position as a research assistant for an ONR project in the fall of 1948. During the ensuing academic year, Raiffa and his fellow student Gerald

Thompson worked on their first research problem: finding efficient algorithms for solving the two-person zero-sum game. Their result, the "double-description method," was discovered independently of George Dantzig's work on the simplex method for linear programming, and had the advantage that it did not require labor-intensive recalculation of the program's "objective function" at each step of the procedure.[109] Then in 1950–51, Raiffa tackled the problem of solving the two-person non-zero-sum game, something that he felt remained wide open despite von Neumann and Morgenstern's theory. His first instinct—typical of a mathematician at work—was to sort through "all qualitatively different 2×2 bi-matrix games" looking for interesting cases that might lead the way to a more general solution.[110] One particularly obstinate example looked something like figure 3.3. In his notes, he suggested that this situation might be like that faced by two wheat farmers in the depression years of the 1930s, "each of whom is better off with full production even though this would depress prices to the detriment of all." In this scenario, the first row and column would correspond with a strategy of lower production, and the second would correspond with a strategy of higher production.[111]

How should one analyze such a situation? Looking at the multiple approaches Raiffa tried reveals most clearly the problems involved in arriving at a satisfactory sense of "solutions" for games. Von Neumann and Morgenstern's stable-set solution suggested that the players might form a coalition, choose strategies (r_1, c_1), and somehow divide the ten points that the coalition received—but, as Shapley would note, this provided little guidance on the value of the game or the players' optimal course of action. The "minimax strategy," originally proposed in the context of the two-person zero-sum game and generalized beyond that context by Nash, had the advantage of providing usable advice, even if the advice

	c_1	c_2
r_1	(5, 5)	(−10, 10)
r_2	(10, −10)	(0, 0)

FIGURE 3.3. One of seventy-eight qualitatively different two-player non-zero-sum games. After Howard Raiffa, "Game Theory at the University of Michigan," in *Toward a History of Game Theory*, ed. E Roy Weintraub (Durham, NC: Duke University Press, 1992), 165–75.

was nonoptimal: the two players should choose (r_2, c_2) and settle for the certainty of breaking even. But was it not possible to chart some middle course between these two polar opposites by viewing the game from a different methodological perspective? One possibility that Raiffa tried was to imagine himself giving advice to *one* of the players (for instance, the row player). In such a situation, he would use any knowledge of the column player's intentions or capabilities to infer the probabilities of the player's choosing c_1 or c_2. With these estimates in hand, it is easy to calculate the probabilities that the row player *should* use r_1 and r_2. Suppose you think the column player will choose strategies c_1 and c_2 with probabilities p_1 and p_2. Then the row player's expected payoff is

$$10p_1 - (5p_1 + 10p_2)\, x$$

where he chooses his strategies r_1 and r_2 with probabilities $(x, 1 - x)$. Thus the problem of solving the game is transformed back into a traditional maximization problem. By hanging a probability on the column player's actions—in effect, treating both as an unreasoning if regular force of nature—it might be possible to force a valuation of the game, albeit at a cost to the subtleties of game theory's original treatment of intersubjectivity.[112] Still another possibility would be for the game theorist to imagine herself a neutral, well-informed arbitrator, called in to mediate between the two players and advise them *jointly* on how to apportion the gains inherent in the game. Inspired by a lecture by industrial relations specialist William Haber in the spring of 1951, Raiffa asked: "Imagine that two of my game-theoretic friends had to play an abstract two-person non-zero-sum game; suppose that, in fear that they would fare poorly, they approached me to arbitrate a solution: to propose a binding resolution of the game. What principles of efficiency and fairness would I want to impose?"[113] Sketching out some rules, he managed to prove that strategies existed that satisfied the rules even if he could not calculate any specific strategies.[114]

Von Neumann and Morgenstern, Shapley, Nash, and Raiffa worked on a common problem, although the results diverged substantially, and to this day, there remains little agreement among practitioners as to what constitutes a "solution" to multiplayer and non-zero-sum games. It is tempting, therefore, as some commentators have done, to ascribe this divergence to differences in individual personality and worldview on the part of the theorists. Thus Martin Shubik, in a memorable turn

of phrase, suggested that one's choice of solution concept almost constituted an act of projection of self onto society. To truly comprehend the
equilibrium concept, you had to be in the presence of Nash himself: "It's
a game and it's played alone."[115] Similarly, Sylvia Nasar, in her biography
of Nash, has proposed that von Neumann and Morgenstern's choice of a
"cooperative" solution concept reflected a more European worldview in
which collusion and cooperation are more prevalent than in individualistic America.[116]

At the same time, the choice of game-theoretic solutions also plainly
hung on a number of epistemological and methodological considerations
that went beyond the personalities involved. Consider the dedication of
Shapley, Nash, and Raiffa, to the cause of calculating expected values
for games, or their inclination toward building their solutions on principles of individual rationality, rather than the social conventions or norms
that were the ultimate arbiters of social outcomes in von Neumann and
Morgenstern's theory. This feature of their work makes sense for a number of reasons, most notably the broader agendas of military worth analysis and logistics research. At the same time, the solution concepts discussed here are certainly idiosyncratic—although not eccentric. Von
Neumann and Morgenstern built a theory of social behavior on a number of assumptions about the ability of players to form coalitions and to
make side payments, which Nash then rejected in an attempt to ground
game theory more firmly in assumptions of rational individual behavior.
Meanwhile, Raiffa's work imagined the game theorist as a trusted coun-
selor offering advice to individuals trying to make strategic decisions individually, or as a neutral arbitrator trying to jointly advise two parties
somehow stuck in a game situation. Correspondingly, Nash's noncooperative theory has proven an alluring talisman for latter-day economists
committed to the same brand of methodological individualism found in
neo-Darwinian evolutionary biology, while Raiffa's vision for game theory subsequently formed the basis for the prescriptive analysis of bargaining and arbitration situations. Such an approach underlies the pedagogy of institutes devoted to "decision analysis," programs that train
decision experts to take the role of trusted advisors or arbitrators in conflict mediations. Both Shapley and Shubik, meanwhile, would retain cooperative game theory as the foundation of their game-theoretic interventions into economic modeling in the 1960s and beyond.[117]

While the new solution concepts of the later 1940s and early 1950s
possess certain features that make sense in light of the mathematicians'

relationship with the postwar military, this context was only incidentally responsible for the subsequent success of the results that emerged in this period. Certainly, RAND and the ONR provided a labor pool of mathematicians with exposure to the theory, and among whom solution concepts could proliferate. But despite certain points of congruity between the military research agendas laid out in this chapter on the one hand, and the solution concepts of Shapley, Nash, and Raiffa on the other, the theories of n-player games that emerged in the later 1940s and 1950s could only have created at least as many problems as they solved. The two-person zero-sum game beckoned toward a general theory of optimization or maximization, and its ability to compute expected values for duels and gambits held out hope for developing the mathematics necessary for Weaver's military worth analysis. But the new solution concepts raised a number of awkward questions. Who is the game theorist? An advisor to one of the players? An impartial observer or "umpire" considering how best to arbitrate between two parties? Or a social scientist, matching solutions of games to "social" phenomena, such as the formation of institutions and different patterns of information? Such questions could only have been destabilizing to the persona of the operations researcher or military worth analyst, even as they suggested alternative uses for the theory.

A Theory Remade

The late 1940s and early 1950s proved to be the glory days of game theory at RAND and at Princeton, an era of expanding budgets and optimism as mathematicians and operations researchers found an audience for their work in the postwar era. However, within a few short years game theory per se appeared to be losing pride of place in the mathematicians' armamentarium—at least, in some quarters. Consider RAND: in 1949, in a year when the corporation's staff and consultants produced at least twenty-five papers on game theory, its mathematics department could state in its annual report that it continued to regard the theory of games as the guiding inspiration for its research.[118] But certainly by the later 1950s, statements by RAND staff also seemed to confirm that game theory had fallen from favor. "For our purposes, game theory has been quite disappointing," remarked Charles Hitch to a reporter from *Harper's Magazine* in 1960.[119] Throughout the 1960s and beyond, a number of prominent RAND staffers would express similar sentiments of disappointment and disenchantment—albeit in somewhat vague terms— with the theory.[120]

It is hard to know how to interpret such changes of sentiment. Broader changes enveloping RAND, traceable to the sudden emergence of a "hot" war in Korea in 1950, or to domestic anticommunism during the McCarthy years, might have played a role in moving game theory out of its previously central position in the corporation's research program. For whatever reason, after the summer of 1952, the kind of free-wheeling summer conferences that brought together academics and RAND staffers around game theory and related techniques became rarer. Also at this time, some prominent members of the mathematics division fell prey to the paranoia of the times. In 1952 J. C. C. McKinsey, a prolific game theorist and author of the first textbook of game theory, was fired from RAND: as an open homosexual, he was apparently deemed a security risk. Two years later, Richard Bellman, by then one of RAND's top mathematicians and the creator of "dynamic programming," was confronted with old charges that he had offered to rent his apartment to a communist spy. Faced with the prospect of losing his job, he had to sell his house to reduce the risk of defaulting on the mortgage. A similar experience befell Herman Kahn, a physicist working at RAND on contracts from the Atomic Energy Commission.[121]

But it seems more likely that the "disappointment" referenced by Hitch in 1960 was only relative to the rather extraordinary hopes invested in the theory in 1945. Weaver, Flood, and the early staff of the RAND military worth division saw game theory's "powerful analysis of competitive processes" as the centerpiece of a kind of evaluative thought-and-practice that could help synthesize the diverse forms of expertise needed to assess the adoption and use of military technologies. Similarly, Dantzig and Rees saw in game theory and mathematical programming a framework for bringing some kind of coherent order to the diverse edicts and rules that had previously governed the military's budgeting process. Yet over time, other techniques—"systems analysis," war-gaming, simulations, "futures studies," the "Delphi method" of forecasting, or Herman Kahn's "scenarios"—emerged for accomplishing these tasks that possessed more flexibility and permitted the inclusion of much greater richness of detail. In each case, these techniques provided exactly what Flood et al. had thought game theory offered—a coherent system of rules and procedures for disciplining the process of integrating and mobilizing knowledge about the performance of complex weapons systems and military operations.[122]

Consider "systems analysis," RAND's signature technique of technology assessment, as it was articulated at the height of its influence in the 1960s. In the words of its chief expositor, E. S. Quade, systems analysis "is not a method or technique; nor is it a fixed set of techniques." Rather, it is "a systematic approach to helping a decision-maker choose a course of action by investigating his full problem, searching out objectives and alternatives, and comparing them in light of their consequences, using an appropriate framework—in so far as possible analytic—to bring expert judgment and intuition to bear on the problem."[123] In such pronouncements, it is hard not to hear echoes of Weaver's speech before the assembled worthies of American social science at the Hotel New Yorker some two decades earlier. But now game theory was relegated to the status of a minor adjunct to the broader systems analysis "approach." Thus, as longtime RAND game theorist Norman Dalkey would note in his contribution to Quade's edited volume, while it is natural to describe "battles" using "the framework of the theory of games," the direct applicability of the theory was highly limited; the study of game theory, the author suggested, was only a prelude or warmup to the much more intricate process of "modeling" complex conflict systems by computer or simulation, which in turn was but one input to the broader analysis.[124] And of course, even systems analysis itself hardly proved up to managing the kind of conflicts over weapons development and national security strategy that had arisen in the context of the B-29 study, as the troubled history of Edwin Paxson's bombing systems analysis in the later 1940s suggests.[125]

Away from the briefing rooms, white papers, and dense technical reports that were the typical products of the RAND analyst's craft, the face of game theory in academia and in public life also shifted as the 1950s proceeded. ONR-sponsored research into "logistics," like Tucker's program at Princeton, continued to support game theory by training graduate students and contributing papers and editorial support for Princeton's multivolume *Contributions to the Theory of Games*—the prime outlet for high-end theoretical contributions to the field. And as operations research began to professionalize in this decade, it acquired the trappings of academic success in the form of societies, journals, and a university curriculum that would form a key intellectual habitat for the theory. But here again, it is hard to miss the differential survival of the various facets of *Theory of Games*. While utility theory would capture the interest of economists and behavioral scientists in the 1950s, other components of game theory, like the theory of multiplayer games, or the extensive form of the game, would receive relatively less attention. Meanwhile, the two-person zero-sum theory won what may best be described as a pyrrhic victory: the recognition of a close connection between such games and linear programs helped keep the theory alive at a critical time, but interest in programming rapidly came to overshadow that given to games.[126] The essential mathematics of the theory survived, becoming a special case in the general theory of optimization, but the resulting work was all but unrecognizable as "game theory."

Here, the underlying failure was surely one of metaphor. As other commentators have noted, "planning" and "programming" may serve as significant cultural decoders of the postwar era in a way that "games" does not. The multiple meanings of "programming"—programming computers and programming human activities—proved critical in joining advances in electronic computing with theories of social organization and human behavior during the 1940s and 1950s.[127] Furthermore, the borderline sensational attention given by the business press to the disturbing possibility of "Robot Planning" based on computerized programming methods being developed by the Defense Production Administration during the late 1940s and 1950s speaks to a broader confluence of meanings that proved far more powerful (certainly, outside of the mathematical community) than the metaphor of "game" ever could be.[128] As a result, a young graduate student in economics, mathematics, or operations research in the 1950s would likely have encountered a very different brand of game theory than that appearing in von Neumann and Morgenstern's book just a few short years earlier. Especially in textbooks of operations research, presentations of the two-person zero-sum game soon came to be treated almost as a sterile warm-up exercise in advance of the main event: the study of programming and related optimization problems. In this context at least, the two-person zero-sum game thus lingered on primarily as a vestigial fragment of the theory's glorious past.[129]

Thus, by 1957, in their survey of the field, R. Duncan Luce and Howard Raiffa would opine that "it is quite possible that ultimately the theory of games will be considered important in mathematics mainly because of its historically significant relations to other parts of mathematics rather than for its own sake."[130] Their conjecture concerning game theory's future place in mathematics (much less the social sciences) may have been premature, but their historical insight was fundamentally sound. Game theory owed its initial survival after 1944 to a concurrence of interests and individuals unique to that time and place: operations researchers and wartime mathematical consultants trying to find a place for their expertise in the postwar order, and a military with a number of very practical (if not necessarily mathematical) problems to solve. Game theory offered nothing like a complete solution to those problems, not by a long shot. But it did solve an essentially social problem for its practitioners and their military patrons—that of providing a coherent intellectual framework capable of spanning the administrative nitty-gritty of weapons evaluation and the quotidian realities of Air Force budgeting practices on the one hand, and the chalkboards and textbooks of Ivy League campuses on the other. It is small wonder that the mathematicians who emerged from the war loved this thing that they called "game theory." But in many ways, their game theory was a far different thing from von Neumann and Morgenstern's 1944 creation. Ultimately, the intellectual and institutional alliances they cultivated would actualize only one of many possibilities latent within the mathematical grab-bag that was *Theory of Games*. In the 1950s, mathematicians and social scientists funded by the Ford Foundation and the Social Science Research Council would rework game theory as part of a framework for the study of decision-making and information-processing behavior, while the 1960s saw the emergence of game theory as a general framework for studying human conflict and cooperation in the context of "conflict resolution" and "peace research." But in the 1940s and early 1950s, these vistas had not yet been sighted.

CHAPTER FOUR

Game Theory and Practice in the Postwar Human Sciences

A t RAND, the Pentagon, and the "logistics research projects" inaugurated by the Navy at various universities, game theory found its first durable institutional and disciplinary foothold by the late 1940s. The mathematics of the two-person zero-sum game helped link together a number of practical activities undertaken in the service of military patrons, and a research program in applied mathematics capable of sustaining some academic interest. However, this success fell short of the lofty ambition expressed by von Neumann and Morgenstern in 1944 and recognized by many of their book's enthusiastic reviewers: to provide a positive theory of human individual and collective behavior, on par with the achievements of classical physics. The work ongoing at RAND and elsewhere might help to satisfy these ambitions to the extent that military practices of planning and budgeting were beginning to conform to the theory's rational ideal. But beyond the specialized and secretive domain of military technology assessment, logistics, and programming lay the vastness of the postwar social and behavioral sciences awaiting the revolution anticipated by game theory's creators.

This chapter explores some of the earliest attempts to connect game theory to the study of actual humans, as individuals and in groups. It simultaneously documents game theory's first tentative steps away from the rarified community of mathematicians, mathematical economists, and operations researchers who were most responsible for its development before the late 1940s, and into broad areas of the human sciences, especially those research traditions associated with interdisciplinary agenda of "behavioral science." Some of the pioneers of this process were mathematicians at RAND and at Princeton, whose experimental studies of game-playing behavior are the focus of the next section. In addition, at least a few mathematicians like R. Duncan Luce and Howard Raiffa, authors of the landmark 1957 text *Games and Decisions*, were involved in concerted efforts to communicate game theory to scholars in the social and behavioral sciences, providing something of a "supply side" stimulus to applications of the theory. Their work was associated with outbreaks of somewhat idiosyncratic research programs in a range of fields, from psychophysics to political science, that can seem perplexing when viewed in isolation, and many of which turned into dead ends and are now quite forgotten. Throughout the 1950s, game theory would crop up in association with a motley and difficult-to-classify collection of publications in these fields: studies of teamwork and information flow in "small groups," mathematical models of learning, and developments in the theory of psychophysical scaling and the measurement of attitudes and values, to name a few.[1]

If there is a common theme that runs through these appearances of game-theoretic ideas, it stems from a problem that had already emerged in the context of mathematicians' attempts to develop the theory of the non-zero-sum game: decision-makers have private minds that are unavailable for direct consultation either by some independent "game theorist" or by their fellow game players. Of course, a powerful current of thought in the pre–World War II American human sciences held that "mind" was *terra incognita*, so that the only proper object of study was observed behavior or actions. This was certainly true for the brand of behaviorism then regnant in American psychology, with its lab rats and stimulus-response studies, but the spirit carried over into other fields as well. Thus the young economist Paul Samuelson, under the influence of Percy Bridgman's "operationalism," had sought in the late 1930s to remake the economic theory of utility, an inherently subjective quantity, in terms of "revealed preference."[2] Similar intellectual currents— operationalism and logical positivism—would inspire Harvard's S. Smith Stevens to develop psychophysics, with its longstanding interest in perception and sensation, along similar lines.[3] The "cognitive revolution" of the 1950s, associated with the work of George A. Miller, Herbert Simon, and Jerome Bruner, among others, gradually began to make discussion of mind and mental processes respectable once again. But more broadly, the problem of how to read minds emerged as a preoccupation of many postwar social and behavioral scientists.[4] After all, a number of questions they hoped to answer were as pressing as they were sensitive. How do you feel about one-world government? What are your attitudes toward communism? Have you had a romantic encounter with a person of the same sex?[5]

Thus, while the theory of games was connected with a problem facing mathematicians and social and behavioral scientists alike, members of these disciplines viewed the theory's problems and promise in quite different ways. Mathematicians, economists, and those who used game theory to develop "models" of human behavior tended to work from the inside to the outside, so to speak: from assumptions about the internal cognition, values, and beliefs of game players to social outcomes. By contrast, many postwar social and behavioral scientists tended to do the reverse, working from observed behavior back to mind—to metrics of attitudes, beliefs, values, intelligence, or personality. So whereas in the first instance the inscrutability of human subjects represented a barrier to developing a positive "theory" of games, in the latter, game theory could suggest new problems and techniques for experimentation and measurement to study mind via behavior. This close coupling of theory and methodology has been noticed by a number of historians of the human sciences in recent years: for example, the way that "normative rules of data collection . . . prescribed for psychologists" get turned into a "theory" that imagines that "the brain acts as if it were a scientist following . . . rules of method."[6] Conversely, the practice of "theorizing" is closely connected with debates over the methodological and organizational aspects of research programs in the human sciences: "theory" inspires method.[7] Such ambiguities would mark the story of game theory's movement back and forth between mathematics and the social and behavioral sciences in the 1950s. In this context game theory was at once a traditional "model" or "theory" of human behavior (albeit one that required substantial correction and augmentation to have any predictive or descriptive value), and a set of "tools" or "techniques" for notating and manipulating data, clarifying concepts, and suggesting novel experiments and investigations. A group of undergraduates taking a phony IQ test or trying to determine which of a set of isosceles triangles is most aesthetically pleasing, or a flight crew learning to work together under the stresses to be expected in outer space—these were the kinds of scenarios that could be captured in the new game-theoretic idiom for the human sciences during the 1940s and 1950s.

Experiments among the Mathematicians

The notion that game theory might eventually establish itself as a positive theory of human individual and collective behavior was as old as the theory itself. In his very first paper on games, in 1928, von Neumann had suggested that he would soon write an article applying the theory to specific examples of play in card games. "The agreement of the results with the well-known rules of thumb of the games (e.g., proof of the necessity of a 'bluff' in poker) may be regarded as an empirical corroboration of the results of our theory," von Neumann wrote.[8] The paper did not appear, although von Neumann apparently included numerical explorations of poker and bluffing in his lectures on games before 1944.[9] And among the postwar generation of game theorists, experimentation—in the broadest sense of a fluid interplay between theoretical language and human interaction—was quite deeply ingrained in their mathematical culture. Among other things, the game theorists at RAND and elsewhere were frequent game players, taking up chess, Kriegsspiel, poker, bridge, or indeed inventing their own games and playing them against one another. Hence it was natural that the first empirical studies of game-playing behavior would emerge from this group of people.

The game-playing culture among the game theorists was rich and varied. At Princeton, afternoon teas in the mathematics department were accompanied by board games, from chess to the new game called "Hex" or "Nash," introduced by John Nash shortly after his arrival at Princeton in 1948.[10] Lunch hours at RAND were famous for their lively games of Kriegsspiel, essentially a game of chess in which the two players cannot see the moves of the other. These began in earnest in the spring of 1949 and continued strong thereafter. The RAND mathematics division even organized distance Kriegsspiel tournaments between members of different Air Force research and development efforts—and kept scrupulous track of who won.[11] Outside the halls of academia and RAND, at least some of the early game theorists were habitual card players. Before consulting and government work took over his schedule, Merrill Flood commuted between Princeton and New York to play bridge with his fellow Princeton PhDs, the statistician Frederick Mosteller and the mathematician Clifford Mendel. Flood was skilled enough that he participated in the 1945 national bridge tournament. Shortly afterward, in early 1946, he wrote to Mendel that "I am quite interested in bridge, and am working on the mathematical theory of bridge along the general lines proposed by von Neumann in his recent book on games. Since it is impossible for me to do this as a hobby I have made it a part of my business [as a mathematical consultant], and I hope and expect to find some way to commercialize (an awful word) the work so that I will at least not lose money on it."[12] Flood's hope of creating a profitable bridge "system" that he could hawk to his fellow players was never realized, most likely because his continued work for the military prevented the postwar penury that he had feared. His draft paper on the topic was apparently never published.[13]

Card games were an important feature of work and play alike for the younger Princeton game theorists. In his first stint at RAND in the summer of 1948, Richard Bellman devised a new card game that combined features of poker and blackjack. Although he exploited the mathematically straightforward structure of the game to identify solutions, there is no indication that he observed plays of the game.[14] A year later, when John Nash was polishing his noncooperative equilibrium solution concept for presentation as a thesis, his advisor insisted that he produce an example wherein he applied his new solution concept to a specific card game. The example he chose was a poker game for three players, suitably modified to limit the available strategies to a computationally feasible number.[15] Over the next few years, as Nash spent his summers at RAND, he and his Princeton colleagues refined a number of such simple games. There is also evidence that they played these games against one another—in part as sport, and in part to critique their game-theoretic analyses. In a note to Morgenstern from August of 1952, Lloyd Shapley described a four-person game called "So Long Sucker."

By the taxonomy of game theory, it was a four-player game with perfect information and side-payments forbidden. At the outset each player received six chips of a single color, unique to themselves, that they bid in turn onto the table or onto other piles of chips. Laying a second consecutive chip of the same color onto a pile allowed a player to "capture" all the other chips in that pile. The game continued (with various provisions for "rescues" and team-formation among the players) until all but one player had to leave the game for lack of chips. Shapley noted that the game was widely played among his acquaintances, with varied results.[16] The principal response appears to have been rage from most of the losers; the winner typically betrayed his coalition partner in the closing stages of the game.[17]

It is unclear just what the mathematicians learned from these experiences. The relationship among theory, experiment, and observation found in accounts of their research could only have been the stuff of nightmares for any midcentury philosopher of science or methodologist of the social sciences. In such a fluid environment, where the theorists were often the experimental subjects and the observers, it is impossible to separate the performance of particular varieties of game-theoretic rationality from the observation of untutored "human behavior," or experimental controls from the constraints of mental calculations or learned patterns of behavior shared by members of this rarefied mathematical community. In fact, the game theorists seemed not simply indifferent to these distinctions: they almost went out of their way to confuse them.

Consider, for example, one of the first informal experimental studies of bargaining that Flood carried out in June of 1949, involving his attempts to buy a car from fellow RAND employee "HK" who was just then moving back east with his family. Assuming HK wished to obtain the "best" price and Flood the "lowest price" given their alternatives, how should the two agree on the terms of exchange at some price in between these upper and lower bounds? This is the classic problem of how two bargaining parties should divide a surplus, perhaps the most basic economic problem to which game theory had been applied. While it is hard to separate Flood's account of the episode from his *ex post facto* analysis, it appears that the two men approached the situation through a formal mathematical lens from the start. The first problem was to determine HK's "best" price p and Flood's "lowest" s, in itself not an easy thing since both men recognized that it involved estimating a randomly distributed variable, perhaps by sampling the prices offered and asked at local used auto dealerships. As a result, "neither party relished the task of estimating his price," and therefore "both attached some additional value [d] to completion of the deal without need for further searching."[18] Fortunately, in this case, the need for additional searching was eliminated "by agreeing that a used-car dealer, who was well-known to both parties would be asked in confidence to state his selling and buying price for the car in 'as-is' condition." Hence the joint gain from the transaction would be a quantity $g = p - s + 2d$, reflecting both the money saved by cutting out the car dealer, and the effort saved by both men in agreeing to the deal without extensive searching.[19] Von Neumann and Morgenstern's theory of the non-zero-sum game had suggested the players would somehow divide this quantity but it gave no guidance as to how. Flood suggested a satisfying solution would be for the two to "split the difference" even as he acknowledged there was no logical reason why they should do so and that pretty much any other division might be satisfactory. In the end, however, the inconclusive circularity of their negotiations and calculations were cut off by the intervention of an event unforeseen by the original analytic framework: HK decided to drive the Buick in question back to the east coast rather than sell it.

Given the time-consuming and inconclusive nature of this encounter, Flood and his colleagues also went in search of individuals whose untutored mental processes might generate more decisive results. The (female) clerical workers about RAND provided one obvious population whose relative lack of mathematical acculturation might have seemed promising. Flood tried out a variant of this bargaining dilemma (the stakes being suitably lowered) with a couple of the RAND secretaries in October of 1949. To the first, he offered a choice between receiving $0.50 or finding a way to divide $1.50 by bargaining with the second. Since the first secretary is assured of $0.50 and should veto any division of the $1.50 that gives her signigicantly less than that amount, Flood assumed that the secretaries would divide the surplus dollar (and as in the earlier experiment he assumed that in the interests of fairness they would "split the difference"). Yet the secretaries did something unexpected: they split the whole $1.50 evenly, leaving the first secretary with less than she might have obtained. The result suggested that the game-theoretic characterization of rationality could easily be trumped by the friendship between the two secretaries. In Flood's words, "The main lesson from this limited experiment is that the social relationship between the subjects can have a controlling influence on their choices." Social forces were at work on the secretaries; by contrast, Flood said little about how the same considerations might apply to understanding the game-playing behavior of his fellow mathematicians.[20]

Children, mathematically naïve yet intelligent, seemed another particularly promising population of subjects for testing theories of games. Flood and his colleagues seem to have met them on their own turf, playing games like "rock-paper-scissors" or "matching pennies" with their youngsters at home. For such games, the theory of the two-person zero-sum game suggested the players should choose their move at random, yet this aspect of the theory raised some serious methodological problems that continue to trouble game theorists down to the present day.[21] The data provided by a single play of the game was ambiguous: the player's choice of move in this one instance could just as well have been guided by a deterministic strategy or a random one. Thus the only way to assess a player's strategy would be to repeat the game enough times to gather statistically significant data. But this experimental tack only raised further difficulties: since people have memories and may seek to learn about the strategy of their opponent based on earlier plays of the game, the players in subsequent rounds are in effect slightly different people than they were in earlier rounds of play. This phenomenon seems to have been particularly noticeable when the mathematicians played such matching games against their children, who were only just learning the rules and strategies involved. "Some of my friends insist they can do very well against their children in matching pennies, or its variants, and others have relatively little success and claim that their children go quickly to a random method of play," Flood wrote. In either case, compared to the von Neumann–Morgenstern theory, "a better theory for the two-person zero-sum case will offer a solution for this dynamic problem good enough for me to use at least against a child of average intelligence. After all, one pays a good deal of attention in real life conflict situations to the identification of flaws in an opponent's habits of action that can be turned to one's advantage in subsequent plays."[22]

The year 1949–50 saw a strong growth in interest in experimentation on games among the mathematicians at RAND, even though such pursuits seem not to have been truly central to their work at the time. Flood's work during the summer and fall of 1949, for instance, seems to have focused on logistical problems that were the centerpiece of mathematical work at RAND, rather than game theory per se.[23] Yet by the next spring, a number of experiments involving games were beginning to leave a paper trail. The major stimulus for these studies appears to have been the advent of John Nash's equilibrium concept for non-zero-sum games, which had appeared in print the preceding fall. While Nash's new solution concept eliminated the coalitions that proved so troublesome for the von Neumann–Morgenstern theory, its apparent generalization of the minimax decision principle to non-zero-sum situations met with substantial skepticism. Tucker had urged Nash to provide an example of a game that demonstrated the applicability of his idea; in a similar vein, sometime during that winter while working at RAND, Flood and his colleague Melvin Dresher came up with several examples of simple games that clarified the distinction between the von Neumann–Morgenstern solution to non-zero-sum and multiplayer games, and the Nash solution.

The most famous of these games is associated with the story of the "Prisoner's Dilemma" (PD). An early version of this story, titled "A Two-Person Dilemma," appears in a memo attributed to Albert Tucker at Stanford, dated May 1950. Its essence is familiar to virtually every student of game theory. Two individuals are brought to jail on suspicion of a common crime, but are held in separate cells and cannot communicate. Both feel confident that the police will let them go if they do not admit to the crime they are suspected of committing. However, both are also told several things by the police. First, if one prisoner "confesses" to the crime while the other does not, the one who confesses will be rewarded while the other will face an extended sentence. Second, if both confess to the crime, both will receive somewhat shorter sentences. In Flood and Tucker's characterization, any uncertainties about the payoffs facing the prisoners are never explicitly dealt with: the story translates readily into a two-person zero-sum game with perfect information.[24]

The memo therefore cast the story in game theory's normal form (see fig. 4.1). Its subsequent analysis of the game is striking. For both players, the pure strategy of "confessing" dominated "not confessing." This meant that both players confessing constituted the game's one and only "equilibrium point." However, despite this "noncooperative solution," von Neumann and Morgenstern's theory had suggested that the players would do better by forming a coalition and not confessing.[25] Hence the dilemma for Tucker and Flood (if not for the prisoners) was that this game was capable of two diametrically opposed solutions, each of which was mathematically valid on its own terms.

But this did not address the larger question of how one would decide between these two logically consistent solutions, and it seems likely that the idea of conducting any extensive experimentation on game-playing

I ⟍ II	confess	not confess
confess	(–1, –1)	(1, –2)
not confess	(–2, 1)	(0, 0)

FIGURE 4.1. After "A Two-Person Dilemma." Box 1, Folder: "Notes, 1929–1967," Merrill Flood Papers.

AA \ JW	1	2
1	(−1, 2)	(1/2, 1)
2	(0, 1/2)	(1, −1)

FIGURE 4.2. An early experimental game. After Merrill M. Flood, "Some Experimental Games," RAND RM-789-1 (June 20, 1952).

behavior was not immediately on the agenda. To a professional mathematician, PD was a stylized example to be deployed at the blackboard rather than the basis for an empirical research program. Yet game investigations had been taking a more rigorously empirical turn at RAND at exactly this time. In January of 1950, Flood and Dresher had tried informal experiments with two of their friends, "AA" and "JW," making the two of them play 100 rounds of the following non-symmetric PD variant (see fig. 4.2).[26] AA chose between the rows and JW between the columns; all payoffs were measured in cents, and no side payments were permitted. Here, the Nash equilibrium occurs when AA chooses 2 and JW chooses 1; the von Neumann–Morgenstern solution would have them form a coalition and choose 1 and 2 respectively, whereupon bargaining would ensue to split the extra cent they received. From looking at their data (and the running journals that the participants kept of their interactions), Flood and Dresher concluded that the players were hardly slaves to the Nash equilibrium (although cooperation was not a given either); further, they neither "split the difference" between their best alternatives to bargaining, nor did they split the entire winnings of their coalition, as the secretaries had done when permitted to communicate freely and implement side payments.[27]

Thus the PD experiments at RAND did little to resolve debates within the Princeton-RAND mathematical community over how to "solve" multiplayer games. Nash himself felt that the RAND experiments proved little, and his criticisms hint at methodological problems that could only have been increasingly apparent to the game experimenters. "The flaw in this experiment as a test of equilibrium point theory is that the experiment really amounts to having the players play one large multi-move game," Nash would write in a letter to Flood. "One cannot just as well think of the thing as a sequence of independent games. . . . There is too much interaction, which is obvious in the results of the experiment." Remarkably, Nash suggested that the game players should change the experimental setup so as to bring their observation in line with his theory, and he called for many different players to enter and exit the game at random while keeping them in the dark as to the choices of their opponents until the game was concluded.[28]

Nevertheless, increasingly formal tests of non-zero-sum game theory took place at RAND from 1950 onward.[29] As the design of Flood and Dresher's early experiments suggests, several key variables (apart from the entries in the game matrix) seemed particularly significant in determining outcomes. "AA" and "JW" were not allowed to make side-payments, although this could be permitted in other setups. So too could the ability of players to make "enforceable" commitments or to issue "threats" to coerce action on the part of other players. "Enforceability" was straightforward enough to standardize in a laboratory setting where the experimenters could have the final say on who got paid what, so that if enforcement of the players' deals was included in the game rules, the use of threats would be redundant. By contrast, communication, more broadly, was all but impossible to control in any systematic way, and it would remain so for virtually every game experiment thereafter, not just those at RAND. While experimenters might ban talking or passing notes in the severest of terms, potentially, all it took was a nod, a smile, a triggered memory of a former interaction, or some less personal social cue to throw off these surprisingly delicate encounters.

By the fall of 1951, RAND was the site of increasingly complex experiments on collective decision-making. As we shall see, studies of organizational behavior per se were becoming important to the research portfolio of the corporation at this time. Yet these studies continued to compare the predictions of game theory with their findings. For example, a set of studies by Flood examined how groups of individuals selected one of several available prizes and bestowed it upon a member of the group. Some game-theoretic system of coalitions and side-payments was potentially applicable to this process, although once again the side-payments were underdetermined. However, Flood also compared the group's choices against those proposed by a "majority rule" algorithm for aggregating individual preferences into collective preferences, an auction, and various principles of fair division. Sometimes these Solomonic procedures appeared to work well, especially if the subjects were "honest" in assessing the utility attached to the various objects. But what Flood dubbed "psychological considerations" could resurface as well in these mixed groups of professional staff and regular employees, game theory sophisticates and novices.[30]

All in all, it is hard not to sense a certain amount of inconclusiveness in the mathematicians' responses to such experiments. If the hope was to confirm (or even reject) particular solution concepts to games, the results were fairly disappointing. Moreover, to all appearances, there was not even consensus within the community of game theorists as to what a proper "test" of game theory might look like. Should one simply tweak the experimental setup until a theory was "verified," as Nash seemed to suggest in his responses to Flood and Dresher's experiments? Or should one start with a given situation—whether in the laboratory or in some simulacrum of "real life"—and seek to understand the forces underlying the behavior of individuals in that environment? Finally, as the evolution of Merrill Flood's game experiments suggests, the nature of the experimental subject was up for grabs as Flood looked to the office secretaries or his children (with their obscure yet more decisive mental processes) as potential experimental subjects when mathematicians, who could mimic game theory's ideal rational calculating agent better than most, proved too troublesome for that purpose. Already by the summer of 1950, then, the mathematicians were beginning to appreciate the epistemological and methodological difficulties involved in bringing the theory of games into contact with actual human behavior.

Modeling the Minds of Cold War Subjects

If there is a dominant story about how game theory found a place in the postwar human sciences, it hinges on a particular mutation in their conception of the human subject during this period. Practical military demands clearly accelerated the development of new pieces of machinery—most prominently, gun control servomechanisms and electronic or electromechanical computers—that were capable of mimicking and extending the calculational and information-processing faculties of the human mind. The emergence of such technologies inspired a number of research programs that variously sought to study and to improve the efficiency of human-machine systems in combat or in business, to model human minds as information-processing devices (a trend often associated with the so-called cognitive revolution in psychology), or to program computers to simulate human mental processes (the focus of "artificial intelligence" research).[31] In its attempt to axiomatize and algorithmize human decision-making processes, game theory as practiced by the RAND mathematicians appears broadly congruent with these lines of development, permitting researchers to model human beings as calculating, rational game players, and to program machines to mimic human game-playing abilities.

The best-known program along these lines was of course cybernetics, the brainchild of MIT mathematician Norbert Wiener, whose seminal work on the subject defined it as "the study of communication and control in the animal and the machine."[32] Cybernetics had its roots in the wartime collaboration between Wiener and electrical engineer Julian Bigelow to produce practical mechanisms for guiding the aim of antiaircraft gunners under contract with the National Defense Research Committee. The apparently self-guided nature of Wiener and Bigelow's "anti-aircraft predictor," which calculated the future trajectory of an enemy aircraft based on its earlier patterns of evasive action, led Wiener to make an escalating series of analogies between this machine and living organisms, in the process generalizing notions of purpose and teleology.[33] By December of 1944 von Neumann would join Wiener in a "Teleological Society" focused on "communication engineering, the engineering of control devices, the mathematics of time series in statistics, and the communication and control aspects of the nervous system," which first met in January of 1945.[34] Von Neumann's interest in this group was most apparently connected with his wartime project of developing computational resources for defense applications, and he hoped to develop analogies between the neurons of the brain and the circuits of computers using the common mathematics of Boolean logic following the fundamental insights of neurophysiologist Warren McCulloch and logician Walter Pitts.[35]

The intellectual reach of cybernetics would subsequently be enhanced by a series of conferences sponsored by the Josiah Macy Jr. Foundation from 1946 to 1953. These naturally attracted mathematicians and engineers, but also individuals from disciplines ranging from neurophysiology to anthropology, psychology to sociology, and physiology to ecology. One concept that gained almost immediate currency among the cybernetics group during this period emerged from "information theory," created by mathematician and electrical engineer Claude Shannon. Shannon's formalization of information, first pub-

lished in 1948, eschewed discussion of meaning or context in communication, defining information in terms of a signal's ability to increase
the likelihood of a receiver's selecting the correct message from a large
number of possible messages. Information was transmitted via cybernetic feedback mechanisms, permitting regulation and control of cells,
organisms, and entire societies.[36] Drawing upon many of these new
ideas, cybernetics-influenced studies of communication and control in
human-machine systems would find support in the years following the
war, especially from military funding agencies. Thus the Office of Naval Research, set up shortly following the war as the Navy's prime R&D
funding arm, supported studies of human interactions with electronic
communication systems at institutions like Harvard's Psycho-Acoustic
Laboratory (PAL). As Paul Edwards has shown, experimenters associated with the PAL took cybernetic metaphors seriously in their research, conceptualizing communication system operators as information processors trying to separate signal from a backdrop of random
noise, much like the electronic coding systems that were the original inspiration for Shannon's information theory.[37]

While game theory often appeared in conjunction with such research
programs, the theory's conception of the human subject frequently existed in tension with cybernetic concepts like homeostasis and information. Von Neumann could give expository lectures on the theory of
games at the first Macy Conference in March of 1946, but Wiener's ver-
dict on the theory as a whole was unsympathetic.[38] Writing in his *Cybernetics*, Wiener would complain that "von Neumann's picture of the
player as a completely intelligent, completely ruthless person is an abstraction and a perversion of the facts." Rather than focusing on the homeostatic mechanisms that were guarantors of order and stability in
the world, von Neumann's theory of multiplayer games instead simply
captured the "welter of betrayal, turncoatism, and deception" that was
"only too true a picture of the higher business life, or the closely related
lives of politics, diplomacy, and war." Yet ultimately cybernetics demonstrated that "even the most brilliant and unprincipled huckster must expect ruin."[39] Thus while game theory would often appear in textbooks
on cybernetics throughout the 1950s and 1960s, it would nestle awkwardly within the cybernetic corpus. The thrust of Wiener's work and
of cybernetics more generally focused on studying dynamic processes
of learning, information transfer, and feedback; von Neumann, by contrast, tended to prefer formal *a priori* axiomatic approaches in studying

the connections between logic, electronic computers, and the structure of the brain.[40]

More broadly, the cybernetic conception of the human subject was but one of several that attracted constituencies within the Cold War study of individual and collective human behavior. Another would be exemplified by the kind of research to emerge from RAND following its 1947 Conference of Social Scientists that helped guide the establishment of a social science program at the corporation.[41] In addition to studies of national security intelligence and of economic conditions, the conference volume devoted significant space to outlining possible "psychological and sociological studies" on a number of key topics, including "aggression and morale," "attitudes and opinions," and "political psychology," which would help guide American policymaking both domestically and internationally. While far from monolithic, the subject that emerges from these pages is less an information-processing cybernetic machine than a murky tangle of values, beliefs, emotions, and attitudes. His inner world would most likely be glimpsed through polls and structured interviews, like the ones that made the name of conference participant Samuel Stouffer, whose Bureau of Applied Social Research at Columbia University conducted the Department of War's massive wartime study of troop morale, attitudes, and indoctrination effectiveness later published as *The American Soldier.* Similar techniques could be employed to assess the success of communist propaganda in nonaligned countries, not to mention the shifting sands of American public opinion on a variety of topics.[42] Or perhaps only the a-rational cultural patterns and rituals of the Cold War subject's social world could be analyzed along the lines of conference participant Ruth Benedict's just-published interpretation of Japanese culture, *The Chrysanthemum and the Sword*, which was originally composed for the Office of War Information.[43]

Related conceptions of the subject would reappear frequently as RAND (alongside many American universities) amassed a portfolio of social and psychological research relevant to national security matters during the later 1940s and 1950s. Thus RAND social scientists would advise military negotiators to the armistice talks ending the Korean War on the "operational code" of their Russian and Korean counterparts, the "concordance of beliefs, values, and perceptions" that reflected their distorted "image of the external world" and "that determined the enemy's political decisions." No matter how biased their beliefs or monstrous their values, getting inside the head of one's enemy was essential.[44] But the United States of the 1950s had its own inscrutable and unpredict-able masses, as suggested by studies of Civil Defense schemes intended to brace Americans for the aftermath of a nuclear attack. In the words of the multi-university consortium of researchers who sought to predict cit-izens' reactions to a nuclear attack on New York City in the early 1950s, looting, rioting, and mob aggression were all possible if the public's fears were not successfully "channeled" to facilitate the "return of the individ-ual to rationality" rather than permitting their progression to full-blown panic.[45]

From the start, the kind of interdisciplinary research in the social and behavioral sciences conducted at the RAND Corporation and elsewhere was closely connected with the rise of another institution that, more than any other, would define and support the field of "behavioral science" during the 1950s: the Ford Foundation. The foundation burst onto the American philanthropic scene with the death of Henry Ford in 1947 and his extraordinary bequest of Ford Motor Company stock. Apart from its sheer size, the foundation was also notable for the way its program of funding captured the zeitgeist of those early Cold War years in Ameri-can intellectual and political culture. Imbued with the spirit of late 1940s internationalism, liberalism, and technocratic meliorism, it combined the idealism of Truman's Point Four program of assistance to the devel-oping world with the national-security oriented technological futurism of military defense agencies like the RAND Corporation.[46] This spirit was perfectly set forth in the foundation's new program of research: a slim 1949 volume, *Report of the Study for the Ford Foundation on Pol-icy and Program*, written by a study committee of eminent academics and research administrators. In it, the committee laid out five major pro-gram funding areas: "The Establishment of Peace," "The Strengthening of Democracy," "The Strengthening of the Economy," "Education in a Democratic Society," and "Individual Behavior and Human Relations."[47]

This last funding area contained within it a series of subpoints that called for research into a range of problems: "the process of learning," "the processes of communications" "the scientific study of group organi-zation, administration, and leadership," and the "scientific study of the causes of personal maladjustment," among others. Finally, the report called for "the scientific study of values which affect the conduct of in-dividuals, including man's beliefs, needs, emotional attitudes, and other motivating forces; the origins, interactions, and consequences of such values; and the methods by which this knowledge may be used by the individual for insight and rational conduct."[48] The knowledge thereby produced, the study committee hoped, would provide a "rational basis for planning and responsible decision-making" in a world where economic productivity and military preparedness were critical to survival, but would also insure the "democratic nature of such planning and control," thereby avoiding the menace to democracy posed by the demands of the state and of industrial society.[49] The scientific study of human behavior, the committee members clearly hoped, would help the free world steer a safe course between the Scylla of social atomism and the Charybdis of totalitarian planning.

While the foundation thus maintained a liberal-democratic line in its public pronouncements—envisioning a subject whose inclinations were open to scientific observation yet who nevertheless remained rational and free—its program in "individual behavior and human relations" was from its inception nevertheless closely tied to the style of social science research practiced at RAND. This connection was cemented by H. Rowan Gaither, a San Francisco lawyer who chaired the Study Committee. When the nonprofit RAND Corporation was established independently of Douglas Aircraft in the spring of 1948, Gaither helped obtain a $100,000 loan from the foundation to get the fledgling corporation started and subsequently took a place on its board of trustees. In 1951, the foundation loaned RAND an additional $900,000 for the maintenance of its working capital, and the following year the loaned funds were converted by the foundation to an outright grant, on the condition that RAND would continue to update the foundation's president (eventually Gaither himself) on research conducted by benefit of the grant.[50] The publications Gaither received present a fascinating sampling of RAND's social scientific research in these days, from studies of political attitudes in East Germany to observations on the behavior of members of the Soviet elite.[51]

In part through Gaither's position astride the Ford Foundation and RAND, RAND social scientists could influence the Ford Foundation's research program in "individual behavior and human relations." With program areas I–IV commanding most of the attention of the foundation's senior officers during the first few years of the foundation's operation, the task of organizing Program V was accomplished largely through Gaither's personal intervention. In the summer and fall of 1951, he met with three advisors to work out the program's priorities: Donald

Marquis, the chair of the University of Michigan psychology department and a member of the Study Committee; the RAND political scientist Hans Speier, who attended the 1946 Conference of Social Scientists and who would go on to direct RAND's Social Science Division; and Program V's first staff member, the sociologist Bernard Berelson. The upshot of their meetings was a more specific plan (approved by the foundation's board in early 1952) for the development of Program V, which was at that point renamed the Behavioral Sciences Program. While the precise boundaries of the "behavioral sciences" were open to question, the term clearly was intended to cover broad swaths of psychology, sociology, anthropology, as well as parts of economics and political science.[52] Overall, the program's goal was "to increase the number of competent behavioral scientists," "to improve the content of the behavioral sciences (including data collection and relations to other fields of knowledge)," "to improve methods of investigation," and "to develop institutional resources." Several cross-cutting areas of application were prioritized including: "political behavior; communication; values and beliefs; individual growth, development, and adjustment; behavior in primary groups and formal organizations; behavioral aspects of the economic system; social classes and minority groups; social restraints on behavior; and social and cultural change."[53]

While the Behavioral Sciences Program would continue in existence for a mere six years, by 1957 it had disbursed nearly $24 million in grants for a wide range of research programs and institutions. These programs would leave their mark on a number of efforts at connecting game theory and the disciplines that were seen as "behavioral sciences" during the 1950s. One major series of grants, initiated in 1953 to stimulate "self-study in the behavioral sciences," would prove significant in establishing channels of cross-disciplinary contact involving game theorists at a number of universities, especially the University of Chicago and the University of Michigan.[54] The founding of the Center for Advanced Study in the Behavioral Sciences at Stanford University with foundation money would provide an institutional waypoint for many of the practitioners of the new, mathematical theory–oriented social science. And finally, a series of Ford-sponsored conferences and summer training sessions would directly promote the adoption and testing of mathematical theories like game theory in a range of disciplines, especially social psychology and economics.

The conception of the subject to emerge from this network of foundations, think-tanks, and university research laboratories connected with
"behavioral science" thus combined a range of perspectives which varied
according to the application in question: from man-as-information pro-
cessor found in studies of man-machine systems at the Psycho-Acoustic
Laboratory, to the inscrutable enemy studied by the RAND social scientists or the liberal-democratic subject of the Ford Foundation. Game theory's conception of the decision-making subject would prove capacious
enough to appeal to representatives of these diverse research traditions.
However, the question of *why* researchers in these fields might actually
find the theory useful requires closer examination of the communities of
scientists who adopted it, and of some of the mathematicians who were
game theory's earliest and most active apostles to these communities.
Until at least the later 1950s, game theory tended to travel along lines of
personal contact with a relatively small number of PhD mathematicians
or statisticians whose social network included receptive members of the
social and behavioral sciences. While there would be a number of these
individuals—Merrill Flood, Harvard statistician Frederick Mosteller,
and University of Chicago mathematical biologist Anatol Rapoport—the
next two sections focus on individuals whose careers epitomized game
theory's career in the social and behavioral sciences during this period:
R. Duncan Luce and Howard Raiffa. Both were PhD mathematicians of
nearly identical vintages; their careers were supported by common patrons, especially the ONR and the Ford Foundation; and both forged
connections between game theory and a set of subfields within the behavioral sciences that depended crucially upon interpreting the theory as
a framework for the generalized study of decision-making by individuals
and groups. Finally, their coauthored 1957 book reviewing game theory
and related mathematical techniques, *Games and Decisions: A Critical
Survey*, would prove a major turning point in game theory's relationship with the social and behavioral sciences by bringing game theory to a
wider audience beyond that reached by von Neumann and Morgenstern's
technically daunting 1944 book.

The Social Network of R. Duncan Luce

In 1946, following an undergraduate degree in aeronautical engineering and a stint in the Navy, R. Duncan Luce returned to the Massachu-

setts Institute of Technology with the intention of pursuing a PhD in mathematics. Luce had been initially drawn toward physics, but had ultimately dropped the field in favor of regions of applied mathematics where one might "have the excitement of working more virgin terrain" beyond the already well-developed mathematics associated with the physical sciences. Cybernetics appeared a promising area given Norbert Wiener's presence in the department, as did economics with the field's already substantial mathematical tradition and the proximity of Paul Samuelson in MIT's economics program; yet after "some early and feeble attempts to write down economic equations," he stumbled upon a then-obscure specialty to which he would ultimately devote much of his career: mathematical psychology.[55] Because MIT had no psychology department at the time, Luce's primary introduction to this world was through MIT's Research Center for Group Dynamics, which had been founded in 1945 by the prominent German émigré social psychologist Kurt Lewin.[56]

In many ways, the problems that were the focus of Lewin's group dynamics seemed to form a natural fit with game theory. Lewin and many of his students—most notably Alex Bavelas and Leon Festinger, the future theorist of "cognitive dissonance"—were practitioners of the subfield of social psychology known as "small group studies."[57] Lewin's signature research in this area from the late 1930s onward focused on developing an experimental understanding of the relationship between personality, motivational orientation, and collective productivity. Thus his famous studies of "democratic" and "authoritarian" leadership styles in groups of fifth- and sixth-graders seemed to suggest that democratic leadership resulted in lowered social tension and greater stability and work productivity.[58] Given the potential applicability of such studies to the problems of social conflict, industrial relations, and national defense, Lewin and his students proved exceptionally successful at finding sponsors for this work in the United States. Among other interested parties, the postwar ONR would come to rely particularly heavily on Lewin's students for their studies of factors influencing leadership and teamwork in small groups, so that by 1950 they would dominate the proceedings of a symposium on work supported by the ONR's Human Relations and Morale Branch.[59]

While the foundational studies of the group dynamics tradition were often qualitative and interpretive, Lewin drew heavily on mathematical concepts in developing his distinctive language for mapping and notating the psychological status of individuals and groups. Some of his earliest work sought to develop general psychological theory by exploring analogies between psychology and the physical sciences. Lewin saw behaviorism, the dominant psychological tradition in the United States from the late nineteenth century onward, as analogous to Newtonian physics: reductionistic, focusing on establishing mechanical relationships between cause and effect, stimulus and response. Given his background in the gestalt tradition, Lewin looked instead to the holism of late nineteenth-century electromagnetism, which explored physical interactions in terms of "fields" that captured the total set of reciprocal relationships between mass- or charge-bearing objects. By analogy, Lewin sought to define a "field theory" in psychology that captured completely both the internal state of an individual as well as external social-psychological relationships. Behavior—"action, thinking, wishing, striking, valuing, achieving, etc."—was defined as a "change of some state of a field in a given unit of time," rather than a mechanical response to an isolable input stimulus.[60] Continuing to develop the analogy to the physics of fields, Lewin experimented with new developments in mathematics—other than the differential equations of classical physics or the statistical techniques of factor analysis then growing popular in psychology—that could formalize the relationships and tensions that made up a psychological field, thereby capturing the total psychological "situation."

Topology, the generalized study of space, closeness, and shape, seemed particularly promising. Lewin's usage of this mathematics in his own work was not particularly formal: his articles are filled with diagrams of psychological fields and "life spaces" with different areas separated by "Jordan curves," a technical term for continuous nonintersecting curves in a plane that completely separate an "interior" region from an "exterior" region.[61] Yet during the 1940s, Lewin's students also began to deploy ideas from topology and set theory more formally. Their principal vehicle for this was the analysis of communication patterns in groups: who talked to whom, and how patterns of communication affected the group's functioning. Such questions had also been a staple of the "sociometry" movement, a systematic approach to the analysis of small groups that was pioneered by the Freudian Jacob Moreno and his students in the 1930s.[62] Nevertheless, the Lewinian concern with formalizing this theory in the language of set theory, topology, and abstract mathematics stimulated his students to forge connections with individuals working in areas of mathematics that had previously received relatively little attention in the social and behavioral sciences.

This situation provided the perfect opening for Luce, the intellectually adrift mathematics graduate student, to find his calling. Not long after his arrival at MIT, a roommate introduced him to some theoretical problems related to the analysis of social networks and the formation of "cliques" within populations that had arisen in a course taught by Leon Festinger, one of Lewin's former students and collaborators. Luce translated these problems into matrix algebra, proving a number of results, and was soon "hooked" on applying mathematics to problems in psychology.[63] The mathematics department was less keen on his new calling, and assigned Luce to a young faculty member to write a dissertation on a topic in abstract algebra. Additionally, upon finishing his degree, his aspirations to "do psychological theory" rather than pure mathematics ran aground on the fact that "at that time departments of psychology hired statisticians, but not mathematicians with absolutely no psychological qualifications."[64] Therefore, since Lewin had died early in 1947 and his Research Center for Group Dynamics relocated to the University of Michigan soon thereafter, it was fortunate that Luce had maintained contact with one of Lewin's students who remained behind: Alex Bavelas, who in 1948 founded a Group Networks Laboratory at MIT with funding from a range of military agencies including the ONR, RAND, and the Army Signal Corps.[65] Given his familiarity with the mathematics of social networks, Luce became Bavelas's in-house math expert, spending three postdoctoral years at MIT between 1950 and 1953.[66]

It is not clear what Luce contributed to Bavelas's research, or whether their project yielded any particularly deep insights into small group behavior. The project's eminently sensible underlying hypothesis was that the patterns of communication between individuals in a group would prove to be correlated with the group's effectiveness in task performance requiring coordinated action. Bavelas's experimental setup controlled the communication patterns available to five subjects by having them sit behind partitions around a common table and by manipulating the slots in the partitions through which they passed paper notes, thereby producing the different communication patterns (see fig. 4.3).

A typical experiment involved each of the subjects being presented with a box of five marbles of different colors; the task for the group was

PINWHEEL　　　　　　　　　　　　**STAR**

(P)　　　　　　　　　　　　**(S AND SF)**

FIGURE 4.3. Possible communication patterns in a small group. From Josiah Macy Jr.,
Lee S. Christie, and R. Duncan Luce, "Coding Noise in Task-Oriented Groups," *Journal
of Abnormal and Social Psychology* 48, no. 3 (1953): 402.

to determine which color of marble all of their boxes held in common.[67]
The laboratory soon tried to automate these experiments via a custom-
ized electromechanical computer, although this pricey gadget was never
actually worked.[68] The collaboration produced reams of data on the
teams' communication patterns as well as Luce's formal mathematical
models, but there was little synergy between the two: experiment or observation of real humans seemed to have little purchase on the spare representation of a small group or society as a set of undifferentiated linked
nodes. Luce's publications during his time at MIT were exercises in pure
graph theory, appearing in mathematical journals with no discussion
of experimental data.[69] Similarly, Bavelas's own publications on social
networks from this period largely sought to establish observed correla-
tions between classic Lewinian concepts such as individual "leadership"
and mathematical measures of that individual's "centrality" (based on a
weighting of the "distance" between nodes in a graph) in different possible group communication patterns.[70]

　　Thus Luce began to explore game theory in his final years at MIT,
finding in its assumptions of self-interest and utility maximization a
somewhat richer picture of human psychology than he had in the com-
binatorics of communication networks. Even so, the group-dynamics
emphasis on connecting patterns of information flow with the functioning of small groups left its mark on Luce's approach to game theory.
Rather than following the well-trodden path of applied mathematicians
at RAND and Princeton toward the mathematics of the two-person
zero-sum game, Luce noted that "those aspects of game theory which
are probably of most interest to the social scientist are the theories
of coalition formation for games with a transferable utility which are

based on the notion of a characteristic function of a game."[71] However, to overcome some criticisms made of the applicability of von Neumann and Morgenstern's theory to the social sciences, Luce proposed "To study n-person games with restrictions imposed on the forming of co-alitions, . . . thus recognizing that the cost of communication among the players during the pregame coalition-forming period is not negligible but rather, in the typical model with large n, likely to be the dominating consideration." The result would eventually be yet another early solution concept for multiplayer games, "Ψ-stability," which Luce introduced in a 1955 paper.[72]

Luce's peculiar use of game theory reflects the centrality of concepts linked to "information" and "communication" in the postwar social and behavioral sciences more broadly, even if there was little agreement as to how best to capture these concepts in mathematical form. Shannon's information theory, which proved so popular among the psychophysicists nearby at Harvard's PAL, focused on understanding how communication between senders and receivers could take place in the presence of random noise; group dynamics studies sought to connect the task performance of groups with the structure of their communication networks. The widespread interest in mathematical theories of information and communication would prove fortunate for Luce's career prospects after 1952–53, as Bavelas's project at MIT began to wind down. Sitting awkwardly astride mathematics and psychology, Luce was only rescued from a job as a technical college mathematics instructor by an offer to head up an ONR–sponsored Behavioral Models Project at Columbia University. Catering to the kind of research being performed at the Group Networks Laboratory and the Harvard group, the project's primary mandate was to produce readable introductions to new mathematical theories emerging from fields like cybernetics, information theory, and game theory for researchers in the social sciences. The project's official director was the statistician (and former ONR administrator) Herbert Solomon, and it was run out of Columbia's Bureau of Applied Social Research.

The research program found at the bureau, epitomized by the work of sociologists like Paul Lazarsfeld, represented an approach to the study of information exchange and related social processes very different from the laboratory studies of group dynamics that Luce had worked on at MIT. Lazarsfeld's career in the United States was closely connected with the Rockefeller Foundation–funded Office of Radio Research, created in 1937 at Princeton, which was charged with examining the impact of radio on American society.[73] When Lazarsfeld moved to Columbia in 1939, the project on radio and media research became the basis for the Bureau of Applied Social Research, formed in 1944 under Lazarsfeld's leadership. Like his Columbia colleague (to 1946) Samuel Stouffer, principal author of *The American Soldier* study and attendee of the RAND Conference of Social Scientists, Lazarsfeld focused his research on the problem of measuring attitudes and opinions among members of the public. Lazarsfeld's signature theoretical technique in this regard was "latent structure analysis," a technique for distilling fundamental attitudes and beliefs out of the reams of statistical data generated by his polls, questionnaires, and interviews. Thus, for example, to measure an individual's relative acceptance of government intervention in the economy, one might ask a series of hypothetical questions to gauge his feelings about specific interventions before aggregating them to create a general attitude index. Whether such techniques were put to work to study American political attitudes, as in Lazarsfeld's study of the 1940 US elections, in *The American Soldier*, or to study opinions and communication systems abroad, the problem of reading minds and mapping the impact of information and communication was the central preoccupation.[74]

Thus, during 1953–54, Luce spent a year working out of the Behavioral Models Project's brownstone rowhouse on the Upper West Side of Manhattan, before heading off to the Center for Advanced Study in the Behavioral Sciences at Stanford for the center's 1954–55 year. Both there and in New York, as part of his job for the ONR, Luce passed much of his time summarizing recent work on mathematical models in psychology and the behavioral sciences. One of his first major products—a long review paper on applications of information theory in the behavioral sciences—would later appear in the edited volume *Developments in Mathematical Psychology* (1960), while others would form the core of the three-volume edited *Handbook of Mathematical Psychology* (1963–65).[75] Yet quite apart from such essentially rapportorial work for the Behavioral Models Project, Luce's time at Columbia and at the Stanford Center for Advanced Study in the Behavioral Sciences clearly marked a continuation of his overall intellectual development away from models of social networks and group interaction. Instead, Luce began to focus on a question that clearly reflected the kind of work ongoing at the Bureau of Applied Social Research: on what logical basis could social and behavioral scientists measure what was on the minds of the individuals they studied? This question was central in Lazarsfeld's search for "latent structures" that underlay the statements and actions of his interviewees. It was also a longstanding issue in laboratory-based psychophysics, the study of the relationship between external stimuli (from sound and light to pressure and temperature) and subjective perception and sensation, a field that increasingly attracted Luce's attention. Thus, in his publications from the later 1950s and 1960s, Luce would seek to develop a general theoretical basis for psychological measurement applicable across the social and behavioral sciences.[76] Yet, as we shall see, Luce's project was far from unique. Already in the early 1950s, while Luce was still at MIT, ideas from game theory, information theory, social psychology, and psychometrics were beginning to coalesce in a major project on "the measurement of values" taking shape at the University of Michigan and at the RAND Corporation with the financial support of the Ford Foundation's behavioral sciences division.

Measuring Values at the University of Michigan

Luce's time at Columbia brought him into contact with Howard Raiffa, another recent mathematics PhD then teaching in the department of mathematical statistics, who also chaired the Behavioral Models Project's steering committee.[77] As we saw earlier, Raiffa's career was in many ways typical of the postwar generation of applied mathematicians who developed close connections with military funding agencies. His 1950 PhD thesis in mathematics—based on an ONR report on approaches to solving non-zero-sum games—gained him an audience with Albert Tucker's group at Princeton and opened the possibility of a successful career at the intersection of operations research, mathematics, and statistics. This career trajectory brought Raiffa to Columbia in the fall of 1952 as a replacement for Abraham Wald, who tragically died in a plane crash in India in 1950. Since, in addition to his familiarity with game theory and topology, Raiffa had spent his first postdoctoral year (1950–51) at Michigan lecturing on Wald's 1950 *magnum opus*, *Statistical Decision Functions*, his appointment at Columbia represented a compromise between the needs of the statistics department and the mathematics department. Yet precisely like Luce, Raiffa's second postdoctoral year at Michigan had provided him with a detour through the behavioral sciences that would leave a lasting mark on his career and on his views on game theory's relevance to the social and behavioral sciences.[78]

This detour was facilitated by some of the first grant applications approved under the newly operational Behavioral Sciences Division of the Ford Foundation. Michigan's connection with the foundation had developed early. In the fall of 1950, along with a number of other universities, it became the recipient of $300,000 of foundation money to provide support "for the further development of university resources for research in individual behavior and human relations" over a period of five years.[79] By the summer of 1951, the funds were being distributed to research programs within the university by application to a Committee on Research Resources in Human Behavior, chaired by Donald Marquis, who, in addition to serving as a senior advisor to the foundation, chaired Michigan's psychology department.[80] While several different research projects attracted funding, an Interdisciplinary Program in the Application of Mathematics to the Behavioral Sciences, led by the psychologist Clyde H. Coombs, would focus principally on general methodological and theoretical innovation. By the end of 1953, the program could report the collaborative involvement of seven faculty members from psychology, philosophy, sociology, economics, and mathematics, as well as over two dozen postdoctoral and predoctoral fellows in these fields. By supporting training and research via stipends and a series of visiting lecturers, the program hoped "to develop students in the behavioral sciences with sufficient training in abstract mathematics to enable them to do rigorous model building and theory construction" and "to stimulate and carry out empirical research related to such theory construction." The program also worked closely with another recipient of Ford money on campus, the Research Seminar in Quantitative Economics, which launched in October of 1951 and catered to economists interested in quantitative methods.[81] Howard Raiffa, the seminar's postdoctoral fellow, simultaneously served as *rapporteur* for the program, taking notes on developments in their regular meetings. As a result, during the 1951–52 year Raiffa immersed himself in the behavioral science literature with an eye toward finding opportunities for "theory construction" of the sort envisioned by the Ford Foundation.[82] The research seminar and the program met regularly to discuss the ongoing projects of their staff members.[83]

By the winter of 1952, discussions in the seminars had refined a theoretical program that seemed capable of uniting the various branches of behavioral science there represented in a common endeavor. As a result, the senior figures in the seminars—Coombs, the economist Lawrence R.

Klein, and the mathematician Robert M. Thrall—approached the foundation directly for additional funding that they hoped would more firmly institutionalize their work. In a grant proposal titled "Proposal for a Research Program in the Measurement of Values" and transmitted to the foundation on February 15, 1952, the three called for startup funds and an annual budget of $83,400 to establish an "experimental laboratory" that would begin operation in time for the 1952–53 academic year. As they noted in the proposal, their research topic, "the measurement of values," was pitched directly at one of the bullet points from the Ford Foundation's 1949 study committee report.[84] The topic also provided an intellectual meeting-point for the diverse researchers connected with the Michigan seminars. Coombs, mentored by experimental psychologist L. L. Thurstone at the University of Chicago, represented a classic tradition of psychophysical research focused on exploring the possibility of measuring subjective responses to stimuli.[85] While his early work had focused on traditional psychophysical topics such as the determination of sensory thresholds and the testing of psychophysical laws governing sensation, starting in the late 1920s Thurstone's interests began to run toward social psychology and the experimental study of values, attitudes, and beliefs—for example, by devising questioning procedures to determine subjects' relative ranking of the severity of crimes, or their attitudes toward individuals of other nationalities. By the spring of 1952 he was also a consultant to the Chicago Quartermaster Depot arranging the design of experiments to measure soldiers' preferences for different combinations of food.[86] These interests were united by a common theoretical focus on exploring the properties of functions that captured the relationship between external stimulae and subjective responses—whether these were the utility functions and indifference curves of economics, or the psychophysical laws relating stimulus and sensation and developed by Gustav Fechner and others.[87] In a similar manner, during the war years Coombs had begun to move away from traditional psychometrical problems of sensation toward more general theoretical explorations of the measurement of subjective quantities (often dubbed "the new psychometrics"), whether via laboratory experimentation, objective or projective testing, or surveys.[88]

Klein, the economist, likewise found the issue of how to measure values central to his discipline. "Since the days of Adam Smith," the grant proposal noted, "economists have been concerned with the problems of value," with "utility" serving as a foundational concept of the discipline, especially the emerging subdiscipline of welfare economics.[89] Although the fortunes of measurable utility had fluctuated significantly in the interwar period, as we have seen the publication of *Theory of Games and Economic Behavior* sparked a reexamination of the concept, especially by the economists at Klein's institutional homeland, the Cowles Commission for Research in Economics at the University of Chicago. There, Jacob Marschak found himself in an exchange of journal articles with Milton Friedman over the extent to which the hypothesis of expected utility maximization could "rationalize" a wide variety of economic activities involving risk, and over the relevance of the new utility theory to classic tenets of welfare economics such as the diminishing marginal utility of money.[90] Game theory was also in part the hook that drew Thrall, the mathematician, into the grant proposal. An expert on the mathematics of linear and nonlinear programming with funding relationships to the ONR, he had also recently been involved in a joint project with Coombs, also sponsored by the ONR, on the "Mathematics of Measurement." While this project was aimed primarily at the measurement of sensation, it also held out the promise of producing "an axiomatic system applicable to psychological values." Such a new system was necessary because "these axioms are not the same as those required for the ordinary operations of arithmetic": subjective values are generally expressed in terms of relative degrees of preference between alternatives rather than absolute magnitudes (much as sensations in response to stimuli are described in relative terms), therefore necessitating the development of a new mathematical framework for their measurement.[91]

Two recent works in particular raised the hopes of Coombs, Klein, and Thrall that such a new axiomatic system for the measurement of values might be at hand. The first was a 1951 monograph by the Cowles Commission research associate Kenneth Arrow, titled *Social Choice and Individual Values*. Arrow—who spent time at the RAND Corporation as a result of a subcontract between the Corporation and the Cowles Commission—was apparently inspired by a question put to him at RAND as to whether nations could be treated as having coherent utility functions for purposes of game-theoretic models of international negotiations.[92] Arrow recognized that this geopolitical question dovetailed with a longstanding debate within welfare economics concerning the possibility of defining the collective good. Political economists in the utilitarian tradition, from Jeremy Bentham onwards, had tended to assume that the values of individuals could be compared, and that individual utilities associated with any given course of action could be summed to determine a collective welfare that could then be maximized.[93] In his work, Arrow cast the problem in the broadest possible terms by inquiring as to whether any logically consistent algorithm, satisfying a number of desirable characteristics (like non-dictatorship), could exist for aggregating the preferences of individuals. In general, Arrow proved that it couldn't. While Arrow's work took a theoretical approach to the problem of preference aggregation, by contrast, a recent article by Harvard statistician Frederick Mosteller and the psychologist Philip Nogee sought to place the measurement of value on a more secure experimental footing. Adopting the definition of utility proposed in von Neumann and Morgenstern's book, their research suggested that experimenters could make measurements to create a reasonable representation of an individual's utility curve by observing the behavior of individuals in a series of simple games, but that those individuals' subsequent behavior in more complicated choice situations could be inconsistent with their previously derived utility functions.[94]

The significance of these two works to Coombs, Klein, and Thrall in 1952 is particularly striking. Arrow's work is typically remembered among political scientists for its profound implications about the foundations of democratic politics, while Mosteller and Nogee's paper was explicitly pitched as a contribution to assessing the empirical adequacy of the von Neumann–Morgenstern theory of utility. For the Michigan group, the critical problem highlighted by these works was not simply that individuals might violate any given set of axioms of rational behavior, thereby somehow falsifying utility theory as a model of human behavior. More important from their perspective were the difficulties Arrow, Mosteller, and Nogee had uncovered for *any* attempt to measure values along the lines envisioned by the Ford Foundation in its research program. Quite apart from what it said about the workings of governments and economies, Arrow's analysis suggested difficulties with the use of statistical aggregation schemes intended to move from data about the preferences of individuals (derived from polling or surveys) and statements about their overall values as individuals or as groups. Such schemes were in effect algorithms for aggregating preferences of the sort that Arrow had proven incompatible with a small number of seemingly desirable conditions on their fairness and the consistency of the results they produced. Likewise, Mosteller and Nogee's work suggested the problem of comparing utility measurements—whether between individuals (the central problem of Arrow's work), or between the same individual at different times and in different situations. In their grant application, Coombs, Klein, and Thrall specifically addressed this point, asserting, "It is felt that the psychological theory of measurement will be useful in situations where interindividual comparisons of utility cannot be assumed."[95] To clarify theoretical concepts and plan an experimental research program, Coombs, Klein, and Thrall called for a "summer seminar," attended by a mixture of mathematicians and behavioral scientists, to "examine the axiomatic bases" of the new mathematical theories of value measurement, "with the immediate objective of planning experiments designed to study their applicability and reasonableness in settings as realistic as possible."[96]

While the foundation did not immediately fund the "experimental laboratory," it went forward with the seminar, which soon expanded its focus from the "measurement of values" to assume its ultimate official title, the University of Michigan Summer Seminar on Design of Experiments on Decision Processes, to be held in the summer of 1952. It is not completely clear why the seminar underwent this change of focus. It was likely due in part to the recognition that many potential invitees to the conference were affiliated with the RAND Corporation, at least in the summers, and that a number of new projects at the corporation were relevant to the topic of the seminar. One of these new projects was the Systems Research Laboratory (SRL), which opened at RAND in the spring of 1952 under the direction of Brandeis psychologist John L. Kennedy. The SRL was set up to take Air Force contracts to support studies of team behavior relevant to "the operation of complex man-machine systems"—in particular, the operation of an "air-defense direction center." In many ways, these studies would represent a more realistic and drastically scaled up version of the note-passing experiments undertaken by Alex Bavelas at MIT only a few years earlier. By the mid-1950s, the project would report on simulations of group communication and learning processes—just getting under way in 1952—carried out using a full-scale model of the air defense center: a crew of thirty to forty men, working for 200 hours, directing some 10,000 flights that defended some 100,000 square miles of territory.[97] This kind of simulation proved fantastically expensive: in 1955 the SRL would spend over $2.8 million.[98]

The 1952 Summer Seminar on Design of Experiments on Decision Processes, Santa Monica

For eight weeks in the summer of 1952, an interdisciplinary group of thirty-seven invitees—nearly twice the original number proposed—met regularly for eight weeks in buildings on the campus of the former Santa Monica City College, which had been vacated pending conversion to a high school. Of the session, a week and a half was devoted to introductory presentations before an interim period during which the attendees were broken down into working groups "for the purpose of delineating specific problems with respect to both theory and experimental design." A number of specific studies were actually drawn up and conducted, both during and immediately following the conference; the session was brought together by presentations of final papers.[99] As Coombs's final report to the Ford Foundation would note, the attendees came from a diverse mix of disciplinary backgrounds including "philosophers (logicians), psychologists, and sociologists whose research in their own fields was variously classified as game theory, utility theory, organization theory, measurement theory, and learning theory."[100] Naturally, given the role of game theory and related areas of mathematics in crystallizing the idea for the conference, many were veterans of RAND's mathematics program like John D. Williams and Merrill Flood who had conducted the first tentative experiments on game theory at RAND in the preceding years. Others, such as Jacob Marschak, represented the Cowles Commission with its ongoing subcontract with RAND. However, still others represented research traditions in the behavioral sciences that were relative newcomers to game theory and related areas of mathematics. In addition to Clyde Coombs, the attendees included a number of experts in learning theory and the study of sensory discrimination. Leon Festinger, the original inspiration for R. Duncan Luce's work on social networks and social psychology, also attended, while experts on organization theory like Herbert Simon and Alan Newell came in their capacity as consultants to RAND's SRL. Von Neumann and Morgenstern were invited to contribute as well, but apparently only Morgenstern could attend.[101]

The proceedings of the summer seminar, ultimately published quite selectively under the title of *Decision Processes* in 1954, make for confusing reading. As with the almost contemporaneous Macy Conferences on cybernetics, they provide a snapshot of a number of disciplines in flux and capture the ghostly outlines of research programs that almost— but not quite—came into existence. The different groups of researchers clearly approached the conference with substantially different skill sets and interests, with a particularly strong division between "those most interested in working out the logical consequences of a theory or a set of axioms, and those most interested in finding out how people acted."[102] Given the preponderance of mathematicians and economists attending, theoretical studies tended to dominate the conference proceedings; however, a number of actual experimental studies appear to have emerged from the seminar's interdisciplinary "working groups," and a few of these were reported in the proceedings. Some of these—for example, the report on "Some Experimental n-Person Games" by a group of the younger game theorists, including John Nash—represented a continuation of the earlier attempts to develop game theory into a general predictive theory of human behavior in game situations. And this presentation to the conference, chronicling the problems involved in testing theories of multiplayer games, drew upon previous experience with the game studies taking place at RAND.[103]

Despite the central role of game theory in bringing the seminar's participants together, the consensus among most of the papers seemed to be that the existing theory of games was of limited value in furthering either the Ford Foundation's goal of "measuring values," or the RAND Corporation's newfound interest in understanding and improving group task performance. Purely theoretical schemes to sum the utility functions of individuals—whether grounded in the theory of games or in some alternative mathematics of aggregation—seemed difficult to justify without resort to a number of ad-hoc assumptions and vague appeals to "reasonableness."[104] Moreover, imposing even a modest number of seemingly reasonable conditions on the methods of measuring and aggregating values could lead to contradictions. For the mathematicians at the seminar, the realm of the logically *impossible* was sharply defined; what might actually *be* remained an ill-defined gray area.

Indeed, apart from "Some Experimental n-Person Games," the other major empirical study to appear in the conference proceedings, conducted by social psychologist Leon Festinger and two of his Stanford colleagues, explicitly assumed from the outset that game theory as it then existed was useless for deriving collective preferences from the preferences of individuals, or predicting how individuals would interact in a group. In this, Festinger quite accurately grasped the message of von Neumann and Morgenstern's theory of multiplayer games. Consider the "simple majority game," the focus of chapter 5 of *Theory of Games*: three players are to divide a fixed pot of money or resources, yet no individual can claim the pot by him- or herself and must therefore enter into a coalition and somehow divide the winnings with another player. In such a game, "a given player has no rational basis for selecting between the other two members to form the initial partnership," so that stable coalitions might *never* form out of sheer collective indecision, or if they did, all three possible coalitions should occur with equal probability. Yet this was "contrary to everyday experience": typically, some coalitions between individuals would form and not others. The reason Festinger et al. postulated for this fact was that truly *social* motives and attributes—motives and attributes that went beyond an individual's desire for a greater share of the game's winnings—were somehow unequally distributed among the experimental subjects. Depending on how the players initially perceived the capabilities and intentions of the others, a dollar received from one coalition partner could be valued more highly than a dollar from another, suggesting that human behavior was influenced by multiple incommensurable measures of value. The resulting experiments thus revealed much about how individuals developed perceptions of their social relationships with one another. Von Neumann and Morgenstern's theory of the three-person zero-sum game, assumed from the outset to be "contrary to everyday experience," nevertheless played a key role in the design and description of the subsequent experiments.[105]

Given these perceived weaknesses of game theory for both the "measurement of values" agenda and for the studies of group problem-solving then in prospect at RAND, many of the contributions to the seminar's proceedings focused on integrating game theory with new theories of communication, information flow, and social structure. Thus Flood and his old bridge partner Mosteller had devoted much of their work the year prior to the seminar integrating game theory and learning theory to form "game-learning theory." Their classic model described the so-called "two-armed bandit," a problem in statistical decision theory that describes the following experimental situation. A payoff is hidden behind one of two closed doors according to some probability distribution; the experimental subject must choose the correct door in order to receive the payoff. If the subject is asked to do this repeatedly, with what probability should he choose each of the two doors so as to maximize the accumulated payoff? This situation corresponded to a game in which the player does not know the interests of his opponent from the outset but must learn them as he plays over time.[106]

Another example of this concern with additional features of social structure and information transfer is found in Jacob Marschak's contribution to the seminar's proceedings, a lengthy treatise introducing an "Economic Theory of Organization and Information." The fundamental project that Marschak proposed involved investigating the properties of different kinds of social organizations that might form, each of which was defined by the different degrees of common interest shared by its members. The "coalitions" of von Neumann and Morgenstern's theory were thus just one form of social organization; other significant organizations were "teams" and "foundations," both of which possessed greater decision-making cohesion than the fractious coalitions. The purpose of such organizations was the gathering and sharing of information among members, and the performance of actions based on this information. Given this understanding of the formation of social groups, Marschak was able to perform some very simple calculations concerning the optimal rate of information gathering and transmission, the best choice of variables, the delegation of tasks, and the like.[107] For Marschak, the exercise constituted a first attempt at developing a mathematical theory of the firm, and he and some of his colleagues at Cowles would begin constructing a general "theory of teams" at this time. In the process, his work quickly became detached from the theoretical idiom laid down in *Theory of Games*, moving instead toward the abstract models of resource allocation, information transfer, and pricing that were the signature ideas of Cowles during the 1950s.[108]

As the "teams" or "organizations" in question got more complex, the phenomena in question began to separate themselves from the mathematics of game theory, or indeed from formal models capable of isolating "values" altogether. Allen Newell, for example, would report to the seminar on ongoing projects at RAND that investigated the function of "organizations" tasked with an air defense mission. Some of the experiments involved simple Bavelas-style exercises in note-passing—for example, four RAND test subjects manning a radar station, two antiaircraft guns, and a supply depot, and tasked with processing information about the location of incoming enemy aircraft and distributing information and ammunition to shoot them down. "This environment was simple enough to allow the subjects to understand it almost perfectly after a few sessions, and thus produce a game-like strategy for solving the problem," as Newell noted. The experiment's simplicity made it a candidate for theoretical analysis, yet it also fell far short of recreating the kind of "spongy" environment "comparable to the real life organization we were trying to simulate." In the SRL Newell had in fact managed to create such an environment, a replica of an air defense center, so that "visiting military men thought the whole outfit operated in a 'realistic' way." This realism was important for RAND's clients, presumably; but unlike the simple note-passing experiment, it gave rise to organization-level "adaptive behavior not taking place on a conscious or intellectual level."[109]

In the end, despite the cast of luminaries the summer seminar had brought together, it could only have seemed like a failure to its organizers at the University of Michigan. Part of the failure was intellectual: as the papers in the conference proceedings seemed to indicate, the goal of finding a coherent theoretical framework for measuring values seemed scarcely closer in the fall of 1952 than it had a year earlier. Ward Edwards, a systems engineer and decision analysis guru, would sum up the problem in his review of *Decision Processes*, by saying that even though "utility" or "subjective value" formed the central concept of the book, Coombs et al. had failed to provide any kind of consistent and general technique for making measurements of utility. As far as Edwards could tell, they exhibited only one actual "method of measurement"—but that method, he suggested, "is based on a very restrictive and unlikely assumption." Not that Edwards felt any particular sense of superiority in making this observation: he too had tried his hand at proposing an experimental methodology for the measurement of utility, and had found it wanting. Although he remained a true believer that utility was (and would continue to be) essential to any plausible account of human decision-making behavior, he nevertheless warned, *"Eventually, someone is going to have to do something about this problem."*[110]

More immediately relevant to the seminar's organizers, the Michigan group would probably not be the ones to "do something" right away, since the experimental laboratory that Coombs et al. had proposed would never be funded. Already in December of 1952, the Ford Foundation's Bernard Berelson would write to Coombs to break the bad news.[111] The conference proceedings that emerged two years later can be read as a sad reminder of a research program that might have been.

From *Theory of Games* to *Games and Decisions*

From the Ford Foundation's perspective, the rejection of the experimental laboratory probably had less to do with the intellectual difficulties exposed by the Summer Seminar and more to do with the success of another project that it had sponsored during the summer of 1952. While Raiffa would coauthor the first paper in the seminar's proceedings—a fascinating meditation on the confusing distinctions between descriptive, predictive, and normative models of human behavior that emerged in the course of the summer—he did not in fact attend the seminar himself.[112] Instead, he spent that summer at Dartmouth College on behalf of a "Program for Furthering the Mathematical Training of Social Scientists," jointly funded by the Social Science Research Council (SSRC) and Ford. The program stemmed from a 1949 symposium, held at a joint meeting of the American Mathematical Society and the Econometric Society and chaired by Jacob Marschak, on the mathematical training of social scientists. Its goal was to develop teaching materials that would introduce social and behavioral scientists to the wealth of new mathematical models that were becoming available.[113] The staff of the program, which in addition to Raiffa included psychologist George Miller and the future Nobel economics laureate Robert Solow among its members, met from June 23 to August 23, 1952, to produce "sets of problems . . . to illustrate the changes in language required for the mathematical representation of a problem" and "a glossary . . . to give the mathematical equivalents of fairly common phrases in the social sciences," thereby easing the adoption of new ideas from game theory and related fields into the social sciences.[114]

The program would gain additional backing in the fall of 1952 when the Ford Foundation granted the SSRC $102,600 over three years to support summer training institutes for social scientists using the materials Raiffa et al. had developed. Subsequently, summer institutes were held at Dartmouth in 1953 and at Stanford and the University of Michigan in 1955.[115] Raiffa, along with a statistician and a psychologist, constituted the teaching staff of the 1953 summer meetings. Convening in New York in April of that year, they set the curriculum of the summer seminar to include:

> sets and relations; axiomatics; probability including stochastic processes; the calculus, including the development of the real number system, differential

and integral calculus of one and several variables, convergences of sequences
and series; matrix theory, linear equations, and quadratic forms; the theory
of games; linear programming; and various mathematical models in the social sciences, including, among others, learning models, and the theory of
choice.[116]

In addition to the regular lecturers, the forty-one attendees—mostly specialists in psychology, sociology, and economics—would be treated to
guest lectures by Leon Festinger, Paul Lazarsfeld, Frederick Mosteller,
and Albert Tucker, among others.[117]

The Behavioral Science Program's shift toward funding the *teaching*
of new mathematical theories emanating from mathematical economics
and operations research, exemplified by Raiffa's summer training programs for social scientists, was directly connected with the demise of the
"experimental laboratory" for the measurement of values proposed by
Coombs, Klein, and Thrall. As Bernard Berelson would note in his rejection letter to Coombs from December of 1952, "With the support of
the Michigan seminar last summer and the grant for mathematical training for behavioral scientists . . . the Division has expressed for the time
being its interest in the application of mathematics to the behavioral
sciences."[118] Meanwhile, back at Columbia during the academic years
1953–1957, pedagogy and the exposition of new mathematical models in
the behavioral sciences remained a key focus of Raiffa's work. With the
arrival of Duncan Luce in 1953, Luce and Raiffa initially set out to fulfill the mandate of the Behavioral Models Project by publishing a series of brief pamphlets on different mathematical techniques relevant to
scholars across the social and behavioral sciences. The two decided to
kick off the process with a concise exposition of the theory of games; the
end result was the 500-plus-page *Games and Decisions* (1957), which reviewed much of the literature on game theory, utility theory, and social
choice theory prior to that date. Much of the material was composed at
Stanford's Center for Advanced Study in the Behavioral Sciences, where
Luce was a fellow from 1953 to 1954, and Raiffa from 1954 to 1955.[119]

Although Luce and Raiffa would claim that "the overall outline [of
Games and Decisions] parallels the original structuring given to the
theory by von Neumann and Morgenstern," their exposition heavily reflected changes in the disciplinary locus of game theory that they had
done so much to bring about.[120] True, the core of their exposition followed the progression of *Theory of Games and Economic Behavior*

from a discussion of the extensive and normal forms of the game, to the two-person zero-sum game, to non-zero-sum games, and finally to general multiplayer games. Yet these traditional areas were augmented by expositions of new bodies of literature that had grown up in connection with the theory since 1944. Some of these new topics plainly owed their presence to game theory's close postwar relationship with operations research: aspects of statistical decision theory, linear programming, methods of solving two-person zero-sum games, and the theory of games with infinite pure strategy sets. Others stemmed from the work of the book's authors in the context of "behavioral science": Luce's "Ψ-stability" and his probabilistic theory of subjective utility scales, and an exposition of Nash's bargaining theory, approached through the lens of Raiffa's "arbitration schemes."

As noteworthy as the specific differences in content between *Theory of Games* and *Games and Decisions* were broader differences in tone and theme. Gone was the bold self-assurance with which von Neumann and Morgenstern proposed what they imagined would develop into a positive "theory" of individual and collective behavior, capable of revolutionizing the breadth of the social sciences. As a description of human behavior, the assumptions of game theory increasingly seemed dubious, or at least contestable. Moreover, the study of multiplayer games was still disintegrating into a swarm of distinct theories, each of which appeared to possess clear validity only in quite specific circumstances. Thus Luce and Raiffa felt that games like the Prisoner's Dilemma sounded "the death knell for the [Nash] equilibrium concept as the principal ingredient of a theory of non-cooperative non-zero-sum games," if the point was to provide "a realistic theory for all possible non-cooperative non-zero-sum games," even as von Neumann and Morgenstern's theory faced similar challenges.[121] While some readers criticized Luce and Raiffa for focusing too much on the theory's descriptive or predictive capabilities without considering the more conceptual, "explanatory" virtues of rational choice theorizing, more shared Herbert Simon's agreement with their "basically pessimistic forecast for game theory." In the absence of a satisfactory theory of *n*-person games, the theory could never have much to contribute "toward the construction of a theory of human economic and social behavior." While "the authors do not offer a satisfactory explanation of why game theory failed to fulfill this initial promise," Simon noted, "my own hypothesis is that the theory rests on a fundamentally wrong view of the human decision-making organism and the nature of rational choice." Simon's preferred description involved actors who displayed "approximate rationality" (akin to what he would later call "bounded rationality") as a way of bringing the ideal of rational action back in line with a descriptive theory of human behavior.[122]

Despite Luce and Raiffa's pessimism about the prospects for the theory of games serving as the basis for a new social science, their exposition of the theory nevertheless held out the possibility of unifying a broad swath of disciplinary terrain around a new way of talking about individual and collective behavior in terms of decision-making and decision processes. This possibility was perfectly reflected in the title— *Games and Decisions*—suggested by Albert Tucker.[123] It was also reflected in the way Luce and Raiffa foregrounded the theory of utility in their book, making it the focus of their second chapter and rescuing it from its position as an appendix to the 1947 edition of *Theory of Games*. Instead of simply supporting the development of a theory of games, utility theory served to introduce a taxonomy of decision-making situations that formed much of the basis for the remainder of the book. These ranged from individual decision-making under certainty (linear programming) and uncertainty (von Neumann–Morgenstern utility), to theories of bargaining and arbitration (game theory), and, finally, to theories of social choice and group decision-making (Kenneth Arrow's social choice theory). As historian of science Hunter Heyck has noted recently, formal characterizations of decision-making, so common during this era, generalized away from the specifics of the thing *doing* the choosing, focusing instead on formalizing the nature of "decision processes" of every kind: human cognition, deliberations in a committee, elections in a democracy. As we shall see in the following chapter, this ability to generalize across levels of analysis, from mind to society, would prove an essential desideratum of theories in academic "behavioral science" during the later 1950s and 1960s.[124]

Thus, in so many ways, *Games and Decisions* served both as an epitaph for an early era of game theory's development and as a harbinger of things to come. The book was dedicated to the memory of von Neumann, who had died shortly before it appeared, yet its publication marked a significant shift in game theory's center of gravity from the one that von Neumann had known. Through Luce and Raiffa's efforts, and through the creation of new networks of patronage by organizations like the Ford Foundation, RAND, and the Office of Naval Research, the mathematics of games sat at the center of a new intellectual terrain—

the study of decision-making—stretching from psychometrics and small group studies to economics and sociology. Moreover, as we will see repeatedly in subsequent chapters, Luce and Raiffa's book would mark a turning point in the history of game theory's spread into the social sciences. While before game theory spread largely along lines of individual contact between mathematicians and particular groups of behavioral scientists, *Games and Decisions* provided a highly readable survey of the mathematics of rational choice that was at last free of the technicalities that generally restricted von Neumann and Morgenstern's 1944 book to a small group of mathematical initiates. As a result of these developments, during the 1960s, game theory would emerge from the laboratory and the technical nitty-gritty involved in studying cognition and behavior to facilitate a sweeping series of debates about the role of human reasoning in managing the Cold War arms race.

CHAPTER FIVE

The Brain and the Bomb

What would keep the world safe in an age of nuclear weapons? So much could go wrong—or at least, that was the lesson of any number of Cold War–inspired novels, films, and short stories, from *Dr. Strangelove* to *War Games*. Armageddon could arrive in the form of a deliberate Soviet attack, overwhelming America's air bases abroad or early warning systems at home; or it could be born of a careless strategic miscalculation. But disaster could also flow from causes that were at once more mundane and terrifying in their intractability: buggy computer programs, poorly soldered electronics, a paranoid wing commander, a lonely watchman in a missile silo, the inscrutable "human factor." And to these threats, what was the answer? Bigger H-bombs, longer-range missiles, more reliable computers, and the steel, concrete, and rebar of civil defense installations were some possibilities; but so too was the recognition that, in the final analysis, wars were both started and prevented by human beings, with all their strengths, frailties, intelligence, and irrationality.

These, in a nutshell, were the issues at the heart of a great series of public debates about the possibility of ending the arms race, the prospects for peace, and the perils of nuclear weapons that played out during the high years of the Cold War. These debates permeated popular culture in the cinema and the press, but they also captured the attention of the day's great intellects, triggering an outpouring of ponderous tomes from university presses, articles in scholarly journals from *Daedalus* and the *Journal of Conflict Resolution* to *Bulletin of the Atomic Scientists* and technical reports to the Arms Control and Disarmament Agency. They were the focus of talks at the Council on Foreign Relations, panel discussions at Berkeley's Institute of International Studies, summer seminars and working groups sponsored by the National Academy of Sciences, or international meetings like the famed Pugwash Conferences, which brought together some of the world's most eminent scientists to ponder the great problems of war and peace in the thermonuclear age. They drew in figures from across the spectrum of academic disciplines and political outlooks, from the likes of Herman Kahn, RAND Corporation strategist and plausible model for Stanley Kubrick's "Dr. Strangelove," to peacenik philosophers like Bertrand Russell; from liberal democrats like Hubert Humphrey to pacifist A. J. Muste; from technically trained "defense intellectuals" in administrative positions at the Pentagon to New Left sociologists like C. Wright Mills.

These debates—often contentious, even acrimonious—form the context in which the idea became widespread that the Cold War between the two superpowers was a *game* in the technical sense of game *theory*, and in which the problem of how to choose rationally in this situation became perhaps *the* central problem of the age. This characterization of the Cold War was not entirely new: for example, John McDonald's early journalistic coverage of game theory for *Fortune* magazine in the late 1940s had invoked von Neumann's coalitional form of multiplayer games to describe the rapidly shifting alliances that marked postwar geopolitics.[1] But the late 1950s and early 1960s saw a burst of new game-theoretic characterizations of the Cold War. The Prisoner's Dilemma game—initially dramatized by the mathematicians as a story of cops and robbers—came to stand in for the arms race, with the superpowers deciding between whether to arm themselves further (confess) or to disarm (not confess). Or consider the game of Chicken, introduced in 1959 by Bertrand Russell to characterize the policies of nuclear "brinksmanship" and "massive retaliation" articulated by US Secretary of State John Foster Dulles. In an image familiar to viewers of *Rebel without a Cause*, Russell painted the superpowers as drag racers speeding toward each other on a one-lane road; the first to swerve stayed alive, but lost the "game." The analogy between Chicken and the predicament of the Cold War powers was further popularized by Herman Kahn in his 1960 blockbuster book, *On Thermonuclear War*; and soon thereafter, this "sport" began to appear in print in game theory's "normal form" matrix.[2] Finally, in a series of high-profile books and articles, Thomas Schelling shoehorned both economic "bargaining" and geopolitics into a game matrix, transforming the arms race into haggling and economic exchange into strategy.[3] As the "rational actor perspective" on international relations pioneered by Schelling and others began to take root in the political science literature during the 1960s, the histories of many key geopolitical events of the Cold War era (from the Cuban missile crisis to Richard Nixon's Operation Giant Lance) would be rewritten through this lens, to the point that post hoc analysis and history could become difficult to distinguish.[4]

The goal of this chapter is not to trace the fortunes of the "game" conception of the Cold War in any exhaustive way, nor is it to assess the role of game theory or "war gaming" in shaping American national security policy.[5] Rather, it is to explore what was at stake in characterizing the Cold War as a "game" in some 1960s-era debates over national security in the nuclear age. It does so by focusing on a community of behavioral scientists that coalesced around the University of Michigan's Mental Health Research Institute (MHRI) in the mid-1950s. The research program pursued by the institute epitomized the kind of ambitious interdisciplinary behavioral science championed by the Ford Foundation in its early years. With support from a mix of private, state, and military agencies, and by emphasizing a combination of theory-building and virtuosic experimental technique, scientists there aimed to create a framework for the unified study of living "systems" at all levels of organization, from neural networks to societies. Indeed, so grand were their ambitions that they turned their sights on the largest scales of human behavior: international politics and macro-level phenomena of war and peace. At Michigan, such ambitions found expression in a program of inquiry *cum* activism associated with the terms "conflict resolution" and "peace research," and a Center for Research on Conflict Resolution (CRCR) that would be copied worldwide.

In this context, game theory helped to facilitate several prominent interventions of behavioral science into discussions about conflict, war and peace in the nuclear age. Not only did the theory play a key role in the conduct of behavioral science at Michigan and elsewhere, inspiring dozens of studies annually probing the psychology of conflict, cooperation, and teamwork in small groups, but game theory's notational conventions would facilitate flows of insights and personnel between the psychological laboratory and other disciplines, from mathematics and philosophy to economics and political science. Game theory thus held out the promise of permitting behavioral scientists and peace researchers to explore the psychological dimensions of conflict and cooperation that underlay interpersonal relations and geopolitics alike. But in the final analysis, attempts to build a scientific approach to war and peace on game theory, and to solve the great problems of the nuclear age, ultimately foundered on the problem of rationality. In the games that seemed most relevant to the problems of the Cold War, like Chicken or Prisoner's Dilemma, did "rationality" reside in calculations of maximization, in the design of less error-prone technological systems, or perhaps in therapeutic interventions of competent behavioral scientists? Correspondingly, who was the voice of reason—mathematicians, "strategists," or members of the various social and behavioral sciences? Against the backdrop of intellectual tensions and personal animosities within the conflict resolution community, increasing campus activism against the Vietnam War, shifts in funding and personnel, and a broader "critique of rationality" in American culture, the ambitious twin programs of behavioral science and of conflict resolution found at the University of Michigan fractured in the later 1960s.

Institutionalizing Behavioral Science at the University of Michigan

The intellectual origins of the Mental Health Research Institute and its style of research lay not at Michigan, but at the University of Chicago. Here, just as at Michigan, the first stirrings of interdisciplinary behavioral science were closely connected with emergence of the field as an object of Ford Foundation largesse. Indeed, the very term "behavioral science" has been credited to a group of Chicago faculty, who began meeting regularly in 1949 to form the Committee on the Behavioral Sciences. James Grier Miller, the leading figure on the committee, would later claim that the term found favor "first, because its neutral character made it acceptable to both social and biological scientists and, second, because we foresaw a possibility of someday seeking to obtain financial support from persons who might confound social science with socialism."[6] Miller's dual explanation for the name rings true. For one thing, Miller had worked to form the committee with the encouragement of Donald Marquis, the University of Michigan psychologist who was then in the process of guiding the Ford Foundation toward the creation of its funding initiative in "Individual Behavior and Human Relations."[7] As a result, Program V at the Ford Foundation and the committee at Chicago worked together closely for a number of years, and Ford money flowed liberally to the university. Like Michigan, Chicago would go on to become one of over a dozen major research universities selected by Ford to receive a $300,000 grant for general support of Research Programs in Individual Behavior and Human Relations in July 1951.[8] Two years later, in February of 1953, an appropriation of $262,500 from the foundation funded Chicago's program of self-study in the behavioral sciences, in parallel with similar self-studies at a number of other universities nationwide.[9]

"Behavioral Science" indeed provided a big tent for the gathered academics. By 1953 the roster of the Committee on the Behavioral Sciences could list eleven faculty members and seven graduate student fellows, with members representing a wide variety of departments. While most came from programs in psychology and physiology, economists, anthropologists, political scientists, and even a historian were also involved.[10] These included *inter alia* psychologists and psychiatrists like Miller, fresh from his wartime work on psychological testing for the Office of Strategic Services and about to emerge as one of the great organizers and fundraisers for research in behavioral science. Ralph Gerard was a neurobiologist who would play a prominent role in the Macy Conferences that launched cybernetics. The economists included Jacob Marschak, director of the Cowles Commission for Research in Economics. Finally, the committee included two "mathematical biologists," Nicholas Rashevsky and Anatol Rapoport. This now-obscure specialty had been represented at Chicago since 1934, when Rashevsky raised a grant from the Rockefeller Foundation to support his work on modeling biophysical phenomena. Initially practiced under the aegis of mathematical biophysics, this was subsequently broadened to a "Committee on Mathematical Biology" in 1947 to signal the committee's interest in modeling a wider range of phenomena, including human relations, on which Rashevsky had just published a book.[11]

From the outset, the group at Chicago accorded a significant place to theory in building a unified science of behavior. Consider, for example, Donald Marquis's remarks at a November 1947 symposium that Miller organized at Chicago to showcase the status of social psychology, and to stimulate roundtable discussions on the possible applicability of psychology to the social problems posed by the advent of the atomic age. Titled "Scientific Methodology in Human Relations," Marquis's essay essentially rehearsed many of the themes that would later run through the Ford Foundation's proposal for research in "Individual Behavior and Human Relations," in whose shaping he would play a critical role. Marquis noted that the prospects for social science and the study of human relations had been hotly debated since the war in venues ranging from the US Senate to "most of the administrative councils of our universities and foundations." These administrators and politicians clearly shared a "recognition that many of the crucial issues of our society today are social issues involving the attitudes and convictions and interactions of people." Moreover, "it is further based on a hope that science might achieve in the social field something of what it has achieved in the physical and biological fields."[12] Marquis then proceeded to lay out what was expected of a mature science of social relations, namely, the ability to formulate, test, and apply theories of human social behavior. Two especially promising new theoretical frameworks for social interaction, he noted, were Rashevsky's *Mathematical Theory of Human Relations* (1947), and von Neumann and Morgenstern's *Theory of Games and Economic Behavior*. At the same time, there remained much to be done in terms of empirically verifying the new theories, comparing them against observations of behavior in controlled situations.[13]

Marquis's mention of Rashevsky's work in the same breath as that of von Neumann and Morgenstern foreshadowed the role that mathematical biology and the theory of games would play as theory-building resources for behavioral science. Individuals associated with the Cowles Commission and Mathematical Biology would emerge as critical figures in the Committee on the Behavioral Sciences' membership. During the fall of 1952 and the spring of 1953, as the Ford-funded self study was getting off the ground, a "theory group" of the Committee on the Behavioral Sciences met almost weekly to discuss promising theoretical frameworks for a unified study of behavior. On January 20, 1953, the group heard a presentation by Rashevsky about his recent work on the mathematics of random neural nets. In a number of recent articles in the *Journal of Mathematical Biophysics*, Rashevsky and his student Rapoport had developed a theory of stochastic networks that described situations in which entities randomly form links. The original inspiration for the theory was neurons in the brain, since Rashevsky's research from the 1930s had generally sought to reduce theories of human behavior to the mathematical properties of their regulatory systems. But in the later 1940s, Rashevsky and Rapoport had adapted their probabilistic network theories to analyze the flow of information in societies and the formation of social structures such as cliques, pecking orders, and the like. For instance, Rapoport had drawn on these theories of stochastic networks to create models of the spread of rumors in an arbitrarily large population. Assuming a single entry point for the rumor and random contacts between members of the populations, Rapoport was able to derive a relationship between the number of contacts held by each member of the population and the final expected percentage of the population that would hear the rumor. As Rashevsky's presentation suggested, models like Rapoport's were significant because they could describe similar phenomena at several different levels of biological organization, the individual brain, and the larger society, thereby suggesting the possibility of a general theory of behavior.[14]

Not long after Rashevsky's presentation, on February 10, 1953, the theory group heard from Jacob Marschak, fresh from the Santa Monica seminar on "decision processes" that had been held the previous summer.[15] Marschak gave the group a summary of developments in game theory that had been on display at the summer seminar. This included an extensive overview of the von Neumann–Morgenstern theory of utility, focusing on clarifying the axioms that described the behavior of individuals under uncertainty.[16] However, Marschak's presentation went beyond individual decision-making to include a very brief treatment of the economic "bargaining problem" using the language of the two-person non-zero-sum game. In Marschak's view, the crucial problem with axiomatic treatments of bargaining, such as John Nash's, was that they invariably introduced assumptions that were not reducible to postulates of individual rationality. The need for such assumptions called for additional modeling of structures at the societal level, such as patterns of information and influence; Marschak would refine and publish some simple models of these patterns in his contribution to the summer seminar's *Decision Processes* volume the following year.[17] The talk concluded with Marschak describing some of the experimental studies of bargaining that he had observed the previous summer.[18]

The presentations by Rashevsky and Marschak reflect the confluence of mathematical biology and game theory–inspired economics at Chicago during this period. Although it is difficult to get a sense of the response to Marschak's presentation from the different members of the theory group, both Rapoport and Rashevsky were plainly aware of the work on game theory, utility theory, and social choice theory going on at Cowles. As Rashevsky's correspondence indicates, faculty from two or-

ganizations scrounged among one another's PhD students to fill their perennially underenrolled graduate seminars and served as guest lecturers
for each other's courses. Marschak, for instance, would present a lecture
in Rashevsky's senior seminar on "Mathematical Theories of Group Behavior" in February of 1955. And even after the Cowles Commission left
for Yale in 1955, Rashevsky would send Marschak the latest edition of
his *Journal of Mathematical Biophysics* as a reminder of the close relationship that had existed between the two programs while they were at
the University of Chicago.[19]

Rashevsky and Marschak's presentations also hint at some of the
themes that would be exhibited by the Chicago Committee on the Behavioral Sciences broader theoretical framework for integrating studies of behavior on all levels of analysis. Often associated with the name
"general systems theory," this framework drew on a wide range of philosophical currents: the organicist philosophy of Alfred North White-
head (of whom Miller was a mentor at Harvard); ideas of information
transfer, regulation, and homeostasis that were emerging from Norbert
Wiener's "cybernetics"; and the holistic philosophy of biology developed
by Ludwig von Bertalanffy since the 1930s. As a number of historians
have shown, such ideas were prevalent among social and biological scientists at the University of Chicago in the immediate postwar period.
The Chicago ecologist Alfred Emerson adopted cybernetic language of
feedback, regulation, and homeostasis to describe the ant societies that
he imagined as functionally integrated "superorganisms," thereby drawing parallels between the organization of the individual organism and
the organization of collections of organisms. Likewise Ralph Gerard's
involvement in the Macy Conferences on cybernetics led him to conceptualize the brain as a self-regulating computing device, similar to the
antiaircraft-fire control devices that Norbert Wiener studied.[20]

Such a theoretical framework appealed to the members of the Committee on the Behavioral Sciences because it provided an exceptionally
attractive template for the organization of interdisciplinary research.
Systems thinking embraced the investigation of analogous structures
("isomorphisms") present at different "levels of organization," whether
the cell, the organ, the organism, a group of organisms, or "society." Rashevsky and Rapoport's work developing the mathematics of "random
networks" was a perfect example: the formation of neural networks and
the spread of information in societies were described by common mathematical structures. And as Miller would point out in a presentation to

the American Psychological Association on the guiding philosophy of the Committee on the Behavioral Sciences, such common mathematical structures enabled the committee's interdisciplinary collaboration, even if the committee members disagreed on almost everything else.[21] As with cybernetics, the language of "models" and "systems" also permitted behavioral scientists to communicate with applied scientists and engineers, as Rapoport noted at the theory group meeting of January 13, 1953.[22] But despite a focus on finding "laws" of behavior that were common across levels of organization, the various strands of systems theory frequently evinced an antireductionist stance. For many, especially Gerard, this permitted the search for isomorphic structures present in different levels of organization without trying to explain the laws of one level of organization in terms of the laws of another. Others tended to focus on the study of emergent properties of systems, characteristics that could not be explained in terms of the properties of the system's components. This perspective dovetails, for example, with Marschak's interest in specifically collective-level models of teamwork and organization, and his emphasis that bargaining problems could not be solved on the basis of assumptions of individual rationality alone. Such a stance—in Marschak's case, against a kind of methodological individualism—had the salutary effect of encouraging interdisciplinarity while simultaneously preserving intellectual boundaries and minimizing disciplinary imperialism.[23]

The Chicago group thus potentially stood on the cusp of a profound conceptual breakthrough, with the perspective of "general systems" facilitating disciplinary convergence around the study of mathematical models, while simultaneously preserving the autonomy of the disciplines involved. As Ford Foundation money flowed to Chicago, in 1953 the university would approve the creation of a new research institute to provide behavioral science with a more secure institutional foothold on campus.[24] But at the same time, a number of obstacles threatened realization of the committee's vision. Some were political, stemming from the toll of McCarthyism on campus. Rapoport would later recall that two members of the Committee on Mathematical Biology were fired for invoking the Fifth Amendment in their loyalty oath hearings. Rapoport, who had briefly joined the Communist Party before the war, was not called before a loyalty committee but nevertheless desired to leave Chicago's "oppressive" atmosphere. In the spring of 1954 he was granted a fellowship at Stanford's Center for Advanced Study in the

Behavioral Sciences; he quit his job and headed west.[25] Moreover, although fundraising for a new institute for behavioral science was ongoing, smaller programs such as the Committee on Mathematical Biology
were under significant financial stress. Rashevsky's office correspondence tells the grim tale of his attempts to raise external funds during
this period from sources as diverse as the Ford Foundation and various
New York advertising agencies, which he hoped might be interested in
the committee's work on rumor spread.[26] He even dispatched Rapoport
to Princeton to meet with von Neumann in a bid to gain outside support
lest the program be threatened with elimination.[27] While Mathematical
Biology ultimately managed to survive at Chicago, Jacob Marschak and
the Cowles Commission did not stay: following a round of negotiations
in 1954 the commission was moved to Yale in 1955, where it remains to
this day.[28]

A new opportunity arose during the 1954–55 year when Miller, ever
the frontman for the Committee on the Behavioral Sciences, was contacted by Raymond Waggoner, director of the Neuropsychiatric Institute
at the University of Michigan Hospital. Waggoner and several of his colleagues had been lobbying state representatives and Michigan's governor to approve a building grant and annual appropriation for an institute
devoted to research into problems of mental health. Their arguments
for the institute—focusing on the costs to the state of untreated mental illness among employees, juvenile delinquency, and other problems—
apparently proved persuasive. In 1955 the state of Michigan appropriated $175,000 to cover the first year of operating costs for a Mental
Health Research Institute (MHRI). In the fall of that year Miller, Rapoport, and Gerard arrived as founding members and began lobbying for
funds to construct a dedicated building. Over the next two years, they
would raise nearly $3,000,000 from a variety of patrons, permitting con-
struction to begin in the summer of 1958.[29]

In terms of personnel and guiding philosophy, the institute effectively
reconstituted in Ann Arbor much of the Chicago Committee on the Behavioral Sciences. In addition to Miller, Gerard, and Rapoport, the historian Robert Crane moved from Chicago to be with his former committee members. Marschak and Rashevsky were offered positions; although
they declined to come, they made appearances as guest lecturers and visiting scholars, and another of Rashevsky's students was added to the faculty in addition to Rapoport. Other high-profile appointments followed.
Merrill Flood, who had done so much to assure game theory a place in

FIGURE 5.1. The Mental Health Research Institute building in 2004. The ideas of general systems theory were expressed in the very structure of the building, with each floor corresponding to a scale of analysis: wet labs in the basement (cell); rooms for interviewing patients on the ground floor (individual); and conference rooms and offices on the top floor (society). Photograph by the author.

postwar operations research, arrived as a professor of industrial engineering and senior research scientist at the institute in the spring of 1959. Alfred Emerson, the Chicago ecologist whose work had been intellectually significant in the development of systems theory, visited to give a talk. The institute also cultivated close connections with other departments and faculty in the university: Donald Marquis and Clyde Coombs in psychology, and Kenneth Boulding in economics, who, like Rapoport and Gerard, had spent the 1954–55 year as a fellow at Stanford's Center for Advanced Study in the Behavioral Sciences.[30] Finally, 1956 saw the creation of the MHRI's in-house journal, *Behavioral Science*, under the editorial direction of Miller, Gerard, Rapoport, Marschak, Marquis, and several other prominent psychologists and social scientists.[31]

The organizational structure of the institute, much like that of the RAND Corporation or CASBS at Stanford, was intended both to enhance the variety of disciplines represented in the organization, and to ensure constant cross-disciplinary interaction. From the beginning, the MHRI's research spanned everything from therapeutic trials and experiments in cellular physiology to observations of group dynamics and anthropological studies of social change. Every level of organization, from cell to individual to society, was represented, and the annual reports specifically summarized research by "level." This facet of Miller's systems theory was even imprinted on the structure of the new multimillion-dollar building that would house the institute from 1959: wet labs and an electronics center were stationed in the basement, rooms for individual and group psychological observation were on the first floor, while conference rooms were located on the top floor. As at the RAND Corporation, whose Santa Monica headquarters was consciously designed so as to maximize the number of interactions between researchers, the MHRI's faculty members were encouraged to lunch together at an adjoining cafeteria, where wide-ranging conversations between all the scientists in residence became the norm.[32] To further integrate across the different levels, institute members presided over a number of cross-level "groups" or "workshops" as well as a weekly general research conference attended by all staff. During the first three years of operation, such cross-level groups included a "mathematical models" theory group (led by Rapoport), a theory group in research methodology, plus regular workshops on "group stress," "individual stress," "microculture," and "linguistics" connected with particular projects being conducted by the institute's staff.[33]

The new institute proved extraordinarily successful at tapping many of the key sources of funding for psychological research available in mid-century America. During its first decade of operation the MHRI secured major grants from the Army and Air Force, the US Public Health Service, the National Research Council, the National Science Foundation, and the National Institutes of Mental Health, all in addition to an annual operating budget supplied by the state of Michigan. During the first five years of the institute's existence, military contracts were a particularly prominent feature of the budget, bringing in just over $215,000 during the years 1955–59 for the study of group performance under stress conditions and the study of team information-processing. This sum almost rivaled that dispensed by the Public Health Service during this period for the other major research project at the institute, experiments on the psychological effects of tranquilizers—new miracle drugs that promised to take the edge off the stress of living in a modern industrial society. Between all of these funding sources, the annual research budget of the in-stitute was approximately $700,000 in the late 1950s, rising steadily to nearly $2,000,000 by the end of the next decade.[34]

Modeling Mutualism, Measuring Teamwork: Game Theory and the Practice of Behavioral Science

The operations of Rapoport's cross-cutting "mathematical models" group, like those of the earlier "theory group" of the Committee on the Behavioral Sciences at Chicago, suggest some of the multiple ways that theory, mathematics, and models functioned inside the MHRI and in behavioral science more generally. Nominally, the aim of both bodies was the articulation of theories of human behavior that were congru-ent with observation, but models simultaneously made new observations and realities—something the institute's researchers seemed to under-stand intuitively. As the MHRI's third annual report noted, the aim of the models group was "the construction of mathematical models for the integration of facts about behavior." Not only did "explaining various problems in terms of models [permit] precise scientific definition and manipulation of those problems" but it helped "point up applications to other problems across levels of behaving systems."[35] In addition, as we saw in the previous chapter, theory and models in the human sciences had a way of shaping the direction of experimental inquiry. The story of game theory's arrival at the institute provides a perfect example of the way that theory, data, and experimental research mutually shaped one another. Game theory found a home in Rapoport's work at the MHRI not so much because it provided testable predictions but because, in comparison to some alternatives, it suggested relatively straightforward experimental procedures for exploring interesting behavioral phenom-ena across different levels of analysis.

While Rapoport had first encountered game theory at Chicago in the 1940s, it was not until his 1954–55 year at the Center for Advanced Study in the Behavioral Sciences that he began to consider more closely the connection between the theory and his own work. The proximate cause of this reconsideration was an encounter there with R. Duncan Luce, who introduced Rapoport to the Prisoner's Dilemma (PD) game, as articulated by Flood and Tucker at RAND just a few years earlier. PD reminded Rapoport of a model of the interaction between two organisms that he had developed and published in 1947 as one of his first sets of articles in the *Bulletin of Mathematical Biophysics*. Building on Rashevsky's earlier models of social interaction in *The Mathematical Biology of Social Behavior* (1947), Rapoport's own model focused on situations in which one organism put forth an "effort" and received a "satisfaction" from its interaction with another. The effort expended and the satisfaction received by one organism depended on the effort that organism expended and on the effort put forth by the other, and vice versa. In such situations, organisms could not in general identify a unique effort that maximized their satisfactions, only a curve of possible maxima as a function of the effort of the other organism. If the two organisms could somehow coordinate their efforts, they *could* maximize their joint satisfaction, which was greater than the sum of their individual satisfactions. However, the most interesting questions in the papers concerned the *stability* of such a joint maximum of satisfaction. Depending on the values of parameters in the organisms' satisfaction functions, a slight variation in effort on the part of one of the organisms could potentially set up a dynamic whereby their relationship either would become wholly parasitic or assume some form of mutualism.[36] Rapoport spotted similar dynamics at work within PD, in which cooperation between the prisoners produced payoffs that exceeded the sum of their payoffs if they acted independently.[37]

Upon arriving at the University of Michigan the following year, Rapoport translated his 1947 model into the notations of game theory.[38] He also sought to connect his continuous-variable model, developed with an eye toward explaining the evolution of parasitism and symbiosis more generally, with psychological experiments on human laboratory subjects.[39] In contrast to the austere theoretical emphasis (and financial conditions) that reigned at the Committee on Mathematical Biology, the relatively plush amenities of the MHRI and the proximity to experimentalists that this institution fostered presented the perfect opportunity for such a project. To allow testing of his continuous-variable models on human subjects, Rapoport began working with the MHRI's in-house instrumentation guru and electrical engineer, Caxton Foster, to design a machine that would allow human subjects to select their "efforts" by turning a set of knobs, while simultaneously viewing their resulting "satisfactions" in real time. A version of the device, based on the Sterling LM-10 analogue computer and nicknamed "Sympar," was tried on pairs of students during the 1955–56 academic year. The preliminary results proved congruent with earlier game experiments at RAND and elsewhere. "Initial tries of the apparatus . . . tend to indicate that when verbal communication is permitted a rapid approach to the social optimum is made, but not otherwise. . . . If [this result] is confirmed, interesting possibilities are offered to control the level of group performance by varying the restrictions on its communication patterns."[40]

The "level of group performance" was of particular interest to Rapoport by the spring of 1956 because of an Air Force research contract he had received shortly after arriving at the MHRI. Rapoport's connection with the military dated at least to the late 1940s, when he had used the mathematics of stochastic networks to model the spread of rumors through a population for an Air Force study called "Project Revere." Directed by a sociologist, the project sought to evaluate methods of communicating with a population (such as aerial leafleting) that could be employed in the event that broadcast media could not be used.[41] The MHRI contract, by contrast, stemmed from an invitation Rapoport received from Wright-Patterson Air Force Base in Ohio to propose a project to study the effects of stress on military personnel: with the space race just around the corner, the Air Force wanted to boost the performance of flight crews operating outside the earth's atmosphere. Rapoport's application was successful, and funding began in the spring of 1956, with the resulting project duly institutionalized within the institute by a weekly seminar devoted to reporting research on "group stress."[42] Rapoport himself emerged as something of a rainmaker for the institute, representing the MHRI at a summer 1957 conference to plan funding for the Air Force's new Office of Scientific Research, which would centralize much of the Air Force's R&D spending. At the MHRI staff meeting preceding the conference, Miller waxed optimistic about the "talks which we have been having with various Air Force officials who are interested in general systems theory," noting that the Air Force was "interested in establishing an institute for advanced study in the behavioral sciences and have contacted us about programs and areas for support which might tie in with their program."[43]

The study of team performance under stress was just beginning to acquire a significant scientific literature in the late 1950s, with far more to come in subsequent decades as the exploration of exotic environments, such as the poles and the deep ocean in addition to space, became a major focus of physiological and psychological research.[44] The kinds of metabolic and mental stresses encountered in outer space were very different from those that Miller's tranquilizers were intended to treat, and few presumably could be easily reproduced in a terrestrial laboratory; hence Rapoport ultimately settled on sleeplessness as the primary "stress" for his research. Nor could the Air Force give Rapoport much initial guidance as to what metrics of "performance" were most relevant to space travel. Instead, the framework of general systems theory, which sought to apply concepts of integration, learning, and homeostasis across multiple levels of organization, appears to have been Rapoport's guiding light. The first measure of performance he tried was the time required to complete an interdependent logical task, one in which team members learned to infer causal connections between different sets of buttons and indicator lights on a machine developed by Miller and his colleagues at Chicago (see fig. 5.2). Such logical tasks—which also mimicked the logic of neural networks in the brain—provided a controlled measure of team

An experiment in progress on the Delta apparatus, used to measure group behavior, supervised by Dr. Anatol Rapoport, at left rear.

FIGURE 5.2. Apparatus for measuring group performance of a logical task (1957). Anatol Rapoport is rear left. From *Mental Health Research Institute Second Annual Report* (1958), 22.

problem-solving and information processing abilities, and were the basis for most of the experiments.[45] However, toward the end of the "group stress" study, Rapoport also tried to measure "teamwork" more directly, by observing the frequency of "cooperation" in a three-way repeated Prisoner's Dilemma game. The game worked in this way: a single "defector" won three points, while the two "cooperators" lost one point; two defectors would each receive two points with the "sucker" losing two; if all three defected they would get nothing, but if they all cooperated they would receive one point each.[46]

Rapoport's decision to make his subjects play the PD *game* as opposed to testing their responses to a Sympar-like machine seems to have stemmed in part from technical problems of laboratory implementation. The machine was not only expensive to build and tricky to program, but its response was easily thrown off by subtle hand tremors on the part of experimental subjects. The PD setup, with its discrete moves and payoffs, was far easier to administer and less dependent on the physical coordination of the subjects.[47] Over the course of two sleepless days, Rapoport's subjects would play three-person PD in eight-hour shifts, with four hours in between for other activities. They were not supposed to communicate with one another, since this would result in potentially unscripted and uncontrolled interactions. It was less clear what effect the three-person setup would have (presumably, it would make the learning curve slower), or whether sleeplessness would reduce the subjects' propensity to cooperate. The results were generally disappointing: the stress appeared to have little effect on cooperative behavior, nor did cooperative behavior arise particularly quickly.[48]

When the Air Force group stress project concluded in 1960, Rapoport continued his research on the repeated PD games by obtaining a series of grants from the National Institutes of Health under the substantially broader rubric of "Studies of Conflict and Cooperation in Small Groups." Providing funding at the rate of over $42,000 per year, these grants continued from the fall of 1960 until the spring of 1967, spanning Rapoport's career at the institute and providing the foundation for much of his psychological research in this period.[49] This research systematically explored the myriad variables that determined the evolution of individuals' game-playing behavior during long runs of repeated two-person PD games. The most obvious of these included the magnitudes of the payoffs (typically, in money that was added or subtracted to the subjects' hourly wage), and the knowledge the players had of the payoff matrix or the actions of their fellow players. Less frequently, these might involve properties inherent in the individual game-players, such as their gender. Communication between the players proved more difficult to study, since it was difficult to establish rigorous experimental controls on face-to-face interactions between players; as a result, for most of Rapoport's PD experiments, communication—by word or gesture—appears to have been specifically banned. The fullest account of these investigations—a true *tour de force* of rigorous experimental methodology—appeared in 1965 as *Prisoner's Dilemma: A Study of Conflict and Cooperation,* coauthored by Rapoport and his assistant Albert Chammah.[50]

The MHRI PD experiments were thus carried out on a far more extensive scale than those at RAND, and for substantially different purposes. The mathematicians had aimed to "test" possible solutions of games like PD; the game was played repeatedly in order to gain a convincing sample size; and the learning that took place over the repetitions was initially a problem to be eliminated rather than an object of study in itself. By contrast, Rapoport's experiments were specifically intended to explore the dynamic processes whereby teams of individuals collectively arrived at cooperative behavior. Moreover, in studying these processes, it was precisely the *lack* of a mathematical theory of how to act in a PD situation that made the game psychologically interesting. As Rapoport would remark in the introduction to his 1965 *magnum opus,* "the potentially rich contributions of game theory to psychology will derive from the failures of game theory rather than from its successes."[51] Games like PD were more like experimental "tools" that could permit rigorous exploration of what Rapoport called "real psychology": phenomena involving "personality, intellect, and moral commitment" that could not be explained in terms of rational means-ends calculation from self-interest alone.[52] In this case at least, mathematics' loss was precisely psychology's gain. And throughout the 1960s, Rapoport ran with this insight, developing an extensive array of techniques for studying virtually every aspect of game-playing behavior: refined methods of handling subjects, standardized recordkeeping materials and rules for gameplay, and automated administration and recording of game experiments. Finally, following up on his 1965 *Prisoner's Dilemma,* Rapoport moved beyond PD in *The 2×2 Game,* which exhaustively examined behavioral reactions to all seventy-eight strategically distinct 2×2 two-person non-zero-sum games.[53]

Rapoport's publications on experimental games, while prolific, were only the leading edge of an avalanche of psychological studies involving games appearing in the late 1950s and early 1960s. The work of Morton Deutsch, a social psychologist and student of Kurt Lewin, represented another significant tradition in this literature. Deutsch apparently encountered PD via Howard Raiffa in the early 1950s, while the former was based at New York University.[54] Building on his research into the social dynamics of interracial housing projects, as well as studies of teamwork done with the support of the Office of Naval Research, Deutsch saw in PD a convenient experimental setup assessing his subjects' tendencies to exhibit trusting and cooperative behavior. But where Rapoport focused on the relatively impersonal dynamics and learning processes at work in long repetitions of the game, Deutsch focused more on the way the characteristics of the individual players affected the way they played. Thus in the studies reported in his classic 1958 paper, "Trust and Suspicion," he tested the impact on the players of various "motivational orientations," verbal instructions as to their objectives during the game as well as the intentions of their opponent.[55] Subsequently, in experimental studies of bargaining situations performed in the relatively plush facilities available at Bell Laboratories, Deutsch studied the effect of verbal threats on the bargaining behavior of individuals trying to solve an elaborate transportation coordination problem similar in essence to the comparatively simple PD game.[56]

The use of "games" in psychological research hit the zenith of its popularity in the mid-1960s, with over 100 results published annually at the peak in outlets from the *Journal of Personality and Social Psychology* to the *Quarterly Journal of Economics* (see fig. 5.3). Michigan and the MHRI were at the crest of this wave: many of the articles on experimental games appeared in the MHRI's in-house journal *Behavioral Science*. Through the work of Rapoport and Deutsch, the repeated PD game in particular would emerge as "the E. Coli of social psychology," a model experimental situation spinning off reams of knowledge about social interaction.[57] Even so, fashions change. By the later 1960s, the popularity of experimental games was on the wane, with the kind of rigoristic research agenda pioneered by Rapoport attracting criticism from those who doubted the ability of his highly stylized and artificial laboratory setups to shed light on real-world social phenomena.[58] Yet then, as a decade earlier, the use of games in social psychological research continued to attract defenders. In contrast to the kind of elaborate and expensive machinery and laboratory setups required to assess "satisfaction func-

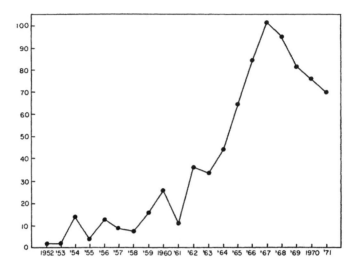

FIGURE 5.3. Reports of experimental games, 1952–1971. From Anatol Rapoport, "Prisoner's Dilemma—Recollections and Observations," in *Game Theory as a Theory of Conflict Resolution,* ed. Anatol Rapoport (Dordrecht: Reidel, 1974), 20.

tions" or observe continuous-time interaction processes, repeated games provided an easy to implement and easy to notate framework that could potentially provide insight into "real psychology." Moreover, Rapoport's conception of games and game theory—as a tool for investigating the psychology of conflict and cooperation—would circulate among a new generation of social and behavioral scientists through his highly readable textbooks, including *Two-Person Game Theory* (1966) and *N-Person Game Theory* (1970).

From Teamwork Studies to Conflict Resolution

All else being equal, one might have expected Rapoport's work on experimental games to produce the typical results of military-funded studies in psychology and organizational behavior: a few academic papers together with dense technical reports, circulated among a small number of scientists and engineers. Yet this would ignore the ambitions of the interdisciplinary social and behavioral sciences in postwar America. It was an age in which science—of which the atomic bomb was the most visible product—was simultaneously feared as a source of new technological monstrosities and hailed as the only force capable of saving humans from further violence. This was true whether one looked for the creation of some new and potent weapons system to deter conflict, or to new international organizations like the United Nations that would somehow save humans from themselves. Out of this environment would emerge several connected areas of scholarship-*cum*-political activism in the late 1950s going variously by the names of "peace research," a "science of peace," or "conflict resolution," as well as the more politically mainstream "arms control."

The possibility that scientists might be able to save mankind from the destructive power of modern weapons, even as they played an outsized role in the creation of those same weapons, had deep roots. Such a message resulted in part from American atomic scientists' own reactions to the advent of nuclear weapons, and specifically to their concern to secure better working conditions and influence over the future direction and use of their research. The June 1945 "Franck Report," written by nuclear scientists at the Metallurgical Laboratory at the University of Chicago, called for international control of nuclear weapons to head off a looming "armaments race" that would unfold as other nations inevitably learned how to produce nuclear weapons on their own. In the aftermath of the report, as well as a petition to the president drafted by prominent nuclear physicist Leo Szilard, Chicago nuclear scientists formed the core supporters of several peace-oriented organizations and institutions: the Emergency Committee of Atomic Scientists, founded by Szilard and Albert Einstein; the *Bulletin of the Atomic Scientists*; and the Pugwash Conferences, which brought together an international assortment of scientists and intellectuals around the issue of arms reduction.[59] Moreover, from an early stage, many embraced the notion that the social and behavioral sciences might serve as foundational disciplines for applying scientific methods to the problem of promoting peace. When James Miller assembled a major symposium on social psychology at the University of Chicago in 1947, the participants' nuclear hopes and fears suffused the proceedings. As Miller would note in his preface to the symposium's proceedings, "The new weapons of mass destruction have dramatized the fact that the present threat to mankind is of a different order of magnitude from any other in history. The issues of international cooperation, effective leadership, and accurate social communication are of unprecedented importance." To address these issues, the conference concluded with a roundtable discussion on the topic of

"Social Psychology and the Atomic Bomb" led by Leo Szilard, in which the participants sought to identify promising research programs in social psychology that would help promote peace.[60]

Behavioral science began to influence the study of international relations at this time. In 1951 Quincy Wright, an eminent professor of international law at Chicago (and brother of geneticist Sewall Wright), penned an essay for an edited volume on the theme of "Research for Peace." For Wright, the most pressing problem in pursuing peace in the age of cold war was to develop a "science of international relations" worthy of the name. As it was presently constituted, international relations did not rise to the level of science, being fractured into fields as diverse as international law, diplomatic and world history, geography and demography, and others. The feature that all of these disciplines possessed in common was that, fundamentally, they were based on the study of human *behavior*. Thus, "a science of international relations should rest on the basic behavioral sciences even though many applications of these sciences to the international field may conveniently be simplified by making assumptions, such as those made in international politics, international law, international organization and international economics, which have a considerable degree of validity under the special conditions in our time." Moreover, a critical first step in developing such a science of international relations was to secure the integration and development of the behavioral sciences in general, so that those interested in a science of peace should first turn their attention to the fundamental questions of behavioral science.[61]

If international lawyers like Wright could call upon emerging "behavioral science" to provide new foundations for a true "science" of international relations, behavioral scientists seized opportunities to proclaim the utility of their knowledge for soothing the international tensions of the Cold War. In a 1955 book, *Towards a Science of Peace*, social psychologist Theodore Lentz of the Attitudes Research Laboratory at Washington University in St. Louis introduced his subject by quoting the first sentence of the United Nations Educational, Scientific, and Cultural Organization's constitution: "Since wars begin in the minds of men, it is in the minds of men that the defenses of peace must be constructed." Such defenses would emerge through a broad program of behavioral research, to be carried out by interdisciplinary teams comprised of psychologists, economists, political scientists, statisticians, and philosophers. Lentz's own work, which focused on measuring public beliefs and attitudes using surveys and psychological scaling techniques, held a prominent place in this research program, which aimed at understanding of the psychological forces that drive wars, especially the attitudes of members of different nations toward one another, and the identification of methods for controlling those attitudes.[62]

Likewise, when a national council of behavioral scientists put out a statement on "National Support for Behavioral Science" in February of 1958, they called attention to international peace as a prime example of an issue that would benefit from their expertise. Just the previous month, President Eisenhower had wrapped up his State of the Union address to Congress by asking for a program of "science for peace," an international initiative to coordinate research on scientific problems relevant to humanity as a whole.[63] The authors of the report readily adapted Eisenhower's rhetoric to suit their needs. "President Eisenhower has asked for a "science for peace,'" they noted. "The issues which can be attacked by behavioral science are the human ones whose solution can guide world affairs along the course from cold war to ultimate peace," yet "there has been almost no systematic research in behavioral science concerning international relations and diplomacy, negotiation, the prevention of war, or the operation of arms control systems. . . . Behavioral scientists could make a specific contribution in this." In particular, behavioral scientists could help to identify and control the causes of war by studying international attitudes, tensions, and stresses induced by technological change and social development.[64]

At the University of Michigan, behavioral science and the problems of war and peace ran together at the Mental Health Research Institute and a closely related organization, the Center for Research on Conflict Resolution (CRCR), which was founded in 1959. The early histories of the two organizations are marked by a number of striking parallels. Among other things, both also owed much to that critical year of 1954–55 that Rapoport, Gerard, and Boulding had spent together at Stanford. For the MHRI, the year proved pivotal in its emergence as an institution; in the history of the CRCR, it is remembered as a moment of intellectual "critical mass," when a group of "brilliant and decent people" came together around a common theoretical framework for discussing international relations. This common framework was provided by the work of the recently deceased British Quaker meteorologist Lewis F. Richardson, whose son was also a fellow at the Stanford Center for Advanced Study in the Behavioral Sciences.[65]

Richardson remains best known for his work creating numerical methods of weather prediction that would subsequently inspire the development of computerized weather and climate prediction models.[66] He was also a Quaker and a pacifist, serving in the Friends Ambulance Unit during World War I and devoting his spare time to developing what he called a "science of foreign politics" that would be quantitative and mathematical. Meteorology, his field of expertise, provided the perfect model for such a "science" since it straddled the border between the purely deterministic sciences, like mechanics and electrodynamics, and a number of newer stochastically oriented branches of science like statistical physics, psychology, and the social sciences. Both meteorology and the human sciences studied phenomena that were random, yet fundamentally regular; in all cases, the hallmark of science would be measurement, even if no "neat simple theory" could encompass all aspects of the phenomena in question.[67]

His central achievement in developing a science of foreign politics—first published in 1919 and refined in multiple publications in subsequent years—was a set of coupled differential equations that expressed the evolution of a nation's "vigor-to-war" (or "warlike strivings," assumed to be a measurable quantity) over time. The rate of change of a nation's vigor-to-war depended on a number of factors—everything from "the defense of that which is held dear" or "national prestige" to pain, fatigue, and "pity for the adversary"—which Richardson captured as terms in his equations.[68] While the equations had many possible solutions, depending on their exact form and parameters, under a reasonable set of assumptions they predicted exponential increases in the vigor-to-war over time if nations built up their defenses and armaments, thereby prompting other nations to respond in kind. Specifically, if a nation's vigor-to-war was proportional to its expenditure on weaponry, Richardson found his model consistent with the arms race that occurred during the lead-up to World War I. Unlike his foray into weather prediction, the fit between the equation solutions and armaments data proved to be quite good; and throughout the 1930s Richardson corresponded with Quincy Wright at the University of Chicago to obtain better data on wars and weapons expenditures to further fine-tune his models.[69]

For those connected with the MHRI's project of building an interdisciplinary "behavioral science," Richardson's work provided a mathematical framework for investigating social conflict that meshed perfectly with their methodological stance: all levels of behavioral analysis, from cells to societies, should be examined concurrently to determine their underlying laws and structures. Rapoport was especially interested in Richardson's equations because they seemed similar to his earlier differential equations modeling the evolution of parasitism and symbiosis, albeit with arms races the result of a self-reinforcing negative cycle rather than the positive feedback-loop that Rapoport imagined for his cooperating organisms. The relevance of such equations to phenomena at the interpersonal as well as international levels seemed a step toward uncovering general laws of behavior that spanned multiple levels of analysis. Arms races, interpersonal dynamics, parasitism, and symbiosis in non-human organisms all could be united by a common theoretical framework.[70] Yet, at least for Rapoport and Boulding, their interest in Richardson's work had deeper personal roots as well. Boulding, a Quaker like Richardson, had been writing antiwar pamphlets since he was a student in England in the 1930s, and had recently completed a book on the economic implications of armament and disarmament.[71] Rapoport, a socialist and one-time member of the Communist Party who nevertheless served in the Air Force during World War II, had been openly critical of American foreign policy since at least the late 1940s.[72]

Back at Michigan in 1955–56, Boulding sought to create an institutional base for the application of behavioral science to the study of social and international conflict. During that winter, he and several other faculty members in sociology, psychology, and journalism joined forces to create a new periodical, the *Journal of Conflict Resolution*.[73] The editors' introduction to the new journal made it clear that conflict resolution represented an extension of behavioral science to the realm of international conflict. While they noted that "up to now the study of international relations has, on the whole, been the preserve of historians and political scientists and of such professionals as lawyers, merchants, diplomatists, and military men," nevertheless, "if intellectual progress is to be made in this area, the study of international relations must be made an interdisciplinary enterprise, drawing its discourse from all the social sciences, and even further." Out of the interdisciplinary social and behavioral sciences, "It is not too much to claim that . . . a general theory of conflict is emerging."[74] Fittingly, one of the earliest issues of the journal was anchored by an extensive exposition of Richardson's mathematical models of conflict written up by Rapoport.[75]

In tandem with the new journal, in 1959, a Center for Research on Conflict Resolution (CRCR) was approved by the university to administer grants to faculty that would sponsor research in the new interdisciplinary field. By the mid-1960s, the center would amass a substantial record of support. Predictably, much of this came from private foundations, like the Carnegie Corporation and the Ford Foundation, which already had a lengthy track record of funding the promotion of international peace. Others, like the National Institutes of Mental Health and the National Science Foundation, were more traditionally academic in their orientation, while the Office of Naval Research and the Air Force Office of Scientific Research demonstrated military interest in the emerging field, even if at least one military official expressed little hope of "convincing responsible persons of its legitimacy and pertinence to military efficacy."[76] With its wide appeal, the center proved to be the first of many such research organizations focused on peace, and something of an institutional template that would be copied internationally. By 1963, when Boulding and his wife began publishing an "International Newsletter on Peace Research," they could list a number of research centers devoted to conflict resolution or peace research. These included (among many others) a Peace Research Institute in Oslo, Norway; a Peace Research Centre in Lancaster, England; and a Conflict Research Society run out of University College, London, and the nearby Tavistock Institute of Human Relations. Many of these organizations would look to Michigan for inspiration, inviting faculty associated with the CRCR to give guest lectures.[77]

Michigan's center drew in a number of faculty closely associated with the Mental Health Research Institute. Boulding, a member of the MHRI's advisory committee, served as director of the center and taught classes on "conflict systems" and international relations for the undergraduate program, all while pouring forth a steady stream of articles on the economics of disarmament.[78] J. David Singer, a political scientist, joined the center in its first year, and was involved in a number of the center's earliest major projects. With the sociologist Robert Angell, he conducted a one-year study on "Value Systems, Foreign Policy, and Soviet-American Co-Existence," and in 1963 he would begin a massive research program aimed at identifying the "correlates of war," statistical indicators that could be used to predict the outbreak of violent conflict. This work, conducted simultaneously under the aegis of the Mental Health Research Institute, continued into the 1970s and resulted in many dissertations, books, and articles.[79] And finally, the social psychologist Marc Pilisuk carried out a number of studies at both the center and the institute—including a collaborative project with Rapoport—that brought insights from laboratory studies of conflict and cooperation to bear on problems of conflict de-escalation and international arms control.[80] The center's projects thus represented applications of classic methods and insights from the behavioral sciences to problems of conflict.[81]

Reasoning in an Arms Race

It was at this point, in the late 1950s and early 1960s, that the metaphor of international conflict as a game suddenly became current, and game matrices became seemingly indispensible for talking about the challenges of nuclear strategy, the possibility of arms control, and the resolution of international conflicts. The particular kind of game deemed most relevant would vary from author to author: nuclear brinksmanship could be discussed by analogy to the "sport" of Chicken; the "arms race" could be compared to the behavior elicited by a Prisoner's Dilemma game; and problems of international conflict and negotiation could be cast as game-theoretic "bargaining." While such analogies may seem natural, they were hardly inevitable. At least within the conflict resolution community, other theoretical frameworks were available for analyzing these problems, as the existence of Richardson's arms-race models suggest. However, as the game matrix's uptake in 1960s-era social psychology suggests, game theory could potentially provide those interested in conflict resolution with a powerful set of "tools" for exploring the influence of higher social-psychological factors, such as "personality, intellect, and moral commitment," in conflict situations on many different levels of analysis.

The functioning of such higher-level social-psychological factors was crucially important in light of some fundamental problems posed by Richardson's equations that were widely recognized, even by many who were unfamiliar with Richardson's work: what might prevent an arms race from escalating into full-blown conflict? And what might limit war once it started? Richardson himself was keenly aware of this issue, for while the dynamical instability of his equations seemed to provide an argument against "balance-of-power" solutions to the international arms race of the 1930s, the equations seemed to offer little guidance on how actually to achieve peace. As he put it, "The equations are merely a description of what people would do if they did not stop to think," if they

remained guided not by "reason" but by the mechanical and reaction-
ary forces of "instinct" and "tradition."[82] The adoption of game theory
by various practitioners of "conflict resolution" was thus closely associated with their interest in probing the social and psychological factors, or
perhaps a higher form of rationality altogether, that might keep conflict
under control.

In this process, two individuals proved most visible during the early
1960s: Rapoport, and the economist and defense intellectual Thomas
Schelling. It is difficult to imagine a more incongruous pair. Schelling
was in many ways the consummate establishment insider, with his Ivy
League academic pedigree, a track record of government appointments,
and frequent visits to RAND. Rapoport, with his immigrant Russian-
Jewish roots, leftist connections, and researcher appointment at the
MHRI, moved in very different circles, despite his working relationship with the Air Force during the late 1950s. Nevertheless, "conflict
resolution" provided a forum in which their paths would cross repeatedly throughout the early 1960s, and game theory provided the common language in which they cast their arguments. But in the end their
disagreements—particularly over the nature of the reasoning needed to
keep the world safe in an age of nuclear weapons—would overwhelm the
ability of game theory to create common ground between them. Their
relationship would also capture in microcosm some of the tensions pulling apart the conflict resolution community during the later 1960s, as the
relatively "big tent" provided by the field in the late 1950s increasingly
proved too crowded to shelter military funding and defense intellectuals
on the one hand, and campus antiwar activists on the other.

Schelling first encountered game theory at Harvard while he was
a student of Carl Kaysen and Wassily Leontief in the later 1940s. Although his doctoral work concerned national income accounting, an initial exposure to the theory as a graduate student interested him in the
problem of how coalition members in a multiplayer game divide their
winnings—a problem of "bargaining" to which the von Neumann–
Morgenstern theory of games had not offered a determinate solution.
Following a series of government posts in the Truman administration, he
took a post teaching international economics at Yale, with the intention
of writing a book on the topic.[83]

The first fruit of this effort was a long article that appeared in print
as an "Essay on Bargaining" in the *American Economic Review* in early
1956.[84] Remarkably, Schelling did not cite any of the growing number

of mathematical explorations of bargaining that followed Nash's seminal papers on the subject, nor did he seem to be aware of the experimental studies of bargaining and non-zero-sum games that had interested
the RAND mathematicians earlier in the decade. In fact, the essay contained no significant use of game theory at all: from the first paragraph,
Schelling took it for granted that the problem of bargaining—how two
or more individuals divide a surplus—admitted no definite mathematical solution. Despite this fact, "there is, however, an outcome; and if we
cannot find it in the logic of the situation we may find it in the tactics
employed. The purpose of this essay is to call attention to an important
class of tactics, of a kind that is peculiarly appropriate to the logic of in-
determinate situations." Specifically, Schelling argued that "the power
to constrain an adversary may depend on the power to bind oneself."[85]
The greatest strength a bargainer may possess was actually a certain
kind of weakness—like that of von Neumann's hypothetical omnipotent
God, capable of creating boulders that He could not lift. Strength lay in
the ability to make irrevocable commitments to a given bargaining demand. The essay therefore explored the concrete details of how this paradox might work itself out in the practical details of economic bargaining and foreign policy alike.

Mathematically innocent though it may have been, the essay nevertheless proved critical in forging connections that would strongly shape
Schelling's subsequent career. It attracted the attention of Joseph Ker-
shaw, deputy director of RAND's economics division, who arranged for
Schelling to spend the summer of 1957 in Santa Monica. Schelling's first
month at RAND was devoted to a program on economic development;
in the next, he was free to explore the work being done by the corporation's various divisions. The visit introduced Schelling to the work of
RAND's most prominent specialists in the dark arts of nuclear strategy:
Bernard Brodie, Herman Kahn, Albert Wohlstetter, and Andrew Marshall, to name a few. He had a chance to visit some of RAND's storied
war-gaming facilities, used to simulate various security crises; he also
began to immerse himself game theory, bypassing von Neumann and
Morgenstern's severely technical tome and spending the fall of 1957 in-
tensively studying Luce and Raiffa's recently released *Games and Decisions*.[86] The "Essay on Bargaining" also introduced Schelling to Kenneth Boulding, who served as a referee for the article. The two quickly
became friends. Not long after, Boulding invited Schelling to a conference to celebrate the launch of the *Journal of Conflict Resolution* (*Con-*

flict Resolution for the first two issues), and Schelling began contributing articles to the new journal.[87] The "Essay on Bargaining" thus positioned Schelling in both the rarefied world of RAND-style nuclear strategy and the fledgling fields of conflict resolution and peace research.

Schelling's first contribution to *Conflict Resolution*, published in the journal's inaugural issue under the title "Bargaining, Communication, and Limited War," nicely illustrates Schelling's newfound connections. The piece drew on Schelling's work at RAND the previous summer, which applied his treatment of bargaining to the problem of how combatants establish norms of conduct to limit the severity of the conflict. In an age when John Foster Dulles's "massive retaliation" remained the official national security policy line, "limited war" was nevertheless a topic of investigation at RAND, and Schelling had spent time the previous summer developing game simulations relevant to the topic.[88] The question Schelling asked was how limits to conflict might be established in battlefield situations where "communication is incomplete or impossible" via a process of "tacit bargaining." Like businesses with "limited competition" bargaining to split a surplus, drivers "jockeying in a traffic jam," a homeowner "getting along with a neighbor that one does not speak to," or a man who has lost his wife in a city and needs to guess the best place to wait for her, combatants in wartime had interests that were not totally opposed; hence they had some aims in common that could be achieved by coordinated action. The identification of situational "cues" and coordinating "focal points"—for example, the 38th parallel in the Korean War—could provide insight into ways that combatants might communicate tacitly to limit conflict.[89]

A second, much longer article in the *Journal*, titled "The Strategy of Conflict: Prospectus for a Reorientation of Game Theory," appeared in the fall of 1958, just as Schelling was returning to RAND for a year's sabbatical.[90] This latter work represented a broad critique of what Schelling called "traditional game theory," which indiscriminately emphasized "methods and concepts that proved successful in studying the strategy of pure conflict" while providing little guidance on problems of bargaining, coordination, limited war, or negotiation.[91] This was, after all, the brand of game theory that Schelling would have encountered in the operations research literature, with its emphasis on theories of mathematical optimization; or in Santa Monica, where in the summer of 1957 Herman Kahn had been circulating drafts of a proposed survey of RAND-style systems analysis, titled *Military Planning in an Uncertain World*, which featured elementary expositions of games like matching pennies or the old standby of "noisy duels."[92] However the alternative to "traditional game theory," with its focus on minimax strategies and pure conflict, was neither von Neumann and Morgenstern's coalitional-form analysis of the non-zero-sum game, nor even Nash's axiomatic treatment of bargaining, but Schelling's own social-psychological analysis of the bargaining tactics, cues, and norms that might produce coordinated action in a non-zero-sum situation. In practice, this meant stripping game theory down to the point that it had almost no content at all, whether normative, predictive, or descriptive. Indeed, just about all that remained of "game theory" in Schelling's article was the game matrix for notating payoffs to players—although this notational system would serve a crucial role in tying together the diverse mix of examples Schelling considered. And so it would serve throughout 1958–59 as Schelling, continuing to write prolifically while at RAND, produced much of the material that (together with his articles in the *Journal of Conflict Resolution*) would later be packaged as *The Strategy of Conflict* (1960).[93]

This work attracted Rapoport's attention and critique. In response, Rapoport drafted an MHRI preprint, "Critiques of Game Theory," in which he started to assess the potential of game theory to contribute to the agenda of conflict resolution.[94] In Rapoport's assessment, game theory was certainly "unfit as a *descriptive* theory of behavior in game-like situations," although the observation of human game-playing behavior could yield interesting insights for the behavioral scientist, as his ongoing research at the MHRI was demonstrating.[95] However, Rapoport felt that Schelling was perhaps too quick to dismiss the normative or ethical implications of the theory's characterization of "rationality." At least "on the level of the two-person zero-sum game," a normative conception of rationality "could be defined in a way that will enjoy widespread acceptance," namely, as maximization of expected payoff. However, in non-zero-sum games such as PD, a simple maximizing conception of "rationality" became paradoxical: as Rapoport asked, "Can we accept the definition of rationality (based on 'doing the best for one self') which leads to a result which definitely is not the best that each player can do for himself?" By exhibiting such contradictions and paradoxes, game theory might open the door to the identification of new norms of rationality that pertained to the players as a group rather than to the players as individuals. Game theory thus had normative implications, whether or not they were "adequate" for making good choices.[96]

Similar themes would run through Rapoport's critical examination of game theory in his 1960 book, *Fights, Games, and Debates*—an instant classic in the conflict resolution literature, often reviewed alongside the work of Schelling. Game theory, Rapoport argued, is not directly applicable to problems of conflict or its resolution; rather, it "stimulates us to think *about* conflict in a novel way" and it "leads to some genuine impasses . . . where its axiomatic base is shown to be insufficient for dealing even theoretically with certain types of conflict situations." Since "these impasses set up tensions in the minds of people who care," those people "must therefore look around for other frameworks into which conflict situations can be cast," for example, an empathetic dialogue of value systems.[97]

Rapoport's interest in the normative implications of game theory for conflict resolution would be reinforced by the national security issues that emerged in the wake of the 1960 presidential campaign. Seizing on national concern provoked by the Soviet launch of the Sputnik satellite, John Kennedy charged that the Eisenhower administration was concealing a sizeable "missile gap" between the United States and the Soviet Union. Other prominent RAND affiliates echoed similar fears in their public comments. Albert Wohlstetter's 1959 *Foreign Affairs* article, "The Delicate Balance of Terror," argued that America lacked a credible deterrent to Soviet aggression.[98] Herman Kahn likewise went public with similar worries about the credibility of America's deterrence posture in his sensational *On Thermonuclear War*, released alongside the books of Schelling and Rapoport during the election year of 1960. Invoking Bertrand Russell's analogy of the Cold War as a game of Chicken as well as the kinds of bargaining problems that were the focus of Schelling's work, Kahn explored the logic of credibility, commitment, and deterrence in such situations, with an eye toward identifying tactics that would increase America's bargaining power. These might include the construction of a "doomsday machine" that would destroy the planet when activated by a Soviet provocation, and robust civil defense programs to convince the Soviet Union that the United States was prepared to go through with fighting a nuclear war if the need arose.[99]

RAND's national profile grew substantially as a result of such works, and Kennedy's election (and his selection of Robert McNamara as secretary of defense) brought RAND staffers and their ideas into power at the Pentagon. A former operations researcher for the Army Air Forces during World War II, McNamara tapped Schelling's former boss, longtime RAND economics division chief Charles Hitch, to serve as assistant secretary of defense and Department of Defense comptroller. Many of Hitch's staff likewise came from RAND, including Alain Enthoven, a systems analyst, and later Daniel Ellsberg, a former student of Carl Kaysen who had distinguished himself for his early work on game theory and utility theory.[100] While Schelling did not join the administration, he eventually played an advisory role in establishing the "hotline" that would at last provide direct communication between Moscow and Washington. Kahn, proving something of a lightning rod in the aftermath of his book, departed RAND to found a think-tank of his own, the Hudson Institute, in 1961.[101]

Thus while Schelling could sit on the editorial board of the *Journal of Conflict Resolution* alongside Boulding and Rapoport, his social world brought him far closer to the actual corridors of power in Washington than the other two men would ever get. Likewise the intellectual distance between Schelling's exploration of the tactics of bargaining (not to mention Kahn's suggestion of a "doomsday machine" as the ultimate bargaining tactic in the great global game of Chicken) and Rapoport's interest in uncovering new ethical frameworks applicable in non-zero-sum situations only grew throughout the early 1960s. From 1961 onward, Rapoport would begin to voice criticism of a group he darkly identified as "the strategists," "the 'voices of reason' increasingly heard today" on the great problems of foreign policy that simply "recommend for the most part a more expert and systematic exploitation of strategic thinking," a replacement of the previous policies of "brinksmanship" with more calculated measures. Rapoport identified such "strategic thinking" with the narrow and individualistic conception of rationality that had proven inadequate in many non-zero-sum games; and given that the "arms race" between the United States and Soviet Union might be realistically characterized as a PD game, this suggested the total inadequacy of strategic thinking for solving problems of international conflict. Instead, Rapoport argued that "what is required is a leap into another conception of rationality, analogous to passing into another conceptual framework," a leap "not unlike Kant's conception of moral choice, according to which one chooses the course that would be to one's advantage if everyone else also chose it."[102]

Rapoport's belief in the possibility of a revolutionary transition to this new "conception of rationality" was reinforced by a finding that emerged during some of the final PD experiments performed under the Air Force "group stress" contract during the fall of 1961. While his sleep-deprived subjects were forbidden to exchange "word or gesture of any kind" (lest their unscripted interactions contaminate the experimental results), during the short breaks between shifts they were permitted to sit near one another while sipping coffee in the institute cafeteria. Intriguingly, despite the complete ban on communication, the likelihood of cooperation rose substantially after the first such break. Rapoport puzzled over this finding for some time, but in the end felt this additional cooperation could only be explained if, as he put it in an autobiographical article, "the mere physical contact between the players induces at least some thoughts related to the fact that *one is playing against someone like self, who is likely to have the same thoughts.*"[103] Through personal contact, game players might be induced to form a kind of enlightened consciousness that had expanded to include others in addition to oneself; this simultaneously suggested that behavioral science, via some group therapist or conflict resolution expert, might intervene in conflict situations to help induce the shift in "framework" that would make a truly contradiction-free rationality possible.

Thus, from the early 1960s onward, Rapoport's scientific and activist agendas built rapidly upon each other. Responding to the vexed politics surrounding the ongoing disarmament conference between the superpowers in the spring of 1962, he would team up to coauthor articles in *The Nation* with MHRI colleague J. David Singer decrying the "unholy alliance of the empire-building physicists, their collaborators in the Atomic Energy Commission, the first-strikers in the Pentagon, the Red-hating politicians, and the sensationalist press" for helping to scuttle the United States' fragile nuclear testing moratorium the previous fall.[104] On the pages of the *Journal of Conflict Resolution*, he simultaneously reported on the use of "Formal Games as Probing Tools for Investigating Behavior Motivated by Trust and Suspicion," summarizing the latest results of his experiments on PD-playing behavior that might identify the processes by which the "leap" to a new form of rationality might take place and by which cooperation might emerge from conflict. In calling attention to the significance of "psychological considerations" for analyzing the "world situation," he also wove together his peace activism with ivory tower disciplinary politics, noting that "it is unfortunate that so much ingenuity is directed at the invention of elaborate mechanical safeguards and at purely strategic analyses" for solving the security dilemmas of the nuclear age, while "so little effort is undertaken with a view toward understanding what makes people behave as they do."[105] Enlightened minds assisted by psychological insight, not better machines or the calculations of the "strategists," would be key to resolving the international tensions of the Cold War.

The event that at last brought Rapoport's long-simmering antagonism toward Schelling and the "strategists" into full view was the 1964 publication of Rapoport's *Strategy and Conscience*. The title itself marked it as a swipe at Schelling's *Strategy of Conflict*, but the villain of the work was plainly Herman Kahn, even if Kahn was never identified by name. Rapoport thus introduced the book with a story of how a "strategist," lecturing at Michigan to a standing-room-only crowd on the topic of "Defense and Strategy in the Nuclear Age," sketched out some scenarios of how a nuclear war might be won: lives lost, cities annihilated, victory secured. Despite "the lecturer's ghoulish jokes," and resulting "ripples of nervous laughter" in the room, the speaker maintained the appearance of being "objective and rational." Yet overcome by "a wave of repugnance" at this performance, during the question-and-answer period Rapoport asked the speaker how he would "defend himself if at some future time he were a co-defendant in a genocide trial." Academic decorum was breached: a number of Rapoport's colleagues thought the question "inappropriate" for failing to keep "the discussion within the mode of reasoned argument and within the sphere circumscribed by the subject of the discussion." In particular, questions of "morality" were judged impertinent in such a conversation.[106]

The aim of *Strategy and Conscience* was to meet the arguments of the "strategists" on their own terms. As Rapoport noted in the book's opening pages, "I have no chance of being listened to . . . so long as I keep insisting that moral issues are central" to strategic problems. However, "the sincere and competent strategic thinker . . . can be reached with 'rational' arguments because he is involved with what he believes to be rational procedures. Indeed, he takes pride in the 'rationality' of strategic thinking, in which he invests his professional competence."[107] Therefore—as he had time and again in the preceding years—he invited his readers to consider the most general problem facing defense planners in what he imagined to be the strategists' own game-theoretic idiom: should they escalate their arms purchases, or reduce them? The game capturing this situation possessed two strategies, "Cooperate" (arm themselves further) or "Defect" (disarm). A reasonable set of payoffs might be captured in the matrix in figure 5.4.

	Cooperate	Defect
Cooperate	(5, 5)	(–10, 10)
Defect	(10, –10)	(–5, –5)

FIGURE 5.4. The arms race becomes a Prisoner's Dilemma. After Anatol Rapoport, *Strategy and Conscience* (New York: Harper and Row, 1964), 49.

As Rapoport noted, this situation was precisely a Prisoner's Dilemma game—a game for which there was no reasonable solution based on naive assumptions of individual maximization alone, as his experimental research at the Mental Health Research Institute seemed to indicate. The same considerations, he hastened to demonstrate, applied to Kahn's preferred formulation of deterrence as a game of Chicken.[108] Rapoport therefore concluded with a plea for psychology. Any approach to the problems of strategy could not rely on the strategists' calculations of rational self-interest alone: it also required the "insight-producing potential" of peace research, psychological techniques that "can induce a reorientation in thinking about international relations," as well as historical research and laboratory experimentation on attitudes and values. The problems of the nuclear age were too important to leave to mere strategy; conscience was essential, on both moral and logical grounds.[109]

Missed Meetings of Minds

The outpouring of responses to *Strategy and Conscience* show quite clearly that Rapoport's line of thought had touched on several major divides in the community of people interested in problems of "conflict resolution." Representatives of the mainline peace movement expressed cautious optimism. Thus A. J. Muste, the prominent pacifist and editor of the journal *Liberation*, wrote in to applaud Rapoport's achievement. Does not a shift from "strategy" to "conscience" point toward an advocacy of unilateral disarmament by the Cold War powers?—he wondered. And he and Rapoport exchanged polite commentary on the book in *Liberation* itself.[110] However, most assessments, like philosopher Sidney Hook's in the *New York Times Book Review*, were sharply critical. Rapoport, Hook recognized, had written a critique not of individual strategists but rather of "the entire mode of their rational thinking."

Yet he wondered, "What does the author offer in its stead? He explicitly rejects the rational approach and the 'problem solving attitude' of the strategist. . . . Appeals to conscience which cannot withstand rational analysis are dangerous guides in a world in which conscience is often the echo of tradition, early training, and cruel social use and wont."[111]

Donald Brennan, a former director of Herman Kahn's Hudson Institute, felt confident that he had demolished Rapoport's argument in his essay review in *Bulletin of the Atomic Scientists*, simply by pointing out that most strategic thinkers did not formally invoke game theory. "Rapoport's reaction to some limited use of game theory by Schelling," he opined, "must have initiated a train of thought so fascinating that he never noticed that most of the community [of strategic thinkers] makes no significant use of it."[112] But beyond this, Brennan leveled the broader charge that Rapoport's writing was "imbued with a kind of surreptitious pacifism" because of his apparent rejection of *any* analytic framework for thinking about national security. Facing the security environment of the Cold War, only two positions were possible. The first, chosen by the "great majority" of Americans, embraced "the idea that national military force may sometimes be used with justice in defense of important national goals or human values," and that "line-drawing problems of degree" provided the only interesting questions. The second, chosen by a small minority of "pacifists," "reject this kind of thinking altogether, and hold that the best overall defense of human values resides in the complete rejection of military force." Since Rapoport questioned the first position, he must inevitably be slouching toward the second.[113]

Rapoport's reply to Brennan attempted to clarify the place of game theory in his argument. The theory may not have formed the explicit basis for strategic thinking, he argued, yet it formed "a contribution to the intellectual climate in which it is taken for granted (especially by those ignorant of game theory) that, by and large, problems engendered by conflict must yield to strategic analysis." More important, as Brennan rightly noted, the point of investigating game theory was to clarify the inherent limitations of *any* theory of rational decision-making in application to questions of strategy. When faced with the possibility of nuclear war, "rationality" lay not in the calculations of self-interest invoked by the strategists, but in the realm of ethics and values. The strategists therefore could not have it both ways: they could not clothe themselves in the aura of mathematical objectivity and "rationality" at the same time.[114]

For his part, Schelling resented being tarred as a "strategist" and "game theorist," especially by a fellow member of the conflict resolution fraternity who used game theory "more than any 'strategist' I know of," in Schelling's opinion. His review of *Strategy and Conscience*, like Brennan's, blasted Rapoport for his sloppy and imprecise treatment of the strategic thinking his book purported to critique.[115] But as Schelling's assessment also made clear, Rapoport was not the first to turn "game theorist" and "strategist" into pejoratives. In the years leading up to the publication of *Strategy and Conscience*, in tandem with RAND's ascendancy at the Pentagon, a number of British operations researchers had criticized what they saw as the overly mathematical approach of American operations research—a development they associated with the influence of game theory.[116] Likewise, reacting to the publications of Wohlstetter, Kahn, Schelling, and Henry Kissinger, as well as near-disasters like the Cuban Missile Crisis in the fall of 1962, "exposés" linking game theory with the formation of American nuclear-strategic doctrine proliferated in 1963–64 as Rapoport's own book went to press. New Left sociologist C. Wright Mills had already identified "crackpot realism" as the mindset of the American military establishment in his 1956 book, *The Power Elite*; a 1963 book by sociologist Irving Louis Horowitz, titled *The War Game*, likewise probed the folkways of a group he dubbed the "New Civilian Militarists," connecting their flawed and inhumane strategic thought with the techniques of "game theory" and "gaming."[117] More such books would soon follow, like Philip Green's *Deadly Logic* (1966), and Norman Moss's journalistic *Men Who Play God* (1968).[118] What had begun as a disagreement between Rapoport and Schelling over the implications of game theory for the development of conflict resolution had now merged with a much broader set of battles over defense policy playing out between various factions within the scientific community, New Left and pro-Vietnam war liberals, activists and establishment.

As a result, for all the panels at meetings of learned societies and exchanges of book reviews that would be devoted to assessing Rapoport's work and the issues it raised, it is difficult to find evidence that the issues at stake were ever resolved, or even really addressed directly.[119] If anything, the ongoing debates exposed the depth of polarization that existed within the interdisciplinary community of scholars and activists interested in problems of conflict that spanned the Ford Foundation, RAND, the Mental Health Research Institute, and similar institutions. Apart from rapidly hardening political lines within the conflict resolution and peace research communities, the central intellectual stumbling block in the debates seemed to be game theory's status as a theory of rational decision-making and its relationship to a "scientific" approach to problems of war and peace. Were the problems of the arms race of a scientific and technical nature? Or was Rapoport correct in suggesting that the superpowers needed to move beyond "strategic" reasoning to a new and enlightened form of rationality that was ultimately grounded in something like empathy? In the wake of the Cuban missile crisis, the United States and the Soviet Union would at last begin to develop a mutual understanding on problems of nuclear arms control, but among the scientists studying problems of war and peace, the divides seem only to have grown.

To some extent, this may simply have reflected a continuation of the combined disciplinary and ethical myopia that behavioral science and conflict resolution had tried so hard to eradicate. Many of the so-called strategists continued their academic research and military consultancies, with defense think-tanks like RAND continuing to bring in contacts for studies of arms control and strategic problems. Especially under the patronage of the US Arms Control and Disarmament Agency (ACDA) from 1961 onwards, arms control became a fairly technical field, the subject of sober, weighty reports by technical experts from a variety of disciplinary backgrounds. Oskar Morgenstern—who had provided an entry into the literature on national security problems with his 1959 book, *The Question of National Defense*—would turn Beltway bandit not long after, emerging as CEO of the Princeton-based consulting company, Mathematica, which among other things would take ACDA contracts. Through Mathematica and various academic institutions, during the latter half of the 1960s, ACDA funded a number of studies to investigate the potential of formal mathematical models for addressing the central problems that had been raised within conflict resolution over the preceding decade: how to design safer weapons systems, how to limit war and de-escalate arms races, and how to make decisions in the absence of firm knowledge about the intentions of one's opponent.[120] Writing for a report to ACDA, the Hungarian-born economist John Harsanyi—whose essays on bargaining in the 1950s had brought him into contact with both Morgenstern and Rapoport—would formally model the uncertain beliefs the game players had about the intentions of their opponents. But while mathematics might help rationalize the empathetic meeting of minds that Rapoport felt he had observed, as we will see in subsequent chapters, it still could not quite reduce it to a matter of rational calculation from axioms of self-interest.[121]

By contrast, whether by design or by historical accident, critics of nuclear-strategic thinking increasingly moved away from the academic contemplation of "peace research" and in the direction of "conscience-driven" activism. There is no better example of this move than Rapoport's subsequent career. In his autobiography, Rapoport recalled: "Hard as I tried during my first years at Michigan to confine criticisms of strategic thinking to the strategists' own ground, the moral and emotional issues could not be avoided when the US launched air strikes against North Vietnam, initiating ten years of horrendous destruction of a Third World country." In the spring of 1965, Rapoport became a faculty sponsor of the first "teach-in" against the Vietnam War, and began to take part in protests and other antiwar activities. Finally in early 1968, much as he had years earlier at Chicago during the height of the McCarthy era, Rapoport left behind what he saw as an increasingly oppressive environment at Michigan to take up a series of prestigious academic appointments in Europe and Canada.[122]

Rapoport's departure from Michigan coincided with the collapse of the wider interdisciplinary project of "behavioral science" and "conflict resolution" that he and his colleagues had brought with them to Michigan in 1955. Both Miller and Boulding had left for other universities in 1967.[123] With Miller gone, the systems approach quickly withered at the MHRI. As its annual reports attest, after a series of interim directors, the institute was finally consolidated in the early 1970s with a much narrower focus on biochemical research. Meanwhile, in the absence of Boulding, the CRCR seemed to be at a loss for direction and funding. By the end of the decade, it came to be seen primarily as a hotbed of campus radicalism and political organizing, especially after serving as the headquarters for a 1970 strike run by the Black Action movement. This last episode featured prominently in the press coverage of the center's eventual closing in 1971, at which time the *Journal of Conflict Resolution* moved to Yale.[124]

Reflecting on these developments, in 1972 J. David Singer could write to a friend that "today, the 'peace research' community is a mess." Despite some early successes formulating a scientific study of peace, after about 1966, peace research was wholly overtaken by the stormy politics of the antiwar movement. Responding to revelations of social scientists' complicity in US counterinsurgency programs like "Project Camelot," or the high profile of operations researchers at the Pentagon and in military consultancies, a new generation of scholars had apparently decided that "science was bad and action/rhetoric was good."[125] Thus it was that the old convictions—about the power of scientific method and theoretical rigor to triumph over both disciplinary boundaries and power politics—that had initially sustained "behavioral science" and, with it, peace research and conflict resolution, at last reached their limits. Perhaps the activists of the 1960s had learned Rapoport's message too well.

Yet game theory would survive the disciplinary and political tumults of the later 1960s, even if many of the grander projects that had sustained it—like the MHRI's interdisciplinary brand of behavioral science research—did not. As we shall see in the next chapter, game theory's association with the arms race would do much to facilitate its transit from psychology into new and wide-open terrain in evolutionary biology. Here, evolutionary biologists would do much to overcome the central stumbling block in the debates over game theory and the arms race, namely, the relationship between game theory and rationality. Their solution, ironically, was to remove the notion of rationality from the theory entirely, thus replacing processes of human reasoning with blind historical processes as the guarantor of reasonable game "solutions."

CHAPTER SIX

Game Theory without Rationality

In the fall of 1973, a war unfolded in an article on the pages of the journal *Nature*. The combatants were animals that had enjoyed only a fleeting existence in a computer program, encoded on punchcards and run on an IBM mainframe at the University of Sussex many months before. "Hawks" pursued a strategy of "total war," fighting tooth and nail against all challengers. Other participants in the conflict adopted more nuanced strategies of "limited war," "retaliating" and "escalating" the conflict if attacked, but generally using "conventional tactics" against the hawk's "dangerous" tactics. When the electronic dust cleared and a body count was performed, the outcome seemed decisive, if counterintuitive: the practitioners of "limited war" had survived as the fittest. The Cold War had been fought in nature, and the hawks had lost.

This article, "The Logic of Animal Conflict," by John Maynard Smith and George R. Price, is often acclaimed as the starting point of a new theoretical tradition in the life sciences, evolutionary game theory, which rapidly established itself at the heart of evolutionary biology. Three years later, when Richard Dawkins published his landmark exposition of the emerging "neo-Darwinian" orthodoxy on evolution in *The Selfish Gene*, his introduction hailed Maynard Smith and Price for introducing "one of the most important advances in evolutionary theory since Darwin." In the pages that followed, Dawkins articulated a vision of the evolutionary process that characterized organisms as cybernetic "survival machines," guided only by their genes, whose reproductive success was determined by their ability to win games against other organisms.[1] By the early 1980s "evolutionary game theory," incorporating the work of Maynard Smith, Price, and several other biologists, was made official in textbooks, conference proceedings, and an extensive associated bibliography. Game theory came to be invoked in explanations of highly diverse biological phenomena, from lopsided sex ratios in obscure species of wasps to territoriality in spiders.[2]

The success of Maynard Smith, Price, and others in establishing "evolutionary game theory" is particularly striking given that several earlier attempts to introduce game theory to evolutionary biology failed to create lasting research traditions—at least under the rubric of "game theory." The most prominent of these unsuccessful attempts was undertaken by the American biologist Richard Lewontin and his colleagues, who initially thought that game theory might play a central role in developing a unified "population biology," a theory of population-level phenomena based on the new techniques of mathematical optimization emerging from applied mathematics and operations research. Yet by the mid-1960s Lewontin was working at cross-purposes to two significant trends in the histories of game theory *and* of evolutionary biology during this period. As we saw in the previous chapter, game theory's connection with behavioral science and with debates over arms control shifted its center of gravity, making it less a mathematical theory of optimization or decision-making (as it had been for military-funded mathematicians and operations researchers) and more an open-ended framework for investigating the interplay between individual self-interest and cooperation in a variety of settings. In tandem, during this decade, long-standing characterizations of the evolutionary process that emphasized cooperation and explained adaptations in terms of their benefit to populations or "groups" would be challenged by an emerging neo-Darwinian orthodoxy that focused attention on the evolutionary role of competing genetic individuals. In this context, game theory provided a framework for reconciling what neo-Darwinian evolutionary theorists saw as the inherent "selfishness" of genes with a number of apparently cooperative adaptations observed in nature, rather than simply accepting these adaptations as normal features of life.[3]

In the larger arc of this history, the emergence of evolutionary game theory is significant for several reasons. Before the later 1950s, game theory tended to travel along lines of direct contact with a small group of military-funded applied mathematicians and economists who were the theory's first adopters. By contrast, evolutionary game theory began as a relatively "indigenous" tradition with weak ties to Cold War military patronage, and whose similarities to other game-theoretic research programs were recognized largely in retrospect. But even if they were only recognized fully in retrospect, the similarities between biology's theory of games and other game-theoretic research programs unfolding in the 1970s and 1980s are striking. John Maynard Smith's concept of an "evolutionarily stable strategy" (ESS) proved to be closely related to John Nash's "equilibrium" solution concept for noncooperative games, which, as we shall see in chapter 7, was undergoing something of a renaissance in economics and the social sciences in the same period. Likewise, W. D. Hamilton and Robert Trivers's use of repeated Prisoner's Dilemma games to explain the evolution of reciprocal altruism rapidly merged with a broader literature on "supergames" that continued to gnaw away at problems posed by the arms control debates of the 1960s.

The story of game theory's adoption by evolutionary biologists can therefore shed light on a broader set of transformations unfolding in the life sciences and in other disciplines during this period. First, the way that biologists found themselves in dialogue with the conflict resolution literature testifies to the lingering influence in academia of the 1960s-era arms-control debates. Yet, while in some cases the ongoing arms-control debates might inspire biologists to consider adopting game theory for use on particular problems, their influence did not completely determine the outcome of this adoption. Thus Maynard Smith, Hamilton, and Price would reach for game theory and adapt it to their purposes not to shed light on human reasoning processes or to suggest guidance on how to conduct negotiations (as had the behavioral scientists and peace researchers of the 1950s and 1960s), but precisely because they felt game theory's spare depiction of rationality captured what they saw as the blind, asocial self-interest of the genes that were the prime actors in evolution. And from this starting point, using the calculations and computer simulations that were becoming *de rigeur* in 1960s population genetics, they could deduce the possibility of altruistic and "social" behavior so often observed in field biology or in the social sciences, thereby setting all of biology (and potentially the social sciences as well) on a foundation of genetics. Thus one of the most remarkable legacies of game theory's encounter with biology was the creation of a new "game theory without rationality" that, ironically, a number of economists and political scientists would subsequently adopt to free their models from unrealistically demanding assumptions about the cognitive abilities of *human* social actors. The emergence of evolutionary game theory was thus part of an interconnected set of conceptual and methodological shifts—toward a

spare methodological individualism and reductionism—that would ultimately bring portions of biology and the social sciences together in the
1980s around an impoverished conception of rationality without reasoning, choice without choosing.

Programming the Lives of Ants and Plants

In the spring of 1964 some of the fastest-rising stars in American biology vacationed together at the house of Robert Helmer MacArthur in
the lush countryside around Marlboro, Vermont. MacArthur had been
a graduate student in applied mathematics before he went to work with
the legendary limnologist and ecologist George Evelyn Hutchinson at
Yale. Following Hutchinson's lead in applying mathematical models to
understand the diversity and distribution of animal populations, he rapidly produced several brief but brilliant papers that drew heavily upon
the new mathematical techniques emerging from operations research
and cybernetics that he had studied as a graduate student. For example, a 1955 article, "Fluctuations of Animal Populations and a Measure
of Community Stability," used ideas from information theory to characterize the stability of ecological communities.[4] By his mid-thirties, Mac-
Arthur was a professor of ecology at the University of Pennsylvania.

MacArthur was at the center of a loose group of young, theoretically
inclined biologists and ecologists. Many of these were guests at Mac-
Arthur's house in Marlboro that summer, and thus E. O. Wilson would
later refer to them as the "Marlboro Circle." Egbert Leigh was a young
mathematician and later a research scientist at the Smithsonian Tropical
Research Institute. Richard Levins had recently received his PhD from
Columbia, producing a dissertation that explored many facets of contemporary ecology using the mathematical tools of decision theory and
convex analysis.[5] Richard Lewontin, at twenty-eight an assistant professor of biology at the University of Rochester and former PhD student of
Theodosius Dobzhansky, was rapidly making a name as someone who
used new mathematical techniques to bridge different areas of biology.
E. O. Wilson, the future creator of "sociobiology" in the 1970s, was a
relative mathematical neophyte, although he prepared himself by studying mathematics assiduously in the odd moments when he was not in the
field on an entomological expedition. Although they were not present
that June, Lawrence Slobodkin, formerly a student of Hutchinson's at

Yale and now professor of ecology at the University of Michigan, and Leigh van Valen, a paleontologist at the University of Chicago, were also connected with the group.

It is debatable how coherent a group this Marlboro Circle actually was. Its members were drawn from many different areas of biology, and while several of them would form long-lasting collaborations (for instance, between MacArthur and Wilson on the theory of island biogeography and between Lewontin and Levins on various aspects of theoretical biology), on the whole their associations appear to have been fleeting. But hovering in the background on those summer evenings was a common disciplinary problem: how to reassert the relevance of "macro-biology"—studies of organisms and groups of organisms in natural environments—in the face of an increasingly hegemonic "microbiology" focused on molecular biology and genetics. A methodological shift toward theoretical rigor and mathematical sophistication seemed a plausible solution. Both population genetics (the study of changing gene frequencies within populations) and population ecology (the study of population structure and demographics in relation to environment) possessed rich mathematical traditions, but communication between these fields seemed stalled. Bringing them together to form a unified "population biology," grounded in rigorous mathematical principles, would provide an answer by field biologists to the rising power of molecular biology. According to Wilson, after two days of discussions this group agreed to a plan that would help realize their grandest hopes for a new synthesis in biology: they would pool their work in a series of groundbreaking theoretical papers, all published under a collective pseudonym. In so doing, they consciously imitated the famed cabal of French mathematicians who, between the 1930s and the 1970s, published a grand synthesis of all modern mathematics under the name "Nicolas Bourbaki." Bourbaki was an obscure nineteenth-century French general; the Marlboro Circle, true to its own inspirations in applied mathematics, chose the name "George Maximin." According to Wilson, his surname was a reference to the "the point of the greatest minimum in optimization theory."[6]

But despite these sanguine early discussions Maximin never took on any more solid attributes, not even a name in print. By the fall of 1964 his constituent authors were beginning to drift in different directions, although many of them continued to collaborate and publish together for years after. But his brief existence reflects the fact that a number of biologists were well aware of game theory's existence over a decade before the publication of Maynard Smith and Price's article. In fact, two members of the group around MacArthur had explicitly employed "game theory" in published work prior to 1964, when they were apparently willing to use its characteristic "maximin" concept as a symbol of their hopes for the future of their work. In 1961 Lewontin had published an article on "The Theory of Games and Evolutionary Biology," and in early 1964 Slobodkin had published "The Strategy of Evolution," which also consciously used ideas from game theory.[7] Other members of the group, such as Leigh and MacArthur, had advanced training in applied mathematics, including game theory, while Levins had studied statistical decision theory (out of Luce and Raiffa's classic *Games and Decisions*) at Columbia in the late 1950s. In short, in 1964 there was every reason to expect that evolutionary game theory would grow from the work of this perfectly positioned group of people. Yet it did not. There is little evidence that their work led to a broad and sustained program of research on "games," and there is quite a lot of evidence that their authors found the game theory of their day peculiarly lacking as a tool with which to mathematize biology.[8]

An understanding of why this was so requires examination of the modeling choices of these theorists, especially their focus on the "population" as the fundamental unit of analysis for theory-building. By the early 1960s, a wide swath of the biological sciences focused attention on populations of organisms as a fundamental unit of analysis, but there was little agreement on how to characterize this entity. The models of population genetics developed by R. A. Fisher in the 1930s depicted populations as thermodynamic ensembles: like billions of molecules colliding in a gas, individuals mated randomly, genes assorted randomly; evolution was characterized by aggregate changes in gene frequency over time. By contrast, researchers in fields such as ecology often assumed that populations possessed features more commonly associated with human societies, such as social structures and learning, or with organisms, such as homeostasis, functional integration, and regulation. Representatives of the Chicago School of animal ecology, for example, focused primarily on analyzing interactions between populations of organisms and a changing physical environment. Here, animal ecologist Warder Clyde Allee and entomologist Alfred Emerson drew on the work of Sewell Wright and their own observations of animal sociality to produce an interpretation of evolution that stressed cooperation, integration, and community. Like many other American biologists at the time, they were consciously responding to "social Darwinism," which appeared to justify constant violence and war both within and between societies.[9] True to the Chicago biology department's traditional orientation toward physiology rather than genetics, these scientists frequently turned to physiological and cybernetic metaphors, taken from the work of Walter Cannon and Ralph Gerard, to analyze populations as functionally integrated wholes. Evolution became characterized as a process tending toward increased "homeostasis," "the regulation, control, and maintenance of conditions for optimal existence," made possible by increased specialization of social roles combined with more efficient group integration. While "competition" still existed, it was carefully regulated to the amount "optimal" for the benefit of the group.[10]

The nature of populations was not simply an academic problem. In both Britain and the United States, the Cold War era saw renewed and intensified concern with overpopulation and its perceived consequences in terms of resource depletion, environmental degradation, third world poverty, the spread of communism, and the increased likelihood of war. Popular books such as William Vogt's *Road to Survival* and Fairfield Osborne's *Our Plundered Planet* (both published in 1948) painted a picture of a bleak future in which uncontrolled population growth would lead to environmental breakdown and widespread suffering. Other works, such as a widely circulated 1954 neo-Malthusian pamphlet titled *The Population Bomb*, were still more alarmist, likening the threat of unchecked population growth in developing countries to that posed by nuclear weapons. While its authors asserted that they were "interested naturally in the humanitarian aspects of the problem" of overpopulation and famine, they were "also apprehensive of the use which the Communists make of hungry people in their drive to conquer the earth." The piece ended with an appeal to readers to support population-control programs, a call subsequently taken up by the ecologist Paul Ehrlich in his 1968 book *The Population Bomb*, and by Garrett Hardin in his widely influential essay, "The Tragedy of the Commons." In response to such concerns, government policymakers and leaders of private foundations aggressively promoted scientific research and technology transfer to promote population health and nutrition in the hope that this would prevent communist domination of the third world. The major American philanthropic organizations, the Rockefeller and Ford foundations, were both spurred to action by the population issue, resulting in funding for agricultural innovation as well as population planning programs for the developing world.[11] The associated high-profile debates over how population densities can be stabilized would ultimately highlight significant ambiguities in the way scientists had characterized populations. The result was a series of controversies pitting population geneticists, who tended to privilege the role of individual evolutionary advantage in producing population densities, against a diverse collection of anthropologists, ethologists, and ecologists who maintained a social or organismal conception of populations.[12]

The event most responsible for crystallizing these tensions over the nature of "population" latent within the small world of British biology was the 1962 publication of a book, titled *Animal Dispersion in Relation to Social Behaviour*, by V. C. Wynne-Edwards of the University of Aberdeen. Here, Wynne-Edwards sought to explain the apparent ability of wild animals to maintain optimal population densities so that they do not overuse their food resources. Drawing on a wide range of literature in ethology and ecology, as well as his own studies of birds in Scotland, Wynne-Edwards argued that successful species have evolved cooperative regulatory apparatuses that keep their densities balanced close to the "optimum number" for a given environment. Specifically, populations behave like engineering feedback devices or cybernetic machines that have the goal of reaching "homeostasis" with the environment. An elaborate system of "conventional competitions," that is, ritualized displays, threats, and conflicts not directly over resources, set up social hierarchies and "pecking orders." Animals lower in the social hierarchy would voluntarily forgo mating and establishing territories "for the common good," thereby providing negative feedback to stabilize population distribution in an optimal manner. Because these social conventions were not always evolutionarily advantageous for individuals, they could arise only as a result of natural selection acting on groups of organisms, or "group selection," a term that Wynne-Edwards coined.[13]

While Wynne-Edwards made frequent use of organismal metaphors in describing animal populations, he owed much of his analytic framework to a 1922 book by the Oxford demographer Alexander Carr-Saunders, titled *The Population Problem*. This book represented an eclectic mix of intellectual traditions: Malthusian demography, Darwinian evolutionary biology, history, and anthropology, to name a few. Carr-Saunders thus drew on a wide range of evidence to argue that humans maintain their populations close to an "optimum number" via social customs and practices, from the restriction of sexual relations to infanticide. Populations might deviate from this optimum number for a variety of reasons: the outbreak of wars or epidemics, migrations, or even simple changes in the "desirable" number of humans. But ultimately, Carr-Saunders argued that war and bloodshed were not inevitable, since cultural change—especially conventions surrounding marriage and the status of women in society—would provide a flexible and humane way of stabilizing population size.[14]

Wynne-Edwards envisioned *Animal Dispersion* doing for the literature on animal ecology and behavior what Carr-Saunders had done for literature relevant to the study of human populations: providing a synthetic overview that established the connection between social behavior and reproductive rates. Moreover, like Carr-Saunders's book, *Animal Dispersion* was released into an environment in which the connection among population, conflict, and social order had become a significant matter of public debate. The reviewers of Wynne-Edwards's book immediately perceived its relevance to Cold War fears of overpopulation, environmental crisis, and war, and in the process *Animal Dispersion* became a media sensation. In Aberdeen the *Press and Journal* treated the book to a multipage review, covered by an enormous photograph of crowds gathering on a street in China. "There are too many of us," the caption stated. "Millions of people all over the world have not enough to eat. Why? Because man has lost the power which all the higher animals have to limit their numbers at a level of genuine subsistence—not a poverty or starvation level."[15] The author concluded that Man needs to reacquire the "wisdom of the animal," to establish social checks on reproduction and ensure that humanity is not reduced to miserable starvation. Other reviewers instantly saw the connection between Wynne-Edwards's book and the Cold War specter of violent conflict over resources. As *The Guardian* noted, Wynne-Edwards has shown that the animal world is not characterized by individual competition, by "nature red in tooth and claw," but by genteel interactions "more in the nature of a sport." In the context of a looming world population crisis, it was urgent that humans curb their unnaturally aggressive and selfish instincts and return to these more natural, benign forms of sociality.[16] Wynne-Edwards could only have been pleased by the reading of his work that circulated in the press: a consultant to Britain's Nature Conservancy and proponent of wildlife protection in Scotland, he saw the argument of *Animal Dispersion* as a first step toward the formation of a naturalistic ethic of conservation based on social customs that regulate population growth and resource use. And although it is difficult to gauge the depth of his feelings about the global problem of human overpopulation with any precision, he did save a copy of *The Population Bomb* pamphlet for his files.[17]

To be sure, many of the scientists associated with the Marlboro Circle were not "group selectionists" in any conscious or overt way. MacArthur had studied ornithology with David Lack, one of Wynne-Edwards's chief critics, and clearly shared Lack's suspicion of group selection. In a letter to Lack in 1962, MacArthur suggested that they should keep a low profile in responding to Wynne-Edwards's *Animal Dispersion*, since the best way to overcome bad science was to ignore it.[18] Yet even if they did not necessarily think of themselves as advocating "group selection," MacArthur and his colleagues often shared a focus on the "population" as the fundamental unit of analysis in biology, basing their mathematical analyses on parameters such as population density and other averages that stood in for the population as a whole.[19] Such a stance makes sense given these biologists' commitment to developing a science of populations capable of uniting population genetics and population ecology, as we have seen. But a focus on populations also permitted them to adopt time-honored mathematical frameworks of population ecology, demographics, and statistics—not to mention new techniques arising from operations research.

Thus, the "population" formed a significant unit of analysis in the theoretical work of the Marlboro Circle. E. O. Wilson, an entomologist by training and intimately familiar with Alfred Emerson's work on ants and termites, readily adopted Emerson's ideas about population-level "optimization" and group selection. Beginning in 1963, he inaugurated a theoretical research program based on "ergonomics," the "quantitative study of the distribution of work, performance, and efficiency in social insects."[20] As he put it, the "matter of the presence or absence of a given caste [of ants], together with its relative abundance when present, should be susceptible to some form of optimization theory, provided we are able to assume selection at the colony level," and "in fact, colony selection in the advanced social insects does appear to be the one example of group selection that can be accepted unequivocally."[21] In particular, he argued that "colony selection" acts like a military logistical programmer or government planner, "regulating" the different caste proportions $W_1, W_2, \ldots$ to maximize the production of queens. The planner is subject to constraints imposed by the environment's food productivity and the frequency of various uncertain environmental "contingencies" (invasions, damage to the colony) that the castes are produced to deal with. If, in responding to a given contingency, the workers from different castes are substitutable for one another in a fixed ratio (based on their relative efficiencies in responding to these contingencies), it was possible to formulate this planner's problem as a linear program. In the case of two contingencies, this is:

Minimize the quantity $W_1 + W_2$ subject to constraints (associated with the two contingencies):

$$W_1 = A + B\,W_2$$
$$W_1 = C + D\,W_2, \text{ where A, B, C, D are constants.}$$

The constants in the model proved difficult to measure directly, so Wilson was hard pressed to predict quantitative caste weights in any specific situation. However, by exploring the properties of the linear program, he was able to establish some general conclusions—for example, caste specialization would increase over time as long as the contingency probabilities stayed constant, and that as castes evolved to be more productive, they would decrease in size.

Over time, Wilson's linear programming models came to be the foundation of most of his interpretive work on ants.[22] However, Wilson, like many members of the Marlboro Circle, did not feel that the lessons of these models were only relevant to such peculiar insect species. Group selection and optimization theory also underlay much of Wilson's synthetic *Sociobiology* (1975), which attempted to explain *all* social behavior, human and animal alike, through evolutionary models.[23] Optimization of population parameters was the fundamental point of evolution: as Wilson would write in the 1970s, "biologists view natural selection as an optimizing process virtually by definition" so that "optimization arguments are the foundation upon which a great deal of theoretical biology now rests."[24]

Richard Lewontin likewise approached the problem of adaptation via group selection and optimization in his 1961 article, "Evolution and the Theory of Games."[25] In this paper, as in later discussions of the Marlboro Circle, Lewontin addressed what he considered to be a major gap in evolutionary theory. Extant models in population genetics focused primarily on gene frequency changes *within* a population, while population ecology had developed a nonevolutionary understanding of the relationship between populations of organisms and their environment.

Game theory seemed promising as a "new mathematics" that might connect the two traditions to explain macrolevel evolutionary changes like speciation and extinction that were caused by a fluctuating environment. Lewontin therefore suggested that a species constantly making decisions about its collective genotype in the face of an uncertain environment is effectively playing a zero-sum "game against nature." The environment first makes its move by choosing from a number of possible "states of nature." The population responds by selecting a "strategy" from the set of possible genotypes G_1, G_2, . . . with associated fitnesses S_1, S_2, . . . , or by selecting some weighted combination of the pure genotypes, thereby obtaining a fitness $W_1 S_1 + W_2 S_2 + . . .$, where the W_i's represent the proportion of the population possessing the ith genotype.[26] This strategy—in biological terms, a polymorphism—is the equivalent of game theory's "mixed strategy," with the probabilities reinterpreted as proportions. Now imagine a plant species trying to choose the optimal number of seeds to produce in an arid environment with uncertain rainfall. In this case, Lewontin argued that it would do best to select a reproductive strategy according to "the maximin criterion of optimality." In other words, the plants would attempt to maximize the minimum possible reproductive utility payoff (in terms of collective reproductive success), an assumption that Lewontin justified by noting that the environment is "capricious," like a clever poker-player, constantly changing and lacking in statistical regularities. Given knowledge of the particular reproductive fitness values for the various possible genotypes (corresponding to numbers of seeds produced), it is possible to calculate the frequencies of these genotypes in the population by seeking their values at the minimax.

Of course, in treating minimaxing as "optimization," Lewontin was transforming the original game-theoretic concept, since the minimax, it will be recalled, was originally conceived as an alternative to conventional optimization theory. But, as we have seen in earlier chapters, Lewontin was far from the first person to do this. As John von Neumann and George Dantzig demonstrated in 1947 while Dantzig was analyzing military logistics, the problem of solving a two-person, zero-sum game is mathematically equivalent to solving a linear program. Lewontin's tableau—plant populations facing "capricious" weather and planning their collective genotype in response—can easily be tweaked to look like Wilson's linear programming model of ant colonies that regulate their caste ratios to manage the contingencies of their environment. Of course,

Lewontin's mathematical idiom was slightly different from Wilson's, since he imagined the population faced with choice under uncertainty as participating in a "game against nature." In formulating this model he relied on game-theoretic analyses of individual decision-making behavior, especially Luce and Raiffa's *Games and Decisions* (1957). Throughout the article, Lewontin explicitly borrowed Luce and Raiffa's notation, and demonstrated that he was fluent in the concepts and techniques they presented.[27] Lewontin was thus working squarely in a tradition of game theory that was prevalent in America in the late 1950s.

These facts, however, only serve to sharpen the question of why Lewontin's ambitions of establishing game theory as a new mathematics of evolution were not realized. Perhaps the problem facing Lewontin (and Slobodkin, whose 1964 article on the "Strategy of Evolution" used game theory in much the same way Lewontin did) was that he had learned the game theory of the day too well. To the intended audience of Luce and Raiffa's book—social and behavioral scientists—game theory was about making rational choices by choosing an optimal strategy when faced with an uncertain opponent. But, especially for a biologist who focused on the analysis of *populations*, this interpretation was fraught with difficulties. It is problematic enough to talk about "choice" and "rationality" in individual (nonhuman) animals, but could one use these terms in reference to animal populations? Do populations possess coherent preferences and utilities that can be optimized? Lewontin himself admitted this difficulty in the final paragraph of his paper. "The final element needed in a game theoretical approach to evolution is a *mechanism*," he argued. "For intra-population genetics, the maximization of $\overline{W}$ [Fisher's average fitness] is a good principle because, in fact, the processes of natural selection provide the mechanism for this maximization. Does a similar mechanism exist for '*maximinization*' of the probability of survival?"[28] Lewontin's suggestion was to make an analogy between the survival and extinction of populations and the life and death of individual organisms, but this simply transferred the problem instead of solving it. Who are the true players in this game, and what properties should be expected of them? Do they have coherent preference structures and the ability to implement their choices (like the military logistical planners) once they are made?

Faced with such problematic questions, it is perhaps not surprising that the invocation of "games" would retreat, giving way to the language

of "optimization," "programming," and "regulation" that fit more naturally with the conceptual framework of population biology. Lewontin
himself rapidly dropped game theory after his first paper on the topic,
concluding that it offered little that could further his goal of achieving
a unified biology of populations. When Richard Levins, a member of
the Marlboro Circle and close colleague of Lewontin, published a book
(based on his doctoral dissertation) on the mathematics of evolution in
changing environments in 1968, many of Lewontin's ideas about lin-
early weighted genotypes and optimality conditions remained. However,
"game theory" received only a passing mention—and that was to suggest that games were not a particularly realistic description of nature.
Thus, while the existence of George Maximin suggests that game theory was available to biologists, the relationship was a fleeting one. As it
had with the postwar operations researchers, game theory dissolved into
the general study of optimization—into scattered vectors, matrices, and
inequalities that bore no apparent resemblance to games—or at least,
games that members of the Marlboro Circle might recognize.[29]

Bridging Chalkboard and Field, Group Selection and "Individual Advantage"

In contrast to the members of the Marlboro Circle, for whom populations were the key theoretical unit of analysis and group selection was
a plausible mechanism for the optimization of population-level parameters, "evolutionary game theory" would emerge primarily from the
work of critics of group selection. Despite the immense popular appeal
of *Animal Dispersion*, Wynne-Edwards's argument attracted immediate
criticism from a number of corners of the scientific community. Among
other things, the book directly challenged the work of Wynne-Edwards's
colleague, the Oxford ornithologist David Lack. Less than a decade
previously, Lack had tackled Wynne-Edwards's central question in his
own book, *The Natural Regulation of Animal Numbers* (1954). Lack's
key finding—that clutch size for a given bird species was relatively constant over a variety of conditions—suggested that population regulation
depended less on social interaction (or indeed, any so-called density-
dependent regulatory process) than on the innate fertility of the species
in question. Of course, as soon as *Animal Dispersion* appeared, Lack

rallied a far-flung network of academic allies to defend his earlier find-ings. In addition to his ornithological colleagues, Lack reached out to prominent geneticists and evolutionary theorists such as J. B. S. Haldane and his star student, John Maynard Smith.[30]

Maynard Smith represented a very different research tradition from Lack: an aircraft engineer turned research scientist at the Galton Lab-oratory at University College, London, in the early 1960s, he was best known for work on the physiology and genetics of *Drosophila*. Yet both shared a shared an interest in understanding the evolutionary basis of fertility, and a conviction that reproductive rates are inherent in the spe-cies in question rather than determined by social interactions between members of that species. Lack came to his conclusions by field obser-vations on the Great Tit (*Parus major*), Maynard Smith from an unex-pected observation about the behavior of his laboratory fruit flies. As he would describe in a 1956 letter to Lack, the behavior involved a funny mating "dance" that males and females performed when they encoun-tered each other. Females would jump back and forth rapidly, this way and that, and then watch as the males tried to mimic their moves at the same rate. The dance would go on until the two had either mated or the female turned away, disappointed by the inability of her counterpart to keep up. What Maynard Smith had noticed was that the slower males tended to be inbred and the faster males outbred; moreover, outbred males tended to be far more fertile than inbred males. So clearly, the females stood to gain from being able to screen out the poor dancers. (Maynard Smith joked that it was good that not all human females did the same, or he would never have found a wife!)[31]

The question was how to explain this phenomenon. "At the time," he later noted, "it was traditional to explain differences in behavior in terms of 'motivation,'" by reference to the organism's conditioning and stimuli. Maynard Smith was dubious: the slowness of the inbred flies, he argued, was not the result of a "choice" or social influence; rather it rep-resented the limits of their genetically determined capabilities. "As it was, I concluded that they did not mate because they could not keep up with the female, no matter how hard they tried." The application of May-nard Smith's insight to Wynne-Edwards's argument in *Animal Disper-sion* was straightforward: limits on animal reproduction were due less to social conditioning and more to inborn limitations.[32]

Not surprisingly, when Wynne-Edwards's book appeared, Lack was quick to invite Maynard Smith to discuss it at an Oxford symposium.

Not long after, as popular interest in Wynne-Edwards's work crested in the fall of 1963, Maynard Smith became the public face of opposition to Wynne-Edwards's thesis. On October 15, BBC radio's signature show, the *Third Programme*, featured an introduction to *Animal Dispersion* narrated by the eminent cellular biologist Michael Swann of the University of Edinburgh.[33] Just a couple of days later, on October 17, a follow-up broadcast on *Third Programme* devoted a lengthy roundtable discussion to the controversy surrounding *Animal Dispersion*, hosted by Swann, and featuring as guests the zoologists Michael Cullen, John Young, and John Maynard Smith. To his guests, Swann posed a critical question that he had identified in his previous broadcast: if the social phenomena described by Wynne-Edwards were in fact real, how could they evolve through natural selection? In reply, Maynard Smith admitted that although *Animal Dispersion* had provoked him to make extensive calculations to see whether the phenomenon was possible, he still had not arrived at a convincing answer. Nevertheless, after exploring a number of possible models, he was quite certain that group selection, however defined, was incompatible with what he called the "selfish business" of natural selection. And by and large the other guests agreed: Wynne-Edwards had succeeded in drawing attention to several poorly understood aspects of animal sociality, but "apart from Maynard Smith's model-making" the book was unlikely to lead to any great breakthroughs in understanding the evolutionary origins of social behavior or the regulation of population numbers.[34]

Maynard Smith persisted in his model-making. The following spring he joined Lack's close colleague at Oxford, Christopher Perrins, in a set of letters to *Nature* that that took Wynne-Edwards to task for his unsophisticated understanding of population genetics and his neglect of ornithological data that seemed to contradict the thesis of *Animal Dispersion*. Maynard Smith's contribution was to derive conditions under which alleles that benefited groups at the expense of their individual carriers might spread. In order for this "group selection" to work, the groups in question had to be strongly isolated; and since such strict conditions were unlikely to be met in natural conditions, group selection was an unlikely mechanism for the evolution of social behavior.[35] In reply to Maynard Smith's mathematical onslaught, Wynne-Edwards could only protest that "the major obstacle to constructive discussion between us really arises from the understandable (though regrettable) differences in outlook and experience between a laboratory geneticist and a field ecologist."[36] There seemed no obvious way to bridge the gap between observations of social interactions between wild animals and blackboard calculations of gene frequencies. Nevertheless, after leaving University College in 1964 for an administrative position at the new University of Sussex, Maynard Smith devoted less time to laboratory genetics and more to searching for new mathematical tools that might allow him to strike a final, decisive blow against Wynne-Edwards's theory.

The intense interest generated by the population crisis of the 1960s provided circumstances under which these two scientists from very different disciplinary backgrounds would be part of a common conversation—even if this conversation revealed more disagreement than agreement. This was especially true three years later when the Royal Society formed a Population Study Group, under the leadership of the pharmacologist Lord Howard Florey, to consider the application of the social and biological sciences to the control of world population levels. Starting in December of 1965, the study group met every few months until Florey's death in 1968, covering such topics as "genetics and population" and the "economics of population growth." Correspondingly, its membership was highly diverse. In addition to high-level administrators from the Family Planning Association and the International Planned Parenthood Federation, the group included the naturalist Wynne-Edwards, the geneticist Maynard Smith, a strong contingent of medical researchers, and several demographers and anthropologists. Once again, "population" was both an object of common concern and a flashpoint of disagreement. "The geneticist's conception of a population, known as the Mendelian population, is not necessarily the same as that of the demographer, geographer, or ecologist," reflected an anthropologist participating in the discussions. Yet the same basic set of problems brought them together. Could organisms somehow "choose" to curtail their reproduction as a result of changing social and cultural influences, or do less malleable facts of their biology doom them to overpopulation, starvation, and war?[37] Wynne-Edwards had proposed one way of approaching the problem, depicting both humans and animals as inherently social beings, albeit ones who had temporarily cast off their natural social structures. Maynard Smith's models had suggested serious problems with the idea of group selection, but a new mathematics of evolution that could explain Wynne-Edwards's findings in terms of individual selfishness remained elusive.

Prisoner's Dilemmas and Reciprocal Altruism

This new mathematics would first emerge in the work of another participant in the meetings of the Royal Society Population Study Group, a young lecturer in zoology named William D. Hamilton who had been invited to attend the study group's meetings by his graduate supervisor from the London School of Economics, the demographer John Hajnal. If the threat of overpopulation and violent conflict hung over proceedings of the Royal Society Population Study Group, none felt it more keenly than Hamilton. Poring over reams of population statistics generated by the United Nations, Hamilton thought he could discern the outlines of future holocausts rooted deep within the evolutionary history of human beings. War, he suspected, was likely to break out in places where distinct populations were experiencing very different rates of increase. Given human biology, such situations would stir "aggressive instincts connected with population perception" and "with the inception of wars." The leadership of the study group was unimpressed with this suggestion, perhaps even hostile to it.[38] But writing over thirty years later in the introduction to his collected works, Hamilton felt vindicated by recent events in East Timor, Kosovo, and Rwanda, where he sensed that competition in population growth rates led inevitably to violence; and he repeated his call for a policy of "disaster acceptance" as an inevitable (if less humane) alternative to abortion or infanticide.[39]

Hamilton is best remembered today as one of the premier evolutionary biologists of the twentieth century, especially for his work on the evolution of altruistic behavior and the theory of kin selection. He was also one of the first biologists to create an enduring program of research around game theory in a series of papers from the mid-1960s. Hamilton had arrived in London in 1961 with hopes of pursuing graduate work on the genetics of altruism and social behavior. The topic had been substantially inspired by his reading of R. A. Fisher's *The Genetical Theory of Natural Selection* while he was an undergraduate at Cambridge. This classic book is best known for its exposition of Fisher's mathematical theory of natural selection and its application to a range of evolutionary phenomena. However, Hamilton was particularly engrossed by the later chapters of the book, which examined human social problems such as poverty and social stratification from the perspective of natural selection and proposed eugenic measures to forestall "racial decay." Such arguments clearly inspired Hamilton to think about the role of evolution in producing human social traits, but their eugenic overtones also imbued Hamilton with an outlook on problems of overpopulation and violence very different from that of Wynne-Edwards and, undoubtedly, the august members of the Population Study Group. In Hamilton's mind, the problem facing humans was not that they had forgotten their natural social instincts: instead, intoxicated by the power of medical science, they had become unnaturally benevolent toward the unhealthy and unfit, thereby encouraging population growth that was both qualitatively and quantitatively maladaptive.[40] This outlook on the population crisis would prove to be centrally connected with Hamilton's line of inquiry in evolutionary theory throughout his career in the 1960s. In effect, Hamilton would invert the central problem of *Animal Dispersion*: rather than asking how organisms could forget their naturally social and cooperative impulses, Hamilton sought to explain why inherently self-interested individuals might behave socially.

Unfortunately for Hamilton, such an intellectual project fell uneasily between the social and biological sciences as they were constituted in the early 1960s. He initially entered the master's program on human demography at the London School of Economics, working with Norman Carrier and then John Hajnal. Given his interests, a more suitable home might have been the Galton Laboratory—the longtime institutional homeland of genetics and eugenics. Yet the laboratory was an institution in transition in the early 1960s: the days of Galton were long past, R. A. Fisher had moved to Cambridge in the 1940s, and J. B. S. Haldane had left for India in 1957. Moreover, after Fisher's departure, Lionel Penrose assumed the Galton Professorship of Genetics and significantly reoriented the institution's mission, changing the name of the in-house journal from *Annals of Eugenics* to *Annals of Human Genetics*, and refocusing its research toward the genetics of disease. From the outset, Hamilton felt that his ideas on the evolution of social behavior met a chilly reception with Penrose, who tried to interest him in a more straightforward project in laboratory genetics before assigning him to another member of the laboratory as an advisee.[41] Hamilton nevertheless persisted in his original project, "trying to prove myself right against all those, including my former professor, Lionel Penrose, who had told me that there was no such thing as genetic altruism or genetic selfishness or aggression, and that, here again, I was on a wild-goose chase."[42]

While his encounter with Penrose clearly came as a disappointment, Hamilton's fascination with the biology of altruism, selfishness, and aggression was in fact nowhere near as marginalized in the intellectual atmosphere at University College as his later recollections suggest. Throughout the 1950s, a number of scientists in the Galton Laboratory had been deeply involved in national and international organizations whose goal was to develop scientific perspectives on the problems of war and violent conflict. During this period, Penrose, a Quaker pacifist, served as secretary of the "Medical Association for the Prevention of War," an international organization of medical professionals whose goals included studying "the causes and results of war." In 1961 and again in 1963, Penrose and his wife Margaret organized conferences on the "Pathogenesis of War," which pursued this goal by mustering perspectives from a broad range of fields including biology, anthropology, and psychology.[43] John Maynard Smith, who knew both Penroses from University College, actually gave a presentation at the First Conference on the Pathogenesis of War in 1961—an exposition of the theory of dominance hierarchies and pecking orders that would feature so prominently in Wynne-Edwards's book just a year later. In bird populations, "a pecking order with ritual rather than real fighting is often found. The winner ceases to attack as soon as surrender is made. The surrender ceremonies are better developed and more obvious in birds than in humans." Despite Maynard Smith's skepticism of the explanation for this phenomenon common within the ethological literature (namely, that ritualization benefits the species as a whole by eliminating needless slaughter), his recommendations for the arms race were strikingly similar to those later found in press reviews of *Animal Dispersion*: humans need to channel their energies in the direction of ritualized activities rather than resorting to actual war.[44]

By the early 1960s Lionel and Margaret Penrose's work on the "Pathogenesis of War" had formed connections with a broader transnational intellectual movement associated with the names "peace research" and "conflict resolution." As we saw in the previous chapter, this movement set down institutional roots in a number of places, most notably the Center for Research on Conflict Resolution at the University of Michigan. Peace research had also developed a core set of intellectual tools for investigating human conflict and cooperation that gave a prominent place to game theory, most notably the work of Thomas Schelling on bargaining and of Anatol Rapoport on the Prisoner's Dilemma game. Rapoport and Schelling's game theoretic studies were widely disseminated as the ideas and institutional forms of conflict resolution spread from the University of Michigan across the world in the early 1960s. By 1962 Penrose was in contact with a number of scientists at the University of Lancaster who had established a "Peace Research Centre" along the lines of the CRCR.[45] This development inspired Penrose and his fellow Quaker and colleague at the Galton Laboratory, Cedric A. B. Smith, to create a reading group of their own that included a number of academics from University College and other nearby universities. Ultimately, Penrose and Smith hoped to create a Peace Research Institute at University College similar to the ones at Michigan and Lancaster.

This goal was articulated in a memo Smith sent to the reading group in August of 1962. Here, Smith suggested two steps that would help catalyze such an institute's establishment at University College. First, he mentioned that he had ordered a number of books as a first contribution to a library that would be open to the community whenever he was in his office. Among the titles in the new collection were Theodore Lentz's *Towards a Science of Peace* and Anatol Rapoport's *Fights, Games, and Debates*—both classics of peace research that called attention to the promise of game theory for understanding conflict. Secondly, drawing on these works, Smith suggested that people involved with the group identify topics in "Conflict Resolution Research" to which they could contribute, and to drum up support for such research among other staff at University College.

In the memo, Smith himself suggested a number of definitions that he hoped would help structure discussion. These included conflict, cooperation, and "altruism," terms that characterized actions by individuals that varied according to the extent to which they helped the individual and harmed or assisted others. Thus competition benefited the individual and harmed others; altruism benefited others and harmed the individual, and cooperation was in the interests of everyone. With these foundational concepts in mind, Smith also called attention to an important theoretical framework for thinking about "conflict and cooperation," namely, game theory. Not only was Smith's presentation of the theory strongly inflected with the language of conflict resolution, but, like Rapoport and the social pychologists, he called attention to the need for the theory to take on board the realities of human biology and psychology—realities that were presumably far more complicated than spare assumptions of rational self-interest.[46]

The formal activities and regular membership of Cedric Smith's peace research group remain generally obscure, but by the mid-1960s it was clearly a thriving enterprise. According to Theodore Lenz, who visited England with his wife in the summer of 1964, Cedric Smith now presided over a "research group [that] numbers about 50, several of whom are connected with the Galton Laboratory or the Tavistock Institute."[47] Many of those involved were Quaker academics who attended meetings at the Friends House on Euston Road, a stone's throw from the University College Union, and a five-minute walk from Cedric Smith's office at the Galton Laboratory off Gower Street. Some of these, such as Adam Curle of the London School of Economics, helped Smith bring Kenneth Boulding to London to lecture on peace research in the summer of 1963. Rapoport himself arrived in London in 1966 to give a series of radio interviews on peace research.[48] Other less-formal activities reveal the scope of the organization's influence and interests. In May of 1963 Cedric Smith gave a well-attended talk on "The Theory of Games and International Relations" at a meeting of the Royal Statistical Society's "General Applications Section."[49] In addition, the group was known well enough among biologists that Wynne-Edwards was present at a meeting of the peace research group in the fall of 1963, shortly after delivering a keynote lecture, drawn from the recently published *Animal Dispersion*, on "Territoriality and Sociality in Animals" at a meeting of the Society for Animal Behavior in London.[50]

The founding of the peace research group coincided closely with the start of Hamilton's involvement with the Galton Laboratory and the period during which he produced his famous papers of 1963 and 1964 that developed the theory of kin selection. Moreover, with Penrose uninterested in Hamilton's project, Hamilton had been assigned to work with Cedric Smith, who would serve as his advisor at the Galton. Smith was clearly excited by Hamilton's thinking on the evolution of altruism—to the point that he would shortly thereafter specifically praise Hamilton's ideas in his 1965 inaugural lecture as Weldon Professor of Biometry. Moreover, at the same time he was laying the foundations for the peace research group, Smith was helping Hamilton review the relevant literature on animal behavior in preparation for Hamilton's famous 1964 paper proposing the theory of kin selection.[51] In this paper, Hamilton sought to explain the evolution of various social behaviors like altruism that had previously been inexplicable in terms of the mathematical theory of natural selection as laid out in R. A. Fisher's work. To explain such behaviors, Hamilton extended Fisher's theory to encompass situations in which behavior affected not only the evolutionary fitness of the individual in question, but also the fitness of other individuals who shared genes in common. In such situations, altruistic behavior could evolve. Hamilton thus sought to reinterpret individual altruism in terms of a deeper and more fundamental selfishness inherent in genes under selection.[52]

Game theory, like altruism, was another of Smith and Hamilton's shared interests, although the extent to which they realized it at the time is unclear. In addition to his involvement with conflict resolution, Smith had been an early reviewer of von Neumann and Morgenstern's *Theory of Games and Economic Behavior,* even writing to von Neumann to point out errors in the first edition.[53] Hamilton recalls first encountering the theory as an undergraduate, but it only made its first brief and qualitative appearance in Hamilton's next major publication, "Extraordinary Sex Ratios," in 1967. R. A. Fisher had argued that species should tend to evolve roughly equal sex ratios, since any individual who produced a surplus of one sex would put the offspring at a reproductive disadvantage. To explore the conditions under which lopsided sex ratios might emerge, Hamilton proposed the following hypothetical model of a situation in which two female wasps parasitize a common food source. If wasps were somehow "ideally gifted," capable of assessing the prevailing sex ratio in the eggs that another female had already laid and using that information to adjust the sex ratio of their own eggs, then Hamilton argued the situation became "very realistically game-like" with each female setting the sex ratio of her eggs strategically in response to the sex ratio of the others. Given this fact, Hamilton suggested that the wasps might "choose" strategies in accordance with game theory's signature "minimax" solution concept for zero-sum games, in which players confronted with uncertainty about the choices of their opponents choose to maximize the minimum payoff they would expect to receive. Hamilton called this strategy an "unbeatable" strategy, since it is unbeaten by any strategy other than itself.[54]

But from the outset Hamilton was concerned about the assumptions behind his model. First and foremost, it seemed unlikely that wasps could possess the kind of cognitive and perceptual abilities needed to be realistic game players. In a letter from early 1968, he reflected on his argument in "Extraordinary Sex Ratios," musing that it seemed incredible that wasps actually realized they were playing a "game" or had any of the perceptual or intellectual capabilities assumed of game-players.[55]

More problematic still was his growing recognition of the difficulties involved in "solving" the kind of games real organisms were likely to play. His analysis of the wasps had implicitly assumed that the minimax solution—the characteristic solution for zero-sum games—was the most relevant one. However, Hamilton had also become familiar with Anatol Rapoport's evolving ideas on how one might solve the non-zero-sum PD game. In 1967 Rapoport would actually claim that such a solution was at hand: if one took into account the fact that humans can develop foresight and expectations, these expectations might change the structure of the PD game to such an extent that the paradox disappears and cooperation becomes rational.[56] But while Hamilton seemed convinced that the PD game was probably widely applicable to evolutionary biology, he nevertheless suspected that Rapoport's solution did not hold up under natural selection. Indeed even human intelligence—although it was ultimately a product of natural selection—probably had little effect on cooperation.[57]

In May of 1969 Hamilton found a natural audience for these evolving thoughts on PD: a high-profile conference held at the Smithsonian Institution in Washington, DC. As described by S. Dillon Ripley, the secretary of the Smithsonian who presided over the proceedings, the conference had been called to review the implications for human affairs of recent popular ethological and anthropological literature. These works, such as Konrad Lorenz's *On Aggression*, Desmond Morris's *The Naked Ape*, Ashley Montagu's *Man and Aggression*, and Richard Ardrey's *The Territorial Imperative* all sought to understand the biological origins of human violence. Given world events then unfolding, such as "wars between people along the Jordan, the Nile, and Niger rivers; assassinations of leaders in Dallas, Memphis, Los Angeles . . . the mixing of student and police blood on the campuses" and "our lack of innate or learned constraints in regard to the biosphere," the question of how "we humans compare with other creatures" was especially timely.[58]

Hamilton's publication from that summer held out little hope for the evolution of altruistic behavior. Considering a number of examples of purportedly group-benefiting behavior in the natural world, he concluded that many of them were instead "mildly selfish," so that, sadly, "the animal in our nature cannot be regarded as a fit custodian for the values of civilized man." He also examined a formal game-theoretic model of interactions between two individuals, each with a "choice" of "strategies": a "normal" gene, and a mutant "selfish gene," with gene frequencies p and q respectively. Here, Hamilton's use of game theory was strikingly different from that of a mathematician or behavioral scientist. Starting from a game matrix listing "payoffs" in units of evolutionary "fitness" or relative reproductive advantage, he derived a difference equation that captured how the gene *frequencies* (rather than the probabilities associated with a "mixed strategy") would evolve over time, assuming random interactions between individuals. In particular, Hamilton examined what would happen to this difference equation if the payoffs in the game were those of the Prisoner's Dilemma, a "controversial problem" that he had studied in Rapoport's publications on game theory. In this case, an initially rare "selfish gene" would nevertheless win out, reflecting the fact that evolution did not select strategies for being the "best" in some abstract sense, but for being better than their nearest competitor.[59]

This conclusion accorded well with his biological intuition, but in the remainder of his paper Hamilton continued to puzzle over the gap between his findings and Rapoport's assessment of human game-playing behavior. Given its roots as a theory of human interaction, game theory seemed to assume that its players could reason and communicate—abilities that allowed humans to perceive a tension between self-interest and the greater good. Nonhumans, by contrast, probably possessed no such cognitive and communicative capabilities; or, more to the point, "by our lofty standards animals are poor liars."[60] So although natural selection "has made us almost all that we are," Hamilton speculated that PD might need to be solved differently depending on whether the players were human or nonhuman, with a "cooperative" solution for humans and a "selfish" solution otherwise.

Thus Hamilton's 1969 paper was far from the final word on the relationship between game theory and evolutionary biology. Even so, the paper would prove influential beyond Hamilton's wildest dreams. Among other things, Hamilton's invocation of Prisoner's Dilemma played a pivotal role in the career of another conference attendee, Robert Trivers, a graduate student working with Harvard primatologist Irven DeVore on the significance of "reciprocation" for building altruistic social relationships in humans and primates. According to Hamilton, Trivers apparently had not thought of applying game theory to reciprocated altruism at the time they encountered each other at the Washington conference in 1969.[61] Their meeting brought together Hamilton's evolutionary application of the Prisoner's Dilemma game with Trivers's interest in situations in which animals interacted repeatedly an indefinite number of times.

As the results of conflict resolution might have suggested, in the case of repeated rounds of Prisoner's Dilemma, cooperative behavior can arise, a phenomenon that Trivers would review in his classic 1971 paper, "The Evolution of Reciprocal Altruism."[62]

Even so, in the late 1960s game-theoretical formulations of reciprocal altruism (not to mention the rigorous application of repeated Prisoner's Dilemma games to biology) remained in the future. Hamilton's musings on the evolution of social behavior, like Wynne-Edwards's, were of a piece with larger cultural concerns of the 1960s about overpopulation and the causes of war and violent conflict. In particular, the interdisciplinary field of "conflict resolution" provided an intellectual space within which thinking on models of conflict and cooperation between individuals could travel between the social and biological sciences. Game theory also fulfilled a key point of Hamilton's (and Maynard Smith's) goal of grounding evolutionary theorizing on assumptions of individual self-interest. At the same time, Hamilton's vision for evolutionary theory could not yet be fully squared with the cognitive-psychological interpretation of game theory and rationality prevalent in conflict resolution; nor indeed could the selfishness of genes be reconciled with many of the features of social behavior documented by Wynne-Edwards earlier in the decade. Organisms had become plausible game players, but Hamilton could not quite yet bridge the gap between nonhuman and human game players, genetic determinism and rational choice, selfish genes and caring groups.

Bringing Gene-Machines to Life

The reconciliation of the apparently social phenomena documented by Wynne-Edwards (and by practitioners of conflict resolution) with the self-interested genes described by Maynard Smith and Hamilton would be the prime accomplishment of John Maynard Smith and George Price's 1973 article, "The Logic of Animal Conflict." The article would do this via two linked theoretical and methodological innovations. First, it made the metaphor of animals as machines literal, reconstituting animals as programmed automata in a computer simulation, and measuring the aggregate result of their interactions through data printouts instead of observing flesh-and-blood creatures in the wild. Second, it built upon Hamilton's earlier work to propose general conditions under which a set of strategies adopted by a population of these simulated organisms would be "evolutionarily stable," "unbeatable" by the field of possible alternatives.

The conceit that organisms are information-processing machines had a long history prior to the mid-twentieth century, but the postwar era saw a profusion of new technologies and theoretical frameworks that proved unprecedented in their ability to blur this boundary between mechanism and life. The wartime development of electromechanical computers, communication systems, and servomechanisms inspired Alan Turing's and John von Neumann's theories of computing and self-reproducing automata, Claude Shannon's theory of information, and Warren McCullough and Norbert Wiener's theory of cybernetics. Drawing on these ideas, geneticists, including Maynard Smith, quickly came to interpret DNA as a repository of coded information, and the organism as an information-processing machine programmed by its genetic material.[63] In adopting this view of life, Maynard Smith was encouraged by his friendship with Donald Michie, one of Britain's first professors of computer science and a wartime colleague of Alan Turing's in the codebreaking operation at Bletchley Park. Avid chess-players, Maynard Smith and Michie in 1961 proposed a design for machines that would mimic human chess-playing abilities, thereby displaying a primitive kind of artificial intelligence. In fact, one of the designs' algorithms, the Smith Two-Move Analyzer (or SToMA), proved capable of besting Maynard Smith in a trial game. If a machine of his own design could beat Maynard Smith in a game of chess, it might not be too difficult to imagine all apparently intelligent social behavior—from dancing among fruit flies to territoriality in birds—could just as easily be produced by game-playing automatons.[64]

Maynard Smith imagined organisms as automata, but it was less clear how these automata were to be programmed and studied in any serious way. This was the contribution of George Price, who would draw on a heady mix of information theory, cybernetics, and computing technologies to create a new and totally general theory of selection and simulation algorithms for studying natural selection inside computers. By the mid-1960s, Price already had gone through at least three scientific careers in some of the quintessential research institutions of the American Cold War military-industrial-academic complex. During World War II, as a PhD student in chemistry, he had worked for the Manhattan Project in the Met Lab at the University of Chicago. Subsequently, he had lectured in chemistry at Harvard, researched transistors at Bell Labs and fluorescence microscopy at the University of Minnesota, survived an unsuccessful stint as a freelance science writer, and finally settled as a computer scientist at IBM's laboratories in Poughkeepsie, New York. Through his stint at Bell Labs, Price developed a keen interest in Claude Shannon's work on information theory and the theories of artificial intelligence and automata being developed by Herbert Simon and others. At IBM he also acquired a facility with the relatively new FORTRAN programming language, which he could apply on IBM's luxuriously powerful computing equipment (in this case, the multimillion-dollar IBM 7094) to create optimization models of the market for high-technology products. However, in the spring of 1967, Price was at last poised to realize his dream of writing about science for a popular audience. A botched medical procedure had handed him a small insurance settlement, allowing him to move to London where he hoped to write a bestselling book on the evolutionary basis of human social customs.[65]

Price never wrote the book. By the end of that fall, he discovered that Desmond Morris's runaway bestseller, *The Naked Ape* (1967), had scooped many of his ideas: for instance, an evolutionary explanation of the hymen that he had submitted in a manuscript to *Science* in 1962. But already Price had moved on to thinking about the evolution of other phenomena such as arachnophobia, visitations by ghosts, human speech, monogamy, and sexual shame. Sometime the next summer, in a visit to one of the libraries in central London that he regularly haunted, Price came across Hamilton's 1964 paper on "The Genetical Evolution of Social Behaviour." By September he was corresponding with Hamilton (who was in Brazil for the 1968–69 academic year), peppering him with questions about how kin selection held up in light of nonrandom mating and sexual selection.[66]

Hamilton's paper had a transformational effect on Price when he discovered it during that summer of 1968. Not entirely believing its results, he set out to re-derive them and, in the process, uncovered a new expression for the rate of gene change in a population; Hamilton's results were a special case. In addition to recovering the theory of kin selection, the equation (known today as the Price equation or the covariance equation) indicated conditions under which organisms could evolve spiteful behavior, as well as situations in which true "group selection" could take place. With Hamilton away, Price visited Hamilton's old advisor, Cedric Smith, who was so impressed by the novelty of the result that he arranged for Price to obtain an honorary appointment at University College and an office in the Galton Laboratory.[67]

Price's new equation, which he came to view as the cornerstone of a "theory of selection" similar to information theory, suggested a new theoretical program for understanding the evolution of social behavior. Price described this in a grant proposal to Britain's Science Research Council, written jointly with Cedric Smith in early 1969. Building on material from *The Naked Ape* and his own proposed book, Price hoped to use his new mathematics of selection to understand the evolution of social behavior in hominids during the Pleistocene. One of the most pronounced features of early man (according to Morris's book) was his proclivity to live in cooperative families and communities. Price felt that neither group selection nor individual selection alone could satisfactorily explain why such social structures would evolve. As a result, with support from the grant, Price hoped to study systems of punishments, retaliation, and reciprocation that altered the incentives facing individuals, thereby making prosocial behavior like altruism and cooperation both individually and collectively beneficial. The task for the theorist was to understand why some such "behavior systems" possess more "evolutionary stability" than others and thereby persist across many generations.[68]

While Price proposed to explore a wide range of behavior systems, especially those surrounding sexual selection and mate choice, like Cedric Smith he was particularly interested in the evolutionary basis of conflict and cooperation, war and peace. Already by February of 1969, Price had produced a paper, titled "Antlers, Intraspecific Combat, and Altruism," which had been provisionally accepted by *Nature*. This paper sought to explain a phenomenon first observed by Darwin: deer antlers, while massive and clearly costly for an animal to develop, are actually better suited for defense than for dealing a fatal blow to adversaries. While such an adaptation might appear to benefit the species at the expense of individuals, Price argued that it could benefit individuals if other members of the species had evolved the ability to recognize and "retaliate" against individuals who engaged in lethal fighting. Part of Price's plan for the grant therefore was to continue research on cooperation and the phenomenon of "limited" animal fights. This last area of work was particularly promising, Price felt, because of the existence of game-theoretical treatments of "limited war" that could be readily adapted to model animal fights. Game theory thus might lead the way to a widely applicable and elegant theory of animal behavior, much as it had for human behavior. Should such a general theory prove elusive, computer simulations of the interactions between individuals might still allow him to obtain realistic models of particular evolutionary situations. Among other things, he proposed to simulate processes of sexual selection over a number of generations—as few as twenty if he ran out of money to pay for run-time on the UCL computers, but potentially as many as one hundred, should the early results not show any interesting selective phenomena.[69] Indeed, Cedric Smith's letter to the Science Research Council supporting Price's grant application requested £50 per year for computation.[70]

Initially the "Antlers" paper was put on the back burner as Price continued to work on developing the general mathematics of selection throughout the summer of 1969. Upon Hamilton's return that fall, Price sent him an extensive manuscript on the topic and the two coordinated the publication of a pair of related papers for submission to *Nature* the following spring.[71] But by the summer of 1970 he had returned to animal combat, urged on by John Maynard Smith, who served as a reviewer for the "Antlers" paper and had recommended its provisional acceptance and suggested revisions. It is unclear what Maynard Smith took from his reading of Price's paper, and no copy of the original is extant.[72] To the end of his life, Maynard Smith would say only that Price had sought to demonstrate that the threat of "retaliation" could encourage limited combat between animals, and that he did not mention game theory. Regardless, not long after reviewing Price's paper, Maynard Smith left for the University of Chicago to spend the year 1969–70 on sabbatical at the Committee on Mathematical Biology, with the intention of applying game-theoretic models to the study of animal combat.[73] Sometime during that year, Maynard Smith formulated the following problem to give to students in his seminar on population genetics. Two animals engage in repeated contests to determine possession of a resource with a given fitness value. The animals have "two 'levels' of fighting available to them" in any given move of the game, "conventional" and "escalated"—terms usually used in discussions of arms races that Maynard Smith likely borrowed from Price's 1969 paper. The two levels are different in that conventional fighting is not intended to seriously injure the opponent, while escalated fighting is. Assuming animals could make different initial moves, as well as respond to the actions of their opponent by "retaliating" or "retreating" in subsequent moves, Maynard Smith concluded that each player had five possible strategies. These he used to create a 5×5 game matrix that listed arbitrary (if reasonable) numerical payoffs for the different strategic combinations.[74]

Compared to formulating the game, finding a solution would prove far more conceptually challenging. Like Hamilton in his exploration of the Prisoner's Dilemma game, Maynard Smith recognized that the players in question could not really be considered "rational" in the traditional game-theoretic sense: hence, he argued, "we are not seeking for [the player's] *optimal* strategy in game theory terms." Instead, the goal was to find what Maynard Smith called an "evolutionarily stable strategy" (ESS), a strategy for which, if it is adopted by most individuals in a population, "there is no alternative strategy which will pay better."[75] In drawing on Price's notion of "evolutionary stability" to replace what he saw as game theory's stronger emphasis on determining "optimal" strategies, Maynard Smith effectively solved a problem that had troubled him since his observations of the dancing fruit flies, and that had been at the core of Hamilton's 1969 paper. Could one realistically assume that organisms were capable of "rationality" or "choice"? The concept of the ESS, a strategy to which "there is no alternative," helped remove both choice and rationality from discussion. At the same time it provided a way to recover the insight, suggested by Price's analogy to human arms races, that the threat of "retaliation" would act to limit combat.[76]

Upon his return to England, Maynard Smith sought to contact Price, with the idea of publishing a joint paper on animal combat. While Maynard Smith would feel confident enough to present his concept of the ESS in a separate paper (eventually published in 1972), Price was determined that conclusive verification of his original "retaliation" hypothesis required better computer simulations of animal combat. Throughout September of 1970 he slogged away, submitting increasingly complex programs to run on University College's IBM 360 Mod 65 and poring over the reams of printouts thereby generated. Progress was agonizingly slow. From his correspondence with Hamilton during this period, it appears that the initial paper on animal combat had begun to spin off all manner of complex models, calculations, and simulations. Some of the simulations appear to have focused on population-level gene frequency dynamics, positing aggregate systems of warfare and combat and tracing their influence on the evolution of fighting styles over multiple generations. Others, which focused on details modeling of fights between pairs of animals, proved incredibly challenging both practically and conceptually. However, toward the end of the month, he could report to Hamilton that he was starting to get back to work on writing. [77] More broadly, by this time Price and Hamilton both were deeply involved in computing, frequently swapping results with one another by letter. Using a series of increasingly intricate simulations, they sought to explain a wide range of biological phenomena, from the evolution of herding and schooling instincts in animals to the preservation of genetic polymorphism in populations.[78]

Yet, for whatever reason, Price never managed to complete either of the papers on animal combat that he referenced in his letter to Hamilton. In part this may have reflected Price's deepening preoccupation with biblical exegesis instead of evolutionary biology. Price had arrived in London an atheist, but he felt that the covariance equation that he discovered in the summer of 1968 could only have come to him by divine intervention. Inspired by this gift of grace, he ran through the streets of Marylebone desperately seeking a church and soon thereafter converted to Christianity. By late 1970 Price was spending much of his time trying to establish the true chronology of the Gospels and searching for codes hidden within the Bible. Heeding Christ's injunction to give what he had to the poor, Price also began tending to the vagrants sleeping on the streets near the St. Pancras train station, with the result that he was soon homeless himself.[79]

Thus, in the end, Maynard Smith seems to have been the force that pushed the 1973 *Nature* paper to completion.[80] The computer simulation that formed the centerpiece of the article was run out at the University of Sussex, although Price's influence on its design was clear. The model involved five behavioral phenotypes—"mouse," "hawk," "bully," "retaliator," and "prober-retaliator"—each of which adopted a different sequence of "conventional" or "dangerous" tactics in response to the moves of its opponent. For instance, the "mouse" phenotype, true to its name, never uses dangerous tactics and always plays conventionally; if attacked dangerously, it will retreat before serious harm is done. By contrast, the "hawk" always plays dangerously until its opponent is defeated. Other strategies represent a mix of these two: the "bully" plays dangerously against conventional tactics, but retreats in the face of repeated danger; a "retaliator" plays conventionally until it is forced to respond to dangerous tactics by using dangerous tactics itself; and the "prober-retaliator" behaves like a mix of the hawk and the retaliator, with strategy chosen by a probability distribution.[81] Having assigned numerical payoffs to the outcomes of losing, winning, and being injured, Maynard Smith and Price used a computer to simulate fights between each possible pair of phenotypes. After 2,000 simulations of each pair, they determined the total score won by each strategy and concluded that either "retaliator" or "prober-retaliator" would prove to be an ESS depending on the initial level of "mouse" in the population, with other strategies appearing only rarely through mutation. The simulation "shows emphatically the superiority, under individual selection, of 'limited war' strategies in comparison with the Hawk Strategy."[82]

In some ways, "The Logic of Animal Conflict" reads like a first report of a new and exciting collaborative research program. Among other things, the conclusion hints that "a more detailed analysis will be published elsewhere." In fact, the paper would serve as an epitaph for the tangled and complicated relationship among Maynard Smith, Hamilton, and Price, with its roots in the debates surrounding overpopulation and aggression in the preceding decade. Little more than a year after it went to press, Price was found dead in a condemned building where he had sought shelter. Hamilton and Maynard Smith were the only two of his scientific colleagues to attend the funeral, held in the classroom of a small church not far from St. Pancras station.[83] Price's departure from the scene left Maynard Smith to pursue his primary focus on the theory of the ESS and mechanisms for achieving equilibrium, eschewing the kind of detailed computer simulations that Price had favored. In a series of papers, Maynard Smith substantially streamlined the ideas from his paper with Price, setting up and analyzing specific, simplified games, such as the "war of attrition" (1974) and the game of "hawks and doves" (1976). This modeling tradition, which took the ESS as its foundational concept, developed principally on the pages of the *Journal of Theoretical Biology*, of which Maynard Smith was an editor.[84] By contrast, aside from an exposition of the covariance equation published in 1975, Hamilton largely moved away from his early work on the evolution of social behavior and focused instead on understanding the evolution of sex and the sources of genetic diversity in populations, eventually quitting London for a position at the University of Michigan.

Even so, by the end of 1973, a new vision for evolutionary biology was coalescing around the mathematics of rational choice. Wynne-Edwards's depiction of populations as social units, characterized by norms of behavior and social structures, had been rejected, as had Lewontin's model of populations maximizing their chance of survival in the face of natural adversity. Instead, there emerged a new conceptual framework for evolution in which selfish gene-machines were the principal actors. A mirage of "social" structure was recovered, however, by the use of game simulations and by the new concept of the evolutionarily stable strategy, which replaced game theoretic notions of optimality or rational choice with those of evolutionary equilibrium and stability.

Game Theory without Rationality

The emergence of evolutionary game theory marks a significant turning point in game theory's history for a number of reasons. In the hands of evolutionary biologists, game theory continued its movement away from the networks of military patronage that had sustained it in the immediate postwar years. Biologists encountered game theory in a range of contexts, most notably in connection with "conflict resolution," that sought to bring together the social, behavioral, and biological sciences to gain insights into the nature of war and peace, conflict and cooperation. As a result, biologists like Maynard Smith, Hamilton, and Price were exposed to a very different kind of game theory than that used by Lewontin in his 1961 article. No longer was game theory primarily a technique of optimization or military worth evaluation, as it had been in the hands of the military-funded applied mathematicians who had dominated game theory prior to the early 1960s, but a flexible set of notations for exploring the logic of conflict and cooperation.

Cross-disciplinary discussions—like the ones associated with conflict resolution or fears of population crises—may have helped to bring a theory of human decision-making to the attention of evolutionary biologists, but this was not sufficient: biologists also needed to actively adopt game theory. As this history suggests, they did so for a number of interconnected reasons. In part, the emergence of evolutionary game theory was bound up with a broader project undertaken by biologists during this period to make mathematical modeling, not the descriptive practices of field observers, the dominant methodology for interpreting macrobiological phenomena of the sort that Wynne-Edwards described in *Animal Dispersion*. This move was most pronounced in the exchanges between Maynard Smith and Wynne-Edwards that lay at the center of the controversy over group selection. For Maynard Smith, the drosophilist and former aircraft engineer, conjuring with the laws of genetics and the rules of arithmetic clearly took precedence over other ways of interacting with and observing animals. This would prove true even in Maynard Smith's observations of sexual selection in his laboratory fruit flies. George Price, the chemist and computer programmer, represents an even more extreme case: for him, the object of study in evolutionary biology was no longer flesh-and-blood animals at all but electronically constituted automata.

For Maynard Smith, Hamilton, and Price (if not for Lewontin and the members of the Marlboro Circle), these methodological choices were deeply connected with a set of assumptions they embraced regarding the nature of the individuals undergoing natural selection. Fundamentally, genes, not social influences, determined individual behavior; genes could not choose, but were selected; genes were blindly selfish, not altruistic. For Hamilton, this selfishness arose out of his eugenic preoccupations, his suspicion that any interpersonal benevolence might simply turn out to be a "higher hypocrisy."[85] For Maynard Smith, the programmed existences of Price's computerized game players simply made real his metaphor of organisms as information-processing machines, guided by their genes alone, with minimal reference to social stimuli. Likewise, Maynard Smith's signature concept of the ESS reinforced his view of organisms as automata by attributing their apparent strategic "choices" to evolutionary "stability" rather than rationally produced "optimality." The difference between this view of the evolutionary process and Lewontin's model of plant species as a coherent entity, trying to maximize their chances of survival in a game against nature, could not be more stark.

Ironically, the end result of game theory's foray into biology thus was a theory of rational choice stripped of anything resembling "rationality" or "choice." In one of a collection of articles that reviewed a decade of growing interest in evolutionary game theory, conflict resolution guru Anatol Rapoport would reflect that the emergence of evolutionary game theory appears "bizarre," for two reasons. First, game theory's roots lay in attempts to formalize "rationality"—a characteristic typically "associated with foresight, prudence, and reasoning ability." Yet, simultaneously, the theory of evolution by natural selection arose in large part with the aim of "*removing* foresight, goal directedness, and other concepts associated with rationality from explanations of the evolutionary process and of its end products." Hence the title of Rapoport's article: "Game Theory without Rationality." Writing in the same volume, John Maynard Smith agreed: "The title of Rapoport's commentary so accurately characterizes evolutionary game theory that I have borrowed it for my own reply."[86]

This fact is significant in evaluating the subsequent spread of game theory *back* into the social sciences from evolutionary biology in the 1980s. As we will see in the final chapter, contacts between economists and Maynard Smith's research program at Sussex would clarify the nature of the ESS, demonstrating that it was a refinement of the noncooperative Nash equilibrium solution concept for games that had just begun to attract a wide following within economic theory in the late 1970s. Hamilton's subsequent move to the University of Michigan brought him into contact with agent-based modelers in political science, most notably Robert Axelrod. Their resulting collaboration influenced such classics as Axelrod's *Evolution of Cooperation* (1984) and sparked a revival of interest in computer simulations of repeated Prisoner's Dilemma games that continues today.[87] More generally, to game theorists tangled in the stringent assumptions of rationality imposed by game solution concepts and equilibrium refinements developed to describe *human* behavior, the focus on evolutionary dynamics and methodology of computerized agent-based models found in evolutionary game theory would prove liberating. The result, over the past three decades, has been a refocusing of game theory, away from the analysis of rational behavior and toward the examination of evolutionary dynamics, from the analysis of optimization to the analysis of equilibrium in human and natural economies alike.

CHAPTER SEVEN

Dreams of a Final Theory

In 1975, a young German graduate student came to join John Maynard Smith's research group at the University of Sussex. This student was Peter Hammerstein, who had recently completed a thesis on mathematical models of bargaining with future Nobel Prize–winning economist Reinhard Selten. Initially Hammerstein kept a low profile; but at length he mustered up the courage to talk to Maynard Smith and ask a question that had been puzzling him. Why hadn't Maynard Smith cited Nash in any of his papers that developed the notion of an evolutionarily stable strategy? Maynard Smith's response was telling: who is Nash? Despite his consultation of Luce and Raiffa's *Games and Decisions*, he had not absorbed Nash's name—or if he did, he never connected it with anything that looked like an ESS. This may be understandable, given Luce and Raiffa's emphasis on Nash's model of bargaining rather than his equilibrium point concept. Nash was also referenced infrequently in the conflict resolution literature that helped port game theory into biology in the 1960s. Finally, though, in the 1970s, the worlds of economists like Hammerstein and evolutionary biologists like Maynard Smith were only beginning to intersect. In any event, Maynard Smith began to cite Nash in modest recognition of a similar idea.[1]

Maynard Smith's acknowledgment may have been modest, but it was also a harbinger of things to come. Within a few years, Hammerstein and Selten would demonstrate the close connection between the ESS and the Nash equilibrium, showing that the ESS is a particular kind of Nash equilibrium in which initially rare mutant "strategies" are allowed to spread through a population. Or to put it in current game theoretic jargon, the ESS is a particular equilibrium refinement.[2] The recognition that the Nash equilibrium solution to games, which had been practically declared dead in the later 1950s, might be the natural condition of life came on the eve of renewed hopes that a final theory of games encompassing the economies of nature and of society alike was just around the corner.

Both game theory in general and Nash's noncooperative equilibrium concept in particular were on the verge of a revival among social scientists at the time that Hammerstein came to work with Maynard Smith. This was most noticeable in economics. Writing in 1984 for an audience of economists, the game theorist Robert Aumann would make his oft-quoted pitch for the Nash equilibrium's central relevance to economic theory, arguing:

> The Nash equilibrium is the embodiment of the idea that economic agents are rational; that they simultaneously act to maximize their utility. If there is any idea that can be considered *the* driving force of economic theory, that is it. Thus in a sense, Nash equilibrium embodies the most important and fundamental idea of economics, that people act in accordance with their incentives.[3]

By that point Aumann's words were not simply a fighting faith: already the language and methodological stance of noncooperative game theory was steadily remaking the fields of industrial organization, financial economics, and many others. By the later 1980s, statements of game theory's interpenetration with economic theory abounded. Not only did game theorists suddenly seem indispensable to a functioning economics department, but now, "bright young theorists tend to think of every problem in game-theoretic terms, including problems that are easier to deal with in other forms."[4] At the center of this theorizing was Nash's equilibrium for noncooperative games, seemingly the central new concept of the field. As one textbook author would write in 1990, "Nowadays one cannot find a field of economics (or of disciplines related to economics, such as finance, accounting, marketing, political science) in which understanding the concept of a Nash equilibrium is not nearly essential to the consumption of the recent literature. . . . The basic notions of noncooperative game theory have become a staple in the diet of students of economics."[5] Perhaps the capstone of Nash-program triumphalism took place in 1994 with the awarding of the Nobel Prize in economics to a trio of game theorists whose work had primarily focused on noncooperative games: John Harsanyi, Reinhard Selten, and John Nash himself.

A number of historians and practitioners of economics have offered explanations for this sudden burst of enthusiasm for noncooperative game theory in their field. A set of new and powerful results by a younger generation of game theorists working from the 1960s onward; the increasing mathematization of the postwar social sciences; the shifting evidentiary demands of American antitrust jurisprudence; and the emergence of serious technical problems with the previously dominant theoretical framework of Walrasian general equilibrium theory: all of these seem like plausible factors. But to focus attention on cataloguing applications within a particular discipline may perhaps miss some of the larger ambitions of game theory's practitioners for their subject, ambitions that often looked beyond economics per se and toward a new kind of theory-building effort that would unite all the social sciences and even evolutionary biology under a shared set of foundational concepts and principles, and that might hold together the descriptive, prescriptive, and predictive aspects of game theory that were in constant danger of drifting apart. It may also downplay some influences and impulses that transcended economics—most notably, the lingering influence of the arms-race debate on game theorists working in the 1960s and 1970s. As we have seen in the case of evolutionary biology, Nash's equilibrium concept was effectively reinvented at the confluence of two intellectual trends: the growing sense among evolutionary biologists that altruism and cooperation were problems to be solved rather than facts of nature; and debates over the possibility of arms control and conflict resolution in a world where binding commitments were impossible. In embracing noncooperative game theory, economics was hardly alone during this period, and this fact suggests the influence of broader intellectual trends and considerations. Certainly, many of the theoretical tools and results that would become canonical to social scientists of the 1980s and 1990s (bargaining under various kinds of uncertainty, results on indefinitely repeated games) were forged in the context of attempts to deal with problems identified during these debates, and these theoretical tools carried over to the academic social sciences. The career paths of individuals like Thomas Schelling, whose analysis of strategy deeply shaped his subsequent work on urban issues and health care policy, put biographical flesh on this story, while shifting sources of patronage for think-tanks like RAND in the 1960s and 1970s suggest the kinds of institutional pressures that helped push intellectual frameworks developed for the study of national security problems into the realm of social policy.[6]

Thus, while the action in this chapter is carried forward substantially by individuals who self-identified primarily as economists, its focus is not specifically on understanding the reception and use of noncooperative game theory in economics during this period. It explores instead a broader set of questions about what was at stake in the Nashification of game theory between the 1950s and the 1980s. It does this by first examining early attempts to build an economic theory of "sparse markets" on the foundation of von Neumann and Morgenstern's theory of multiplayer games, and some critiques and problems that emerged in connection with that program in the 1960s and 1970s. The proponents of the "Nash program" for game theory, most notably John Harsanyi, imagined that their theoretical approach would overcome a number of problems associated with von Neumann and Morgenstern's theory of multiplayer games, thereby laying a more secure foundation for a final, general theory of social behavior. Moreover the central concept of noncooperative game theory, when interpreted as "the embodiment of the idea that economic agents are rational," helped to maintain ties between game theory's status as a description of social behavior and a prescription for rational conduct. A guide for arms-control negotiations, a description of the evolutionary process, a completely general theory of social interaction—this new brand of game theory seemed ubiquitous during this period, held together by Nash's equilibrium concept.

Yet beneath these apparent successes, concerns about game theory's utility as a coherent "theory" for the social sciences, and the relationship between game theory and the rational decision-making that it purported to study, remained never far from the surface. Nash proved the existence of equilibrium strategies for noncooperative games—something that had not been done for the von Neumann—Morgenstern stable set solution—but many of the noncooperative games most interesting to natural and social scientists possessed more than one equilibrium point. As with Tucker, Flood, and the Princeton mathematicians in the 1950s, the search for determinate solutions to such games (and for algorithms to find these solutions) became a central preoccupation of the game-theoretic community, and a key driver behind the explosion of game-theoretic literature that has occurred since the 1970s and has continued to this day. New "solution concepts," whether noncooperative or otherwise, have proliferated, and ad hoc methods for solving games and selecting equilibrium points abound, chewing up pages upon pages in journals like the *International Journal of Game Theory*. The literature is now so enormous and intricately technical that it all but defies efforts at coherent characterization. In the words of economist Christian Schmidt, as of the early 2000s, "game theory seems to have lost its original unity. . . . The question today is whether the name 'game theory' should remain in the singular, or become 'game theories' in the plural."[7]

The proliferation of "game theories" has gone hand in hand with the further disintegration of any straightforward conception of "rationality" and the problematization of game theory's rationality postulates more generally. Whether the aim is to model behavior or guide it, the difficulty is the same: straightforward and generally applicable rules of rationality will not necessarily get you where you want to go. This certainly seems to have been the conclusion of Reinhard Selten, co-winner of the 1994 economics Nobel with Nash and Harsanyi, who nevertheless in the course of his career drew an increasingly sharp distinction between game theory's world of mathematical artifice and the decision-making of actual individuals and organizations, much less the study of how we ought to choose. By contrast, a number of practitioners of game theory have sought to ground the Nash equilibrium concept less in human decision-making processes or uncontestable principles of rationality than in the dynamics of evolutionary processes. Thus the apparent triumph of game theory in the social sciences between the 1970s and the 1990s can be read less as a beginning than as an end—of the dream of a final, unified theory of rational choice spanning the social sciences and policy, and the latest unraveling of long series of debates over rationality that von Neumann and Morgenstern's book provoked.

Competing Programs for Game Theory and Social Science, 1950s–1960s

Despite the attractiveness of game theory to some economists associated with RAND and a few other institutions, the lament commonly heard among the first generation of game theorists concerned the continued indifference of economists to their work. Martin Shubik, a Morgenstern student and one of the pioneers of game theoretic applications in economics, would note even in 1953 that while numerous books and articles had appeared in the eight years since the appearance of *Theory of Games and Economic Behavior*, most of these were of a "mathematical rather than an economic nature," focusing on applications to military problems. While "several universities in the United States and Canada give courses or seminars in game theory in the department of mathematics . . . few departments of economics teach game theory, and only the most cursory of asides or footnotes are accorded it in the textbooks."[8] Nevertheless, the first major applications of game theory in economics appeared pretty much where von Neumann and Morgenstern suggested they might: in thinking about the economics of "sparse markets," situations in which prices are set by interactions among a small number of individuals. While the prime focus of economic theory had always been on the analysis of competitive markets with many buyers and sellers, models of imperfect competition—for example, of a duopoly where two firms set prices for a good—had a long (and contentious) history in economics by this time, attracting the attention of Augustin Cournot, Joseph Bertrand, Francis Edgeworth, and others. Starting in the 1950s, economists began to cast the venerable problem of imperfect competition in the language of game theory, thereby attempting to write game theory into a long established lineage of economic theorizing.[9]

The central figure in this first wave of research was Martin Shubik himself, who, as an economics PhD candidate at Princeton, was closely connected with the mathematicians orbiting Albert Tucker's Fine Hall games seminar. From an early date, Shubik saw a close connection between John Nash's "equilibrium solution" to two-person non-zero-sum games and Cournot's analysis of a duopoly—that is, of a situation in which two firms must decide how to set their prices and production schedules for a market. While the two firms could extract a maximum profit from the market by cooperating to fix quantities produced or prices like a monopoly, independent action would lead to lower prices and profits, albeit not generally as low as in a many-player competitive equilibrium. This latter solution to the duopoly problem Shubik identified with Nash's equilibrium solution for the equivalent game.[10] But like von Neumann and Morgenstern, the champions of a cooperative vision for game theory, Shubik remained skeptical of Nash's solution to such games, just as he joined many economists in their criticism of Cournot's solution to the duopoly problem. As Shubik would note in one of his first papers from 1952, Nash's solution "implies one possible state of society which is logically possible but does not seem to be very relevant to societies as we know them." Specifically, "we can consider the possibility of co-operation. Game theory suggests that joint maximization [of the duopolists' revenue] will be striven for, and there will be a side-payment. Limits can be placed on the size of the side-payment by this economic theory, but the actual side-payment made can be determined only by additional assumptions."[11]

Shubik's recognition that mathematics could uncover the "logical possibilities" of social order, but that the outcomes of interesting social games were underdetermined, dependent on the inclusion of "additional assumptions," was not simply an acknowledgment of the limitations of game theory: it was a central feature of his research program in economics, just as it was for social psychologists of the 1950s and 1960s. The complexity and variety of social institutions inevitably force economists "to depend heavily upon sociological or institutional study," he argued. Game theory provided a new approach to economics "that is not offered elsewhere," not because it could identify clear-cut solutions to games of interest to social scientists, but because "it lays bare the fine structure of competitive situations in terms of the strategical complex of individual motivations working at cross purposes, given specified powers and a specified state of information." This "structure" was perhaps best captured through a game-theoretic taxonomy—akin to the classificatory apparatus of natural history—that could categorize the theories connected with various kinds of economic situations and institutional contexts (see fig. 7.1). As a result, "it would not be unreasonable to call game theory mathematical-institutional economics," a new synthesis of the abstract and mathematical theorizing that was becoming increasingly popular in the postwar

General Theory of Games

Cooperative Games	Semi-Cooperative Games	Non-Cooperative Games
Duopoly, Duopsony Oligopoly Cartel Theory bilateral monopoly	Oligopoly Cartel theory	Monopoly, Monopsony Pure competition Cournot, Bertrand duopoly Macro-economics

FIGURE 7.1. Martin Shubik's game-theoretic taxonomy of economic situations. From Martin Shubik, "The Role of Game Theory in Economics," *Kyklos* 7, no. 2 (1953): 27.

era, and the historically specific studies of the institutionalists, who even
at midcentury remained a force within the American economics profession.[12] This synthesis acknowledged fully that "complexity exists in the
economic world," without "remov[ing] the possibility that there may be a
refined and sophisticated general theory underlying all the special cases
which appear . . . to make up the whole of economic activity."[13]

Shubik's 1959 magnum opus, *Strategy and Market Structure*, sought
to cement game theory's territorial expansion over the study of sparse
markets. The book, which included a laudatory foreword by Morgenstern, reviewed the kind of static noncooperative theory of oligopoly
presaged by Nash's work, but rapidly moved past Nash to lay the foundations for a dynamic theory of oligopoly based on repeated economic
"games of survival," in which firms interact many times sequentially in a
battle for market share. Such dynamical games furthered Shubik's program of "mathematical institutional economics" as they needed to take
into account a number of case-specific parameters reflecting the competitive strength of different firms, the state of information that firms possessed about one another's capabilities, and the credibility of the other's
threats.[14] Yet the book's reviewers, despite their interest in Shubik's project (and in his "games of survival" framework in particular), seemed
uniformly disappointed, and the way in which they were disappointed
reveals a great deal about their expectations, both for game theory and
for economic theory more generally. As one reader put it, "We have
faced [the] oligopoly problem for a long time without making much of it,
but since the publication of von Neumann and Morgenstern's *Theory of
Games and Economic Behaviour* we have been told repeatedly that the
solution lies in game theory. Up till now, there has been plenty of propaganda but not much pay-off."[15] Or as another reader noted, while Shubik's professed agenda in the book was merely "to begin to develop a
unified [game-theoretic] approach to the various theories of competition
and markets," after the buildup of anticipation over the past decade and
a half, "such very modest claims for game theory in economics may now
look rather studied and even a little fishy."[16]

True, several results in the work seemed significant. The repeated
"games of survival" approach suggested some intriguing insights for the
study of industrial organization, especially by suggesting connections
between the propensity of firms to cooperate on the one hand, and the
prevailing interest rate and perceptions about the future value of money
on the other. But on the whole the theory failed to answer some fun-

damental questions that were of longstanding interest to economists. In particular, the "allocation of resources and the distribution of income" had long been the stock questions of economic analysis, to which neo-classical economics had provided answers in the case of competitive markets, yet game theory's treatment of imperfect competition left these unanswered. With his insistence that game theory's axiomatic approach to rational behavior left the final state of a social system underdetermined, Shubik more or less admitted that the theory could not answer such questions.[17] John Harsanyi's assessment of the book in *Econometrica*, despite acknowledging that Shubik's work represented a first serious effort to apply game theory in economics, concentrated its fire on precisely this point. According to Harsanyi, "Shubik's own model does not supply determinate predictions on how the market will be divided between the different duopolists or oligopolists—though to supply such predictions seems to be one of the principal tasks of a theory of duopoly or oligopoly. The model's failure to supply such predictions also greatly restricts its quantitative, empirically testable implications."[18]

Harsanyi's concern with finding "determinate predictions" for the oligopolistic division of a market paralleled his vision for the future development of game theory. In publications from the late 1950s onward, he argued that the goal for the theory should be to do for multiplayer games what von Neumann had done for the two-person zero-sum game: identify "a determinate solution, which is based on a few very natural postulates of rational behaviour, and which defines a unique value (payoff) for the game to each player."[19] It was here that von Neumann and Morgenstern had failed and Nash seemed to light the way: Nash's noncooperative solution concept guaranteed the existence of logically unassailable "equilibrium points" for games, while his model of "bargaining" opened the prospect that cooperative games could be reduced to noncooperative games with unique solutions. Thus while Harsanyi noted that many interesting games possessed multiple equilibrium points, he was nevertheless convinced that a single point could always be chosen. As he put it in 1959, "if a game had several equally admissible solutions, none of them could represent a really stable equilibrium point in bargaining among rational players. For, each player would naturally prefer that particular solution that would yield him the highest payoff among all admissible solutions; and obviously no agreement could be reached on this basis." This "deadlock" could, however, be broken by a further round of bargaining that would result in the selection of a unique solution.[20]

Harsanyi's insistence on the necessity and possibility of finding "determinate predictions" or "determinate solutions" to games reflected in part his ambitions for game theory to serve as a guide for rational conduct in bargaining situations, even if it was not always clear how precisely his analysis should inform decision-making. As we saw briefly in chapter 5, some of his most-lauded work from the 1960s—his analysis of games with incomplete information—appeared in venues like the *Journal of Conflict Resolution* and in reports to the Arms Control and Disarmament Agency.[21] And for organizations like ACDA, in order for the analogy between arms races and games to be at all fruitful, the ability of a game's "solution" to guide behavior was crucial. Again, von Neumann and Morgenstern's solution to the two-person zero-sum game led the way by providing a "satisfactory definition of 'rational behavior,'" even if work remained to be done on calculational techniques for identifying the best, most robust strategies.[22] Despite the perceived gulf between the way game theoretic rationality was defined in zero-sum and non-zero-sum contexts, Harsanyi's treatment of bargaining clearly had at least some prescriptive as well as descriptive content. As Mathematica consultants John Mayberry and Francis M. Sand would note in their summary of Harsanyi's research submitted to the Arms Control and Disarmament Agency, "At this stage of the history of mathematical work in the Social Sciences, it is not possible to define clearly the role of models such as Harsanyi's Bargaining Games in the real situations which they discuss and sometimes describe." On the one hand, "they clearly do *not* describe bargaining behavior in any naive sense, nor are they 'merely' normative, i.e., it is not possible to prove that rational people in a bargaining situation like the one referred to by a model must arrive at the solution prescribed by the model." Even so, the authors continued, "there is ample flexibility in the way these models handle information and preferences to *permit* any person who accepts the postulates of Harsanyi's theory to make use of his model and arrive (jointly with the other persons who participate in the bargaining) at the same solution."[23] The rationality assumptions present in Harsanyi's analyses of bargaining thus seemed general and "flexible" enough that they might gain assent from actors with a wide range of beliefs and values, yet once accepted they would guide their progress toward "the same solution" to their negotiation problems. Uniqueness was essential.

Apart from this normative dimension of the application of Harsanyi's work to arms-control problems, Harsanyi's game-theoretic inclinations also supported a broader program of setting the social sciences on a foundation of austere methodological individualism. As he would argue in an article from the later 1960s, "people's behavior in society can be largely explained as a more or less *rational pursuit* of various personal objectives and interests," without reference to properties of collectives or through the employment of various "functionalist" explanations that had long dominated sociological theory.[24] Invocations of irreducibly "social" determinants of human behavior remained popular "mainly because up to very recently no satisfactory individualistic explanations of social behavior were available." One exception to this could be found in the economic theory of perfect competition, where the postulate of rational individual behavior appeared to imply unique and stable prices. Game theory had been developed to address situations of imperfect competition, situations involving "strategical interdependence between two or more rational individuals," and although it had initially failed to connect rational behavior in such situations with particular outcomes Harsanyi felt optimistic that a perfected theory was imminent.[25] As early as 1965 he felt assured enough to cite a forthcoming book manuscript, titled "Rational Behavior and Bargaining Equilibrium in Games and Social Situations," that would explicate his general theory.[26] A couple years later, in 1967, Harsanyi he would assert that "in recent years . . . the present writer [has] succeeded in developing a new approach to game theory which does yield determinate solutions for all classes of games, and thereby provides a rational-behavior concept rich enough to serve as a basis for a general individualistic theory of social behavior."[27]

Harsanyi's book would take substantially longer to emerge than he had initially hoped, but his impulse to produce a solution concept that worked from axioms of individual rationality to "determinate" solutions for any game nevertheless resonated with the ambitions of the one area of social science that he felt had successfully placed macrosocial predictions upon a foundation of individual rationality: the economic theory of perfectly competitive markets. As we saw in chapter 2, in the 1930s both von Neumann and Abraham Wald had tackled the problem of how to solve the sets of linear inequalities that related aggregate production and consumption in an economy. The related development of linear programming after the war led Tjalling Koopmans to declare that the problem of socialist calculation was on the verge of being solved. However, the economic models to emerge in the 1950s were far more general and

flexible than the earlier input-output and programming models of economic activity, even if their investigation was undergirded by similar areas of mathematics such as the various fixed-point theorems and related
areas of topology. Under reasonable and relatively relaxed assumptions
about the shape of production functions, demand curves, and utility and
profit maximization by individuals and firms, the premier economic theorists of the 1950s like Kenneth Arrow and Gerard Debreu were able
to prove that solutions to such general equilibrium models existed, and
that they were welfare-maximizing in the sense that no individual could
improve their lot without disadvantaging another.[28] Of course, these solutions were not necessarily unique—a problem that general equilibrium analysis shared with the theory of games, and that stemmed from
the underlying nonuniqueness of the fixed points whose existence was
proved by the Brouwer and Kakutani theorems. However, their existence sparked a multidecade program of research to investigate the
properties of different equilibrium points—especially their stability, the
ability to withstand small perturbations—and to develop numerical techniques for their identification. This formidably technical literature was
exemplified by the work of Cowles associate Herbert Scarf and Princeton mathematical-programming guru Harold Kuhn, among others.[29]

It is hard to overstate the allure of such models to postwar economists. Admittedly, game theory represented a much more general theoretical framework than general equilibrium theory, being capable of
modeling a far wider class of situations than those in which economic information could be reduced to a system of market-given prices. But for
those interested in turning economics into a science on par with mathematical physics, the equations of general equilibrium seemed to provide precisely what Shubik's mathematical-institutional economics could
not: a path to obtaining predictions about the macroeconomy based on
the assumption of rational economic conduct by individuals. A focus on
the study of "equilibrium" in particular echoed concepts derived from
nineteenth-century physics that had long inspired economic theorists.[30]
The emphasis on spare and elegant axiomatic rigor reflected the field's
embrace of the new "modern" mathematics of set theory and topology
associated with the theory of games.[31] Moreover, as with the development of mathematical programming at RAND and in the Pentagon,
the computer algorithm remained a key instrument for binding together
both the descriptive and normative aspects of economic rationality. With

postulates of individual decision-making programmed into a computer, the dreams of producing a final theory of society and of engineering a rational economic order seemed within reach.[32]

In this context, despite the theory of games being a much more general framework for analyzing social interaction than models of competitive markets, for much of the 1960s game-theoretic models gained status among economists in no small part from their ability to approximate the behavior of perfectly competitive markets as the number of game players increased.[33] Yet simultaneously, in connecting games with the theories of competitive equilibrium, the empirical, institutional grounding that had been so significant in Shubik's treatment of sparse markets dropped from the picture. By the 1970s, despite his central role in founding "the cooperative game theoretical analysis of general equilibrium theory," Shubik "became disillusioned with its entire apparatus, since it was totally lacking in any institutional detail."[34] Thus, at the close of the 1960s, game theory remained caught between several conflicting impulses and research programs: the "mathematical-institutional" approach of Shubik; the almost Newtonian successes of mathematical economics as epitomized by general equilibrium theory; and the Nash-inspired methodological individualism of Harsanyi allied (on occasion) with the prescriptive agenda of arms-control research.

Noncooperative Game Theory Ascendant

If the writings of its proponents and detractors alike are to be believed, Nash and Harsanyi's vision for game theory and for game theory's role in social science ultimately won out in the 1970s and 1980s. In particular, their work paved the way for game theory to assume a central role in several areas of economics, as the identification of various economic situations with noncooperative games and the search for solutions to such games became a core part of the disciplinary canon of the field. This is certainly the impression one receives from the outpouring of celebratory literature that followed upon the award of the 1994 Nobel Prize in economics to Nash, Harsanyi, and Selten. As the prize citation would argue, "the three laureates constitute a natural combination," their works building upon one another inexorably and inevitably. Nash provided the foundations of noncooperative game theory with his papers from the early 1950s. Harsanyi extended the theory to encompass issues concerning the players' beliefs and information. And finally, in addition to collaborating extensively with Harsanyi on developing a general theory of games, Selten examined the dynamics and stability of noncooperative equilibrium solutions in relation to a number of economic problems, most notably market entry decisions and price sluggishness.[35]

Much commentary on the work of this trio has emphasized their brilliance in reworking game theory into a form that was fundamentally more suited to the social sciences than the one that von Neumann and Morgenstern left behind. In noting economists' enthusiasm for noncooperative game theory, one author went so far as to identify a nineteenth-century "economic approach to games before game theory (and correspondingly without games)" whose development was actually thwarted by von Neumann and Morgenstern's book and its focus on cooperative solutions. Thus the publication of *Theory of Games* "paradoxically blocked or, at least, deviated subsequent informal research undertaken by certain economists for its own sake."[36] Yet even those who did not jettison *Theory of Games* entirely tended to emphasize the genius of the Nobel Prize winners in overcoming significant difficulties left behind by von Neumann and Morgenstern, even if their original vision had been fundamentally sound. Game theory's initial promise was to provide "a common methodology for analyzing strategic interaction within all the social sciences." However, it was only "the conceptual and technical breakthroughs established by Harsanyi, Nash and Selten" that have permitted game theory "to fulfill the high expectations that were expressed . . . immediately after the publication of von Neumann and Morgenstern's *Theory of Games and Economic Behavior.*"[37]

But what these "high expectations" actually were, and how Nash and Harsanyi had succeeded where von Neumann and Morgenstern had failed, has often been less clear. One suggestion, made by Walrasian theorist S. Abu Turab Rizvi, was that game theory emerged in economics to fill the void left after the framework of general equilibrium theory ran into serious technical difficulties in the 1970s. In the middle of this decade, a number of theorists drilled in the mathematical idiom pioneered by Arrow and Debreu in the 1950s discovered that the simple conception of "rational consumers" being defined by consistent preference orderings over commodities simply did not imply the kind of mathematically well-behaved aggregate demand functions that were necessary to produce stable, functioning markets. The axiomatic rigor of general equilibrium theory, in effect, devoured the theory's most prominent results. This prompted theoretically inclined economists to take refuge in the more general framework of noncooperative game theory, with its comparatively solid grounding in individual rationality. The dream of producing a final theory of the economy from a spare depiction of its actors, temporarily disturbed by the findings of the 1970s, moved on to a more general analytic framework that seemed to hold similar promise.[38]

Another possible explanation—recently suggested by historian of economics Nicola Giocoli—focuses especially on the role of Harsanyi's work in producing valuable results in its obvious area of application in economics, namely, imperfect competition. From the 1970s onward, the arguments of the Chicago School of economics increasingly gained favor in the courts in landmark US Supreme Court decisions. The Chicago perspective on antitrust stood older arguments about the power of dominant firms and the possibility of predatory behavior on their heads in effect by arguing that truly anticompetitive behavior—price wars, predation, and other abuses of market power—were in fact economically irrational and hence rarely observed. Market dominance was simply a reflection of a better product at lower cost. It was precisely against this backdrop that arguments began to emerge from game theory in the early 1980s suggesting that a number of anticompetitive practices could be explained in terms of rational, profit-maximizing behavior if there existed some uncertainty on the part of market participants about the capabilities of different firms.[39] Like superpowers using the threat of nuclear Armageddon to wring relatively trivial concessions from their adversaries, firms could potentially take advantage of the uncertainties of their opponents to get what they wanted by threating behavior that could ultimately hurt themselves. Here, the game-theoretic innovations of Selten and especially Harsanyi—initially developed in the context of arms-control debates during the 1960s—proved a perfect fit with the problems at hand, and helped to shift American antitrust policy and jurisprudence by the early 1990s toward a reliance on game theory.[40]

A third suggested explanation for game theory's hold in economics since the 1970s, associated with the historical writings of both Giocoli and the 2008 economics Nobelist Roger Myerson, focuses attention on the centrality of noncooperative game theory to the research program of "mechanism design" that took off in economics during this period. The general equilibrium models of the 1950s had largely taken a particular institution (namely, markets in which everyone always knows what

prices are and adjusts their behavior accordingly) for granted as the defining feature of economic life. Mechanism design theory, by contrast,
imagined the social scientist as an engineer of social institutions, considering the consequences of many different incentive structures on human behavior, and focusing on the ways in which incentives can induce
the sharing of private information and produce better social outcomes.[41]
With the gradual development of theories of rational choice, "it became
natural to think about the application of such rational-choice analysis to
social problems other than production and allocation of material goods.
But nonmarket applications of rational-choice analysis required a more
general analytical framework for doing rational-choice analysis without
the traditional market structures of goods and prices."[42]

The main point here is that the noncooperative strain of game theory (and the Nash Equilibrium in particular) was an essential ingredient of "mechanism design" from the moment of its creation. The foundational thinker on this subject was Leonid Hurwicz, one-time Cowles
Commission staffer and early reviewer of *Theory of Games*. In the 1950s
Hurwicz had established himself as an authority on general equilibrium
theory and on the programming models that appeared to solve the computational problems of resource allocation in a centrally planned economy.[43] However, in a series of papers dating to the early 1970s, Hurwicz
began to consider the general problem of how incentives and information
interacted in political and economic systems, thus opening a window to
a new and flexible discussion of human institutions. In considering novel
ways of organizing an economy or a political system, it was essential not
only "to design process rules which, if followed, would have certain desirable consequences," but also to consider "whether one could expect
the participants to follow the rules" or whether they would deceive other
market participants or a central planner about their preferences and
capabilities. This led Hurwicz to consider, among other things, conditions under which the sharing of accurate economic information might
be made "incentive-compatible," a situation he identified with a Nash
equilibrium in which no participant in an information-gathering process
had an incentive to withhold information or deceive as long as the others did likewise. As mechanism design established itself as a significant
area of research in the later 1970s and 1980s, attracting the interest of a
new generation of theoretically sophisticated economists, noncooperative game theory also grew in influence.[44]

What this literature suggests, then, is that the Nash-Harsanyi-Selten canon was appropriated by various groups of researchers during the 1970s and 1980s: certain disappointed general equilibrium theorists, specialists in industrial organization reacting to the successes of the Chicago School approach to antitrust policy, and mechanism design theorists, to name a few. If one views these developments as part of a common movement, it is not difficult to understand the enthusiasm surrounding the 1994 Nobel Prize. What is less clear, however, is whether this sequence of appropriations constituted a "victory" for the spirit of the noncooperative canon, or whether the appropriators turned to the canon each for their own, very different reasons. The post-1960s history of game theory in fact suggests the latter: that the widespread adoption of noncooperative game theory in the social sciences did not depend on the decisive success of Nash-Harsanyi's agenda for solving games. Indeed, as of the late 1980s at least, the use of noncooperative game theory in economics proved vulnerable to some of the same criticisms made of Shubik's use of cooperative games to study sparse markets.

Thus while Walrasian general equilibrium theory was encountering technical difficulties in the 1970s, by the end of the 1980s at the very latest it should have become clear that Harsanyi's dream of establishing a final theory of games that linked unassailable axioms of individual rationality with the selection of determinate equilibrium points was also becoming hopelessly mired in technical problems of its own. In fact, it appears already to have been in trouble when we left Harsanyi in 1967 triumphantly suggesting that his new unified theory of rational behavior would at last banish functionalism from the social sciences. His "Rational Behavior and Bargaining Equilibrium in Games and Social Situations," already cited as a manuscript in 1965, only appeared in 1976.[45] And even then the theory remained incomplete, dealing fully with only a particular type of games that Harsanyi called "classical games"—games of complete information (in which the players have complete information about the rules of the game and the utilities of their opponents), which are "either fully cooperative or . . . fully non-cooperative," and which "can be adequately represented in normal (e.g., matrix) form." Despite being aware that this class left out many interesting and practically significant real-world game situations, Harsanyi could only gesture to a much broader solution concept that he was developing with Reinhard Selten that would hopefully hold for all games and promise "that at some future date he will produce a third book dealing with the new theory."[46]

The basic problem facing Harsanyi and Selten was again that equilibrium points to a given game are not typically unique, much like the outcomes of games under the von Neumann–Morgenstern theory. In a noncooperative game, while "rational players may be expected to use . . . strategies corresponding to an equilibrium point," since this represented a "best reply" to anything the other players would be expected to play, it was not clear which of several points a rational player would ultimately choose. Selten had begun to tackle this problem in the mid-1960s while studying repeated game models of the dynamic processes by which members of concentrated industries discover prices. In formulating the notion of "subgame perfect" equilibria, Selten noted that some equilibrium points that appear in the normal form of a game become irrational if the game is actually played out in extensive form, since they would require players to make decisions that are irrational in the context of particular subgames.[47] Since Harsanyi's models of bargaining involved a choreographed sequence of steps, they were vulnerable to the same kinds of conflicts between normal and extensive form analysis that troubled Selten. Thus, throughout the 1970s, Selten (often in collaboration with Harsanyi) developed a series of further "equilibrium refinements"—ways of winnowing down the number of possible equilibrium points, often by introducing some additional criterion of "stability." For example, the notion of "trembling hand perfect" equilibria focused on identifying equilibria for which improbable deviations in strategy on the part of the players would not result in a collective shift to different equilibrium points.[48]

But as the literature on equilibrium refinements began to grow rapidly in size and technical complexity throughout the later 1970s and 1980s, the connections between "rational behavior" on the part of individuals and the selection of particular equilibrium points began to fray.[49] Restricting solutions to some kind of equilibrium point could perhaps be justified in terms of an underlying principle of rationality, but the increasing complexity of the refinements must have made game theory less convincing as either a description of behavior, or as a guide to action for would-be game players. In many ways, Harsanyi's promised "third book" that would outline his general theory for solving all possible games represents a kind of *reductio ad absurdum* that proves the point. As with his *Rational Behavior and Bargaining Equilibrium in Games and Social Situations*, which readers awaited for over a decade, the final theory so long promised appeared gradually throughout the early 1980s as a series of intermediary technical reports. When the completed work appeared in 1988 as *A General Theory of Equilibrium Selection in Games*, coauthored with Selten, the 378-page behemoth outlined an equilibrium-selection algorithm of extraordinary complexity. The complexity stemmed from the fact that there were many seemingly reasonable principles of "rationality" or "stability" that could be applied to select a unique equilibrium point, but these principles frequently gave conflicting advice. This required Harsanyi and Selten to identify additional principles capable of resolving the conflicts. The resulting algorithm did indeed select unique solutions for each game, but was so labyrinthine that, as one reviewer noted, "this, surely, is not the way real people interact," nor are the resulting solutions ones that players "by virtue of their rationality, must *necessarily* choose."[50] In a similar vein, another reader would note that "although Harsanyi and Selten see themselves as offering "rational" criteria for selecting equilibria . . . I do not think that many readers will be convinced that other criteria may not be equally valid," nor did they even really attempt to defend this assertion. Nor did they offer much in the way of empirical confirmation of their results. Instead, they were offering "a compilation of their own best judgments" about how to solve games, even if many of their particular choices ultimately seemed "arbitrary."[51]

The uncertainty over how to justify the adoption of particular solution concepts from first principles of rational decision-making did not go unnoticed in the economics literature, despite game theory's apparent success during these years. Consider, for example, the reception of the landmark *Handbook of Industrial Organization* that appeared in 1989 with the aim of surveying the rapidly growing collection of new theoretical approaches to the field since the 1970s. As the many extensive published commentaries on the book noted, the older debate between the Harvard and Chicago approaches to industrial organization (apart from their different attitudes toward government intervention in the economy) had been conducted with a very different set of intellectual tools. While Chicago-style arguments often relied on neoclassical models of general competitive equilibrium or monopolistic markets, the field was by and large "nontheoretical, or even antitheoretical," so that, as George Stigler would remark in 1968, "there is no such thing as industrial organization theory."[52] By the later 1980s, however, the *Handbook* would largely focus on the exposition of theories of industrial organization with a chapter on game theory at its core. In the opinion of its back-ers, game theory represented a kind of middle way between the two intellectual traditions that had dominated the field, promising to combine the "methodological discipline of equilibrium analysis" with the hitherto largely interpretive study of noncompetitive markets. The prospect that the theory held out for analyzing such markets in terms of social equilibrium beyond the subjective perturbations of particular economic actors was thus especially tantalizing.[53]

Yet the *Handbook*'s presentation clearly recognized the incompleteness of the victory that game theory had achieved—at least, as of that date. As the author of the survey chapter on game-theoretic methods noted, in practical problems even small changes in the information structure of a game could dramatically change the set of likely equilibria, so that the "predictive content of game theory" was questionable. As a result, the author felt that the applications discussed left one with a "fairly pessimistic view of the likelihood that game theory can hope to provide a purely formal way of choosing between [equilibrium points]." Thus, while "research on equilibrium refinements is proceeding quite rapidly," "we would not want to base important predictions solely on formal grounds," and practitioners evaluating practical problems of antitrust policy "will need to combine a knowledge of the technical niceties with a sound understanding of the workings of actual markets." The professional judgment of economists (or game theorists) rather than calculation from first principles of rationality would remain the prime ingredient in the study of sparse markets.[54] If the book's authors remained uncertain about the prospects for game theory in economics despite their generally positive outlook, game theory's natural critics—economists associated with the University of Chicago—were even more dismissive. Thus another reader of the book could express intense skepticism due to "the seeming inability of the recent theory to lead to any powerful generalization." For game theory in particular the problem "seems beyond remediation" since it was only capable of generating "almost interminable series of special cases. The conclusions drawn from these cases tend to be very sensitive to the way problems are defined and to the assumptions that follow."[55]

And so thirty years after the publication of Shubik's *Strategy and Market Structure*, many of the same issues debated in the wake of that book (including Harsanyi's own criticisms of Shubik) seemed to be resurfacing in the new economics of industrial organization that sprang from Harsanyi and Selten's brand of game theory. Even those who held that "game-theoretic equilibria have considerable predictive power" and were therefore subject to empirical testing in the social sciences nevertheless had to concede that, even after twenty years of theory development, "the game-theoretic revolution in industrial organization has not yet resulted in much empirical work," so that testing of game theoretic models against data would remain substantially a task for the future. At least we can conclude that the relative paucity of empirical work was not a bar to the adoption of game theory in industrial organization as of the later 1980s.[56] Where justification of particular solutions from first principles of rationality seemed questionable, the choice of an appropriate solution to games was remanded to the discretion and expertise of the individual economist, pending further empirical work to choose between the myriad "special cases" thrown up by the theory, or some future theoretical breakthrough that would set the selection of social equilibria on a more rigorous foundation.

And to at least some economists deeply conversant with developments in the preceding decade, future breakthroughs also seemed unlikely within the existing game-theoretic framework. In a 1991 article Ariel Rubinstein would suggest:

> There exists a widespread myth in game theory, that it is possible to achieve a miraculous prediction regarding the outcome of interaction among human beings using only data on the order of events, combined with a description of the players' preferences over the feasible outcomes of the situation. For forty years, game theory has searched for the grand solution which would accomplish this task. The mystical and vague word "rationality" is used to fuel our hopes of achieving this goal. I fail to see any possibility of this being accomplished.[57]

This was not to say that game theory had no role in the social sciences—far from it, Rubinstein asserted—but that those who hoped to apply the theory would need to think much more carefully about how to interpret the relationship between the theory and the behavior it purported to model. "We are attracted to game theory," Rubinstein noted, "because it deals with the mind," and so his preferred solution seemed to lie in the direction of reintroducing psychology back into the theory, "incorporating psychological elements which distinguish our minds from machines" and thereby rendering game theory "even more exciting and certainly more meaningful."[58] Game theoretic rationality alone was too weak a reed on which to hang the pent-up hopes of social scientists seeking a final theory of society along the lines promised by Walrasian general equilibrium theory, or by the Nash-Harsanyi program of noncooperative game theory.

The Death of a Dream and the Life of a Theory

The Harsanyi program for game theory had tried to weld together the many facets of game theory into a unified whole. Starting from principles of rational conduct with descriptive and prescriptive valences, and calculating through to predictions of collective outcomes, it was at home in discussions of arms-control policy, academic economics or political science, evolutionary biology, and beyond. But with this agenda in trouble by the later 1980s, game theory (and the conception of "rationality" at its heart) fragmented once more, falling back on alternative disciplinary alliances and building itself into new research agendas. Rubinstein's suggested solution to the shortcomings of game theory, typical of a number of social scientists, aimed to boost the theory's predictive power through the injection of additional psychology into game-theoretic models: "laws of perception, bounds on rationality, and the processes of reasoning employed by the players." On the other hand, if this history is any guide, game theory need not be predictive in order to be useful, and an alternative research program might just as easily downplay prediction or description in favor of some other desideratum. This concluding section therefore examines the fragmentation of game theory by focusing on the research agendas of several game theorists whose work epitomizes this trend.

Remarkably, an example of one of the most radical departures from Harsanyi's agenda for game theory and social science emerged in the work of his long-time collaborator and co-winner of the 1994 Nobel Prize in economics, Reinhard Selten. Initially trained as a mathematician at Frankfurt University in the mid-1950s, his fantastic memory facilitated brilliantly lucid expositions of game theory and related topics to the economics faculty, so that his colleagues would remark that "he knows mathematics by heart."[59] Perhaps because of this distinctive cognitive approach to game theory—so unlike the applied mathematicians at RAND who embraced a mindset of rational calculation—or perhaps because of the lingering influence of institutionalism in postwar German economics, Selten felt from the beginning that the mathematics of game theory and the thought processes of most human beings were very different things. In his view, "the hypothesis of human beings guided by purely rational considerations appeared doubtful," and "the way psychologists investigated the problem of human behaviour, i.e. by experiments, seemed more reliable."[60] Herbert Simon's critique of the use of rational actor models in economics, developed in the wake of the 1952 Santa Monica conference on decision processes, was evidently well known to the economists in Frankfurt at the time. However, the fields of "experimental economics," later to be associated especially with the work of American economist Vernon Smith, and "behavioral economics," which represented an incursion of psychological research into economics led by Amos Tversky and Daniel Kahneman, only emerged later. Thus Selten's experimental studies of the behavior of oligopolists, published in the late 1950s and the 1960s, were among the first of their kind.[61] These experiments were often developed in conjunction with game-theoretic models, but they were not wholly intended as "tests" of the theory. Rather, Selten employed them as open-ended opportunities for gathering a wealth of information about the reactions and thought processes of human subjects in a simulated environment.[62]

One event that helped to crystallize Selten's concern over the descriptive adequacy of game theory's conception of rational choice was his 1973 discovery of the so-called chain store paradox, which emerged from his experimental research. In the scenario Selten imagined, a chain store (player A) with branches in twenty towns is faced with twenty local entrepreneurs in those towns, each of whom might borrow money to open a rival store that would depress the chain store's profits. In sequence, these entrepreneurs must decide whether to enter the market or not; if they do, the chain store must decide whether to implement "aggressive" price cuts to defend market share, or a "cooperative" strategy of finding a new market equilibrium with two competitors. "The cooperative response yields higher profits in town k, both for player A and for player k, but the profits of player A in town k are even higher if player k does not establish a second shop. Player k's profits in case of .an aggressive response are such that it is better for him not to establish a second shop if player A responds in this way."[63] Applying strict calculations of advantage to any individual decision (what Selten called "the induction theory") would thus suggest that the entrepreneur enter the market and the chain store respond cooperatively, leading in time to a situation where the chain store shares the market in all twenty towns with a competitor. However, if the chain store acted by what Selten called a "deterrence theory," making an example of any individual competitor, the strategies of the chain store and the entrepreneurs were less clear, as would be the final outcome. While only the induction argument is "logically correct," "nevertheless the deterrence theory is much more convincing," and in general, based on his experience, while "mathematically trained persons recognize the logical validity of the induction argument . . . they refuse to accept it as a guide to practical behavior."[64]

Not long after Selten published his account of the paradox in 1978, a number of game theorists sought ways to "rationalize" the deterrence strategy as part of a larger project during this period to wring deep economic insights into imperfect competition from the analysis of "super-games" that modeled repeated interaction between game players. Thus some of the seminal work of David Kreps and Robert Wilson on "reputation effects" in industry in the early 1980s transformed the chain-store game into a game of incomplete information, finding that a deterrence strategy could prove game-theoretically rational if one introduced uncertainty about payoffs into the game.[65] By contrast, readers seem to have paid relatively little attention to an alternative resolution to the "paradox" that Selten provided in his 1978 article: since "formal reasoning" takes a great deal of time, individuals are more likely to act either on the basis of a boundedly rational "routine" or on the basis of "imagination," the creative envisioning of alternatives. While it was difficult to make such a model of decision-making more formal and mathematical, Selten felt that it better captured the way actual game players might think and behave in the chain store game.[66] This sense was further reinforced by the emotional state that Selten experienced in confronting the paradox he had uncovered. As he would recall later, "The game theoretical analysis of the chain store game by backward induction is very easy and does not put any strain on the cognitive abilities of human beings. . . . Nevertheless the conclusion is behaviorally unacceptable. The author was so worried about this contradiction that he felt three weeks of physical discomfort." An analysis of decision-making thus had to include the fact that, in such problems, a decision-maker might encounter "the motivational limits of his own rationality"—or perhaps, in a modern twist on the Socratic concept of *akrasia*, of knowing what is good and rational but failing to act accordingly. Selten therefore tartly criticized the tendency of those like Kreps and Wilson who "run away from

the problem" by replacing the chain-store game with others offering solutions that were more to their liking.[67]

Thus even as he was poised to win the 1994 Nobel Prize in economics alongside Nash and Harsanyi for his work on game theory, Selten
was inching closer to jettisoning several key assumptions about human
decision-making that lay at the heart of rational choice modeling in economics. By this time, forays of experimental psychologists into economics under the rubric of "behavioral economics" had developed enough
of a following that Selten could note three distinct positions that economists hold "as to the rationality of human decision behavior." Most common were either "naïve rationalism," which holds that individuals actually maximize expected utility, or "anomalism," which suggests that
people possess utility functions but may not always maximize due to
limits on their time and capacity for calculation. The latter position had
gained some following in economics due to the success of psychologists
like Amos Tversky and Daniel Kahneman in documenting the ways in
which experimental subjects often relied on time-saving mental "heuris-
tics" to make choices, thereby potentially producing "biases" that represented deviations from truly rational behavior.[68] However, Selten found
himself increasingly leaning toward a third position, which he dubbed
"constructivist." The constructivist, he wrote,

> believes that people do not have utility functions and do not maximize any-
> thing. Decision procedures are viewed as guided by multiple goals, which are
> not easily comparable. Such procedures seek to avoid tradeoffs among dif-
> ferent goal dimensions, but where this is impossible, preferences need to be
> constructed on an ad hoc basis. This results in highly task specific methods of
> decision making which do not lead themselves to an interpretation as optimi-
> zation procedures, not even approximately.[69]

Selten described his own viewpoint as being "radically constructivist,"
even declining "to use the term 'utility' with its heavy rationalistic con-
notations in a behavioral context."[70] Hence Selten felt comfortable putting a firm dividing line between game theory—which "is for proving
theorems, not for playing games"—and the study of how people actually do (or even should) make decisions. In place of the calculations and
theorem-proving of game theory, Selten has continued to emphasize laboratory experimentation and observation of economic behavior in complex simulated environments.[71] As a result, as one of Selten's former stu-

dents would recall in 2010, "some purely theory oriented game theorists seem to believe that Reinhard Selten is a turncoat, who has lost or (even worse) has left the path to the 'pure and true' cause of game theory."[72]

Admittedly, Selten remains something of an outlier in enforcing so radical a separation between game theory and his understanding of the way that economic actors actually make decisions. But his stance represents one pole of logical possibility in the debate: the notion that game theory's axioms of rational decision-making and the rules that real human beings follow are simply not comparable, and that game theory is a separate, self-contained logical exercise. A less drastic response to the changing fortunes of the drive for a final theory of games can be found in the thinking of British economist Kenneth Binmore during the course of the 1980s, which captured the evolving zeitgeist associated with this heady period in game theory's history. Binmore's early career had followed the well-traveled path from mathematical statistics to mathematical economics, finding him developing mathematics textbooks for students at the London School of Economics (among other things) in the early 1980s.[73] Although he had previously written about aspects of social choice theory, during the 1980s he began publishing prolifically on game theory, both from the perspective of a theorist and experimenter testing game-theoretic models of bargaining and, to a significant extent, as an expositor of recent developments in the field.[74] These expository works almost map the course of a complete romance with the theory, beginning with an initial moment of starstruck courtship, and followed by a long cooling-off period as its strengths and weaknesses came into sharper focus.

Binmore's most effusive expression of affection for the theory, "written at a time when my enthusiasm was at a peak," is found in an "introductory chapter to an ambitious book on game theory" that would ultimately never be written.[75] Much of game theory's promise seemed to lie in the fact that "almost any situation of interest to social scientists is a game" so that game theory held out hope for "a comprehensive analysis of conflict and cooperation in human and animal societies"—thus potentially placing upon the theory a far more ambitious mandate than even von Neumann and Morgenstern had claimed.[76] But even if it might be too much to say that the social sciences were a subset of game theory, Binmore was clear that "unless and until major advances are made in game theory, social sciences are doomed to remain but a poor relation of the physical sciences," and that "it is inconceivable" that debates over the great economic issues of our time "will ever be conducted in a remotely scientific manner without the intervention of game-theoretic ideas."[77] Given this fact, the apparent failure of social scientists to embrace von Neumann and Morgenstern's work from an early date needed some explanation, much of which could be found in the difficulties associated with *applying* the theory, rather than with the bare mathematics of the theory itself.[78] Among other things, the spare description of reality offered by game theory could bypass details of physical reality that constrain the capabilities of decision-makers, or limits on the game player's computational capacity might prevent the solution of particular games of interest. However, the most difficult problem seemed once again to stem from the question of whether game theory's rational *homo economicus* offered a valid description of the way that any given collection of game players behave. "Game theory," Binmore noted, "short-circuits the problem of describing human behavior by assuming that players optimize," but the theory does not explain why we should expect humans to be optimizers. True, education might help make people more rational, but, as Richard Dawkins had shown in *The Selfish Gene*, "gene packages" that maximize on the assumption that others do the same will tend to be selected over those that do not, opening the possibility that our status as *homo economicus* is ensured by nature.[79]

The role of selective processes in ensuring game-theoretically rational behavior was closely connected with Binmore's embrace of the Nash equilibrium as "the fundamental idea of game theory."[80] Not only did "rationality, as understood in game theory, [require] that each agent will perforce select an equilibrium strategy," but equilibria could be produced by a-rational agents as well, so that "evolutive" theories of equilibrium formation are "of greatest significance in so far as applications in the foreseeable future are concerned."[81] But, in contrast to Harsanyi, Binmore seemed reluctant to embrace general principles of game-theoretic rationality beyond the Nash equilibrium condition. No matter the cognitive processes granted to the game-playing agents, the selection of particular equilibria could in general only be explained by reference to a richer understanding of the game *environment* than that typically found in game theory's representations of "games" as matrices or decision trees. Therefore, rather than tilt at "blanket equilibrium definitions intended to be applicable in all abstract games independently of the equilibriating mechanism," a better approach would tailor "equilibrium concepts to the environment."[82] Given this fact, Binmore would find himself asking: why retain a "prescriptive" or rationalistic interpretation of equilibria at all? "The reason is that the freedom of such descriptive evolutionary models from the limitations of prescriptive theory is bought at a heavy price. Descriptive evolutionary models are bound to be highly overspecific, i.e., they will assume vastly more than our knowledge can justify. Moreover, even the simplest of evolutionary processes is likely to generate dynamic systems which are intractable to mathematical analysis."[83] The Nash equilibrium could thus perhaps serve as a last link standing between game theory as a theory of how *to* behave and game theory as a theory of how people *do* behave, and between the elegant and general mathematical superstructure built up by the mathematicians and the mathematically intractable, overspecific "dynamic systems" of the evolutionary biologists.

Even as he retained the Nash equilibrium as a core game-theoretic principle, Binmore's subsequent work reflects the changing sets of practices surrounding game theory as it developed during the 1980s and early 1990s—away from theorem-proving toward the multifarious study of complex evolving systems. This is true if we look to his contributions—in the same vein as philosophers like Brian Skyrms or political scientists like Robert Axelrod—to arrive at a naturalistic account of the origins of human sociality and ethics, drawing substantially upon computer simulations of evolutionary processes or findings in evolutionary biology.[84] Here, as in Shubik's prescient earlier study of imperfect competition, game theory might serve as a generator of taxonomies of social situations and the different equilibrium configurations that they gave rise to: "A game theory without an accompanying classification of games would be like biology without the Linnean scheme," he would write.[85] Meanwhile, in practical applications, like the design of the United Kingdom's 2000 3G bandwidth-spectrum auction, which has been trumpeted as a key victory for applied game theory, principles of rationality (or indeed, any "theory" whatsoever) seem to have been submerged. While an enormous game-theoretical literature on auctions has emerged since the later 1980s, the impact of the intricate auction rules needed to sell the bandwith spectrum in accord with the objectives of the British government were most successfully tested in a simulated laboratory environment.[86] As Binmore would write of his approach, "Since the three stages of an Anglo-Dutch auction are quite complicated, we thought it especially important to test its efficiency in the laboratory" rather than relying on *a priori* calculations.[87]

Although "theoretical" considerations did seem to play a role in the rough-cut analysis of various auction schemes, simulations provided both more richly detailed results and were far more convincing to policy-makers and auction staff. "We think that their experience in playing the roles of bidders within our experimental software had a significant effect in bolstering the confidence of non-economists on the auction team in the workability of the design. (By contrast, mathematical equations have very little persuasive power.)"[88] Whether the aim was the influencing of public policy or the construction of social and ethical theory, a final theory of games and social situations proved in practice quite dispensable, even if some of its principles of rational decision-making continued to hover over the patchwork of modeling and simulating practices that began to grow up in its absence.

The Survival of Game Theory

Thus, fifty years after the publication of von Neumann and Morgenstern's book, and despite the celebratory pronouncements surrounding the 1994 Nobel Prize, it is hard not to be struck by a sense that game theory's history was beginning to repeat itself. Just as von Neumann and Morgenstern's book had done in the 1940s, the results of Harsanyi and others from the 1960s onward raised hopes for something like a general formulation of rationality, and for a "theory" of social behavior derived therefrom. But in both cases, before long, the theory ran up against formidable technical problems and challenges of interpretation. Statements of disappointment or disillusionment appeared, and ambitious book projects were delayed or shelved permanently. And yet remarkably the theory lived on, having formed new sets of disciplinary and institutional alliances via the selective appropriation and reinterpretation of certain pieces of its mathematical machinery, fracturing and moving into new territory.

The responses of Selten, Rubinstein, and Binmore are particularly intriguing in the way they hint at some of the many possible ways this process can play out, and has played out in recent years. Selten's inclination has been perhaps the most radical: to set the theory apart as the province of mathematicians and theorem-provers, while simultaneously embracing in his economic work a model of human decision-making that denies the descriptive validity of the decision-theoretic foundation on which game theory rests. Real humans, in this view, act in accordance with significantly different rules and reasoning processes than the ones laid down by von Neumann and Morgenstern in 1944, or those suggested by the subsequent extensions and refinements of their theory. Yet few others have been willing to go quite that far, preferring to retain certain essential assumptions and mathematical structures from the theory while adding much greater richness of detail, psychological or institutional, to their simulation and modeling endeavors, in pursuit of greater predictive power or verisimilitude. And, as Binmore's comments suggest, such an approach may enhance realism, but at a cost to the prescriptive interpretation of game theory as a general set of precepts for how we should reason and act. Thus the demise of Harsanyi's vision for game theory and for its place in social science, and the emergence of alternative interpretations of game theory (especially the evolutionary perspective on equilibrium selection), marks the unraveling of the conversation that this history has traced: a tangled set of debates about the nature of human rationality, the possibility of reducing rationality to a calculus, and the potential of such a calculus to serve as the basis for a theory of social behavior. And with the unraveling of this conversation, according to a number of commentators, game theory has come to encompass a highly diverse set of practices. The essentials of noncooperative game theory may remain a starting point for "modeling" human interaction in parts of economics and political science, but they now meet reality alongside a host of additional assumptions about human psychology (or details about the environment in which the game is played) that are not readily reduced to widely accepted postulates of rationality. Moreover, game theoretic traditions that were previously considered heterodox or eccentric—such as cooperative game theory—continue to appear in various areas of the social sciences.[89]

As the place of game theory in the social sciences has evolved, game theorists' explicit assessments of the significance of their work have shifted subtly as well. Thus, in reviewing the history of the Nash equilibrium concept in economics, the 2007 economics Nobel laureate Roger Myerson concluded with a review of Sylvia Nasar's biography of John Nash, *A Beautiful Mind*. In a rare critical remark about the work, he focused on the message of the book's final chapter, which sought to demonstrate the significance of Nash's theory for contemporary economics by discussing the central role game theorists played in designing rules for auctions of cellular telephone bandwidth by the US Federal Communications Commission. Such specific "applications" were impressive, yet he concluded that "there is something unsatisfactory in this chapter, because it misses the primary importance of noncooperative game theory as a unifying general structure for economic analysis." More than the theory's relevance to "any one substantive application," he argued, "the practical effects of such general conceptual structures are manifested . . . in the way that scholars make connections between applications, carrying insights from one area into another."[90]

More recently, Herbert Gintis has given a rather harder-edged airing of such ideas in his book, *The Bounds of Reason: Game Theory and the Unification of the Behavioral Sciences* (2009). In Gintis's presentation, game theory comes across as the last, best hope for unifying the study of behavior from sociology and anthropology to economics, political science, and biology, abolishing nonsensical "self-conceptions and dividing lines among the behavioral disciplines" "in the name of science." This unification is possible because, as Gintis puts it, "game theory is a general lexicon that applies to all life forms. Strategic interaction neatly separates living from nonliving entities and defines life itself."[91] It is difficult not to hear in Gintis's remarks echoes of so many projects that have gone before, from the Vienna Circle, von Neumann and Morgenstern, and the Ford Foundation's interdisciplinary behavioral science program, to Harsanyi's ultimate solution concept. Yet in his characterization of game theory as a "lexicon," it is hard not to sense also a retreat from some of the bolder ambitions of game theory's creators or their most ardent postwar followers. Why this lexicon and none other? If Selten's banishment of utilities from the assessment of human decision-making behavior is any guide, large parts of the lexicon may not be as "natural" as they seem.

Any answer to this last question must start with the kind of history related here. Game theory's intellectual roots lie principally in mathematics and its rich stock of theoretical structures, from graphs to subadditive measures. These were originally forged in contexts at some remove from the study of humans, yet the potent brew of metaphors, notations, and results found on the pages of *Theory of Games and Economic Behavior* spread out into the social, behavioral, and biological sciences as a result of a peculiar set of conditions found in postwar America. Clusters of institutions and ideas encouraged a kind of expansive, theory-centric interdisciplinarity, which fed upon and encouraged a wide-ranging set of debates about the nature and potential of rationality in an age of complex technologies and international tensions. In this context, as we have seen, game theory in practice functioned as something like a common language, a facilitator of exchange between groups of individuals with very different working practices and often quite different understandings of the significance and potential of the theory. Game theory, as it has been bequeathed to its latter-day practitioners, provides a heterogeneous collection of tools for notating, speaking, and reasoning within the human sciences. But while these can prove exceptionally useful and ergonomic, if the debates related in this chapter are any guide, they are unlikely to be the only possible ones. They are thus to some extent traditional and conventional—an outgrowth of a particular history and set of practices for the study of reasoning and social interaction.

Acknowledgments

The seeds of this project were sown in the spring of 2000, when I spent a semester teaching decision theory to undergraduates with Daniel Goroff (Harvard Department of Mathematics) and Howard Raiffa (Harvard Business School). Not only did I spend a year prepping for the course by working through the classic literature on utility theory, Bayesian statistics, social choice theory, and game theory, but that experience first put in my head a set of questions that I have followed ever since: about the relationships between mathematics and thought, thinking and doing, the human sciences and mathematics, and whether "rationality" itself might have a history. Despite my subsequent metamorphosis into a historian of science, professors Raiffa and Goroff would remain involved and interested in my work, and I cannot thank them enough for their kindness and support through the years.

A number of teachers and mentors have been extraordinarily generous with their time over the years. Joan Fujimura, Victor Hilts, Gregg Mitman, Lynn Nyhart, Elliott Sober, Richard Staley, and Jeremi Suri of the University of Wisconsin have pride of place in this list. So too do Bruce Masters, Clark Miller, and Shobita Parthasarathy, and the members of the Wesleyan University History Department, the Science in Society Program, and the College of Environment, who have done so much to make Wesleyan the perfect place for completing this work.

It is all but impossible to thank adequately all of my colleagues in the history of science and economics who have read or commented on my work, but I especially want to recognize S. M. Amadae, Roger Backhouse, Lorraine Daston, Neil DeMarchi, Philippe Fontaine, Michael Gordin, Hunter Heyck, David Kaiser, Tinne Kjeldsen, Judy Klein, Rebecca Lemov, Robert Leonard, Harro Maas, Philip Mirowski, Jill

Morawski, Mary Morgan, Joe Rouse, Gil Skillman, Mark Solovey, Thomas Sturm, Will Thomas, and E. Roy Weintraub.

At times I have conferred about the past (and present) of game theory with Robert Axelrod, Gerd Gigerenzer, Melvin Guyer, Olaf Helmer, William Horvath, Jeremy Leighton John, Richard Levins, Richard Lewontin, R. Duncan Luce, Anatol and Gwen Rapoport, Howard Raiffa, Larry Samuelson, Thomas Schelling, Martin Shubik, J. David Singer, John Maynard Smith, and Marina von Neumann Whitman. These people have been unfailingly kind and helpful in answering all manner of questions, and in some cases reading portions of the manuscript or its precursors.

My research has been supported and furthered by numerous grants and fellowships over the years: on several occasions, by the Max Planck Institute for the History of Science in Berlin; by Wesleyan University's Center for the Humanities and the College of Environment; a National Science Foundation Doctoral Dissertation Improvement Award; a Charlotte W. Newcombe Doctoral Dissertation Fellowship; a Vilas Travel Grant, and the William Coleman Dissertation Fellowship (sponsored by the University of Wisconsin Institute for Research in the Humanities). To Mark R. Gordon, formerly of Sanford C. Bernstein & Co., I also owe thanks for inadvertently financing much of my graduate education.

Many archivists and curators have provided me with assistance, always far above and beyond the call of duty. The library staffs at Duke University's Perkins Library and the Library of Congress provided invaluable assistance with the papers of John von Neumann and Oskar Morgenstern. Vivian Arterbery and Ann Horn of the RAND Corporation have been unbelievably generous with their time and energy. Idelle Nissila proved incredibly helpful in navigating the archives of the Ford Foundation. Innumerable librarians from the National Archives provided assistance and illuminating conversations, including Marjorie Ciarlante, Rebecca Livingston, Alan Walker, and Barry Zerby. They also have helped me work through the papers of the NDRC, the Applied Mathematics Panel, the Applied Psychology Panel, the Office of Naval Research, and various records of Air Force R&D policy. At the University of Michigan Bentley Historical Library, Karen Jania and Malgorzya Mus were most helpful in researching the papers of the Mental Health Research Institute, the Center for Research on Conflict Resolution, Kenneth Boulding, J. David Singer, Clyde Hamilton Coombs, and Merrill Flood. More librarians and archivists helped with the writing of chapter 6 than any other chapter. These included the staff of Queens University Special Collections for assistance with the Wynne-Edwards papers; the staff of University College London for assistance with the papers of J. B. S. Haldane and Lionel Penrose; Dr. Jeremy Leighton John of the British Library for assistance with the largely unprocessed papers of William Hamilton and George Price, and Dr. Linda Birch (Alexander Library of Ornithology, Oxford) for assistance with the papers of David Lack. Finally, considerable thanks must go to Jennifer Milligan, Josef Keith, and Marigold Bentley (Religious Society of Friends in Britain) for help uncovering the scope of "peace research" at University College London.

Many dear friends and colleagues who have provided food, company, and moral support without which this book would never have happened: Neil Andrews, Florence Hsia, Libbie Freed, Fred Gibbs, Camilo Quintero, Matt Lavine, Erika Milam, Brent Ruswick, Jonathan Seitz; and especially Scott Burkhardt and Tom Robertson.

The book is dedicated to my wife, Amrys, and to my family.

As always, the customary caveat applies.

Notes

Chapter One

1. See, e.g., Box 42, Folder: "Game Theory Book: Notes and Papers, 1957–1969, n.d.," Oskar Morgenstern Papers (hereafter Morgenstern Papers), David M. Rubenstein Rare Book and Manuscript Library, Duke University.

2. H. Scott Bierman and Luis Fernandez, *Game Theory with Economic Applications* (Reading, MA: Addison-Wesley, 1993), 81.

3. See especially Geoffrey Martin Hodgson, "Behind Methodological Individualism," *Cambridge Journal of Economics* 10 (1986): 211–24; Hodgson, *How Economics Forgot History: The Problem of Historical Specificity in Social Science* (New York: Routledge, 2001); Donald P. Green and Ian Shapiro, *Pathologies of Rational Choice Theory: A Critique of Applications in Political Science* (New Haven, CT: Yale University Press, 1994); Ian Shapiro, *The Flight from Reality in the Human Sciences* (Princeton, NJ: Princeton University Press, 2005).

4. See, e.g., the text of the Royal Swedish Academy's press release for the 1994 prize: http://www.nobelprize.org/nobel_prizes/economics/laureates/1994/press.html.

5. See especially Christian Schmidt, "Game Theory and Economics: An Historical Survey," *Revue d'économie politique* 100, no. 5 (1990): 589–618, for a statement of this position. On the contributions of Nash and Harsanyi in particular, see Roger B. Myerson, "Comments on 'Games with Incomplete Information Played by "Bayesian" Players, I–III': Harsanyi's Games with Incomplete Information," *Management Science* 50, no. 12 (2004): 1818–24; Myerson, "Learning Game Theory from John Harsanyi," *Games and Economic Behavior* 36, no. 1 (2001): 20–25; Nicola Giocoli, "Three Alternative (?) Stories on the Late 20th Century Rise of Game Theory," *Studi e Note di Economia* 2 (2009): 187–210; Giocoli, *Modeling Rational Agents: From Interwar Economics to Early Modern Game Theory* (Northampton, MA: Edward Elgar Publishing, 2003). On the history of game theory more broadly viewed from the perspective of the history

of economics, see Mary Ann Dimand and Robert W. Dimand, *The Foundations of Game Theory* (Lyme, NH: Edward Elgar Publishing, 1997); Dimand and Dimand, *A History of Game Theory*, Routledge Studies in the History of Economics 8 (New York: Routledge, 1996); Christian Schmidt, *Game Theory and Economic Analysis: A Quiet Revolution in Economics* (New York: Routledge, 2002); and Robert J. Leonard, "Reading Cournot, Reading Nash: The Creation and Stabilisation of the Nash Equilibrium," *Economic Journal* 104, no. 424 (1994): 492–511.

6. Biography has therefore proven a popular historiographical genre through which to approach the history of game theory; see, e.g., Philippe Fontaine, "The Homeless Observer: John Harsanyi on Interpersonal Utility Comparisons and Bargaining, 1950–1964," *Journal of the History of Economic Thought* 32, no. 2 (2010): 145–73. See also Sylvia Nasar, *A Beautiful Mind: A Biography of John Forbes Nash, Jr., Winner of the Nobel Prize in Economics, 1994* (New York: Simon and Schuster, 1998); Robert J. Leonard, "From Parlor Games to Social Science: von Neumann, Morgenstern, and the Creation of Game Theory, 1928–1944," *Journal of Economic Literature* 33, no. 2 (1995): 730–61; Robert J. Leonard, "Ethics and the Excluded Middle: Karl Menger and Social Science in Interwar Vienna," *Isis* 89, no. 1 (1998): 1–26; Leonard, "'Between Worlds,' or an Imagined Reminiscence by Oskar Morgenstern about Equilibrium and Mathematics in the 1920s," *Journal of the History of Economic Thought* 26, no. 3 (2004): 527–85; and Leonard, *Von Neumann, Morgenstern, and the Creation of Game Theory: From Chess to Social Science, 1900–1960* (New York: Cambridge University Press, 2010).

7. Peter Galison, "The Americanization of Unity," *Daedalus* 127, no. 1 (1998): 45–71. On new patterns of funding in the postwar social and behavioral sciences and their implications for the organization of research see, e.g., Mark Solovey, "Riding Natural Scientists' Coattails Onto the Endless Frontier: The SSRC and the Quest for Scientific Legitimacy," *Journal of the History of the Behavioral Sciences* 40, no. 4 (2004): 393–422; Hunter Crowther Heyck, "Patrons of the Revolution: Ideals and Institutions in Postwar Behavioral Science," *Isis* 97, no. 3 (2006): 420–46; Philippe Fontaine, "Stabilizing American Society: Kenneth Boulding and the Integration of the Social Sciences, 1943–1980," *Science in Context* 22, no. 2 (2010): 221–65; Jefferson Pooley and Mark Solovey, "Marginal to the Revolution: The Curious Relationship between Economics and the Behavioral Sciences Movement in Mid-Twentieth-Century America," *History of Political Economy* 42 (2010): 199–233; Mark Solovey, *Shaky Foundations: The Politics–Patronage–Social Science Nexus in Cold War America* (New Brunswick, NJ: Rutgers University Press, 2013).

8. See, e.g., Steve J. Heims, *Constructing a Social Science for Postwar America: The Cybernetics Group, 1946–1953* (Cambridge, MA: MIT Press, 1993); Paul N. Edwards, *The Closed World: Computers and the Politics of Discourse*

in Cold War America (Cambridge, MA: MIT Press, 1996); Philip Mirowski, *Machine Dreams: Economics Becomes a Cyborg Science* (Cambridge: Cambridge University Press, 2002); S. M. Amadae, *Rationalizing Capitalist Democracy: The Cold War Origins of Rational Choice Liberalism* (Chicago: University of Chicago Press, 2003); Hunter Crowther-Heyck, *Herbert A. Simon: The Bounds of Reason in Modern America* (Baltimore, MD: Johns Hopkins University Press, 2005); Jamie Cohen-Cole, *The Open Mind: Cold War Politics and the Sciences of Human Nature* (Chicago: University of Chicago Press, 2014); Joel Isaac, *Working Knowledge: Making the Human Sciences from Parsons to Kuhn* (Cambridge, MA: Harvard University Press, 2012).

9. See, e.g., Paul Forman, "Behind Quantum Electronics," *Historical Studies in the Physical and Biological Sciences* 18 (1987): 149–229; see also Stuart Leslie, *The Cold War and American Science: The Military-Industrial-Academic Complex at MIT and Stanford* (New York: Columbia University Press, 1993); Michael Aaron Dennis, "Our First Line of Defense: Two University Laboratories in the Postwar American State," *Isis* 85 (1994): 427–55; and Peter Galison, *Image and Logic: A Material Culture of Microphysics* (Chicago: University of Chicago Press, 1997).

10. See, e.g., Lily Kay, *Who Wrote the Book of Life? A History of the Genetic Code* (Stanford, CA: Stanford University Press, 2000); Edwards, *Closed World*; Chunglin Kwa, "Representations of Nature Mediating between Ecology and Science Policy: The Case of the International Biological Programme," *Social Studies of Science* 17, no. 3 (August 1987): 413–42; Peter Taylor, "Technocratic Optimism: H. T. Odum and the Partial Transformation of Ecological Metaphor after World War II," *Journal of the History of Biology* 21 (1988): 213–44.

11. See Philip Mirowski, "When Games Grow Deadly Serious," in *Economics and National Security: A History of Their Interaction*, ed. Craufurd D. W. Goodwin (Durham, NC: Duke University Press, 1991), 227–60; Mirowski, *Machine Dreams*; William Poundstone, *Prisoner's Dilemma* (New York: Doubleday, 1992); Peter Galison, "The Ontology of the Enemy: Norbert Wiener and the Cybernetic Vision," *Critical Inquiry* 21, no. 1 (1994): 228–66, esp. 231.

12. On this point see especially Mirowski, "When Games Grow Deadly Serious."

13. On "theoretical tools," see David Kaiser, *Drawing Theories Apart: The Dispersion of Feynman Diagrams in Postwar Physics* (Chicago: University of Chicago Press, 2005). The notion that game theory constitutes a set of theoretical tools has been explored in Paul Erickson, "Mathematical Models, Rational Choice, and the Search for Cold War Culture." *Isis* 101, no. 2 (2010): 386–92; and Joel Isaac, "Tool Shock: Technique and Epistemology in the Postwar Social Sciences," *History of Political Economy* 42 (2010): 133–64.

14. Andrew Warwick, "Cambridge Mathematics and Cavendish Physics: Cunningham, Campbell, and Einstein's Relativity, 1905–1911. Part I: The Uses

of Theory," *Studies in the History and Philosophy of Science* 23 (1992): 625–56; Andrew Warwick, *Masters of Theory: The Pursuit of Mathematical Physics in Victorian Cambridge* (Cambridge: Cambridge University Press, 2003); Kaiser, *Drawing Theories Apart*; David Kaiser, "A Ψ Is Just a Ψ? Pedagogy, Practice, and the Reconstitution of General Relativity, 1942–1975," *Studies in the History and Philosophy of Modern Physics* 29 (1998): 321–38.

15. See, e.g., Jean Lave, *Cognition in Practice: Mind, Mathematics and Culture in Everyday Life* (New York: Cambridge University Press, 1988); Jean Lave, "The Values of Quantification," in *Power, Action, and Belief: A New Sociology of Knowledge?*, ed. John Law, Sociological Review Monograph 32 (London: Routledge and Kegan Paul, 1986); Gary Urton, *The Social Life of Numbers: A Quechua Ontology of Numbers and Philosophy of Arithmetic* (Austin: University of Texas Press, 1997); Helen Verran, *Science and an African Logic* (Chicago: University of Chicago Press, 2001).

16. L. Magnani and N. J. Nersessian, ed., *Model-Based Reasoning: Science, Technology, Values* (Dordrecht: Kluwer Academic, 2002); Ursula Klein, *Experiments, Models, Paper Tools: Cultures of Organic Chemistry in the Nineteenth Century* (Stanford, CA: Stanford University Press, 2003); S. de Chadarevian and N. Hopwood, eds., *Models: The Third Dimension of Science* (Stanford, CA: Stanford University Press, 2004); Mary S. Morgan, "Economic Man as Model Man: Ideal Types, Idealization and Caricatures," *Journal of the History of Economic Thought* 28, no. 1 (2006): 1–27; *Science Without Laws: Model Systems, Cases, Exemplary Narratives*, ed. A. Creager, M. Norton Wise, and E. Lunbeck (Durham, NC: Duke University Press, 2007); Mary S. Morgan, *The World in the Model: How Economists Work and Think* (New York: Cambridge University Press, 2012).

17. The game here is based on D. M. Kreps, *Game Theory and Economic Modelling* (Oxford: Oxford University Press, 1990), 11. For an exposition of this line of argumentation—much of which can be traced to von Neumann's very first article on game theory in 1928—see, e.g., Albert W. Tucker, *Game Theory and Programming* (Stillwater: Dept. of Mathematics, Oklahoma Agricultural and Mechanical College, 1955). For another more recent treatment, see Ken Binmore, *Game Theory: A Very Short Introduction* (Oxford: Oxford University Press, 2007), 4.

18. Martin Shubik, "Game Theory and Operations Research: Some Musings 50 Years Later," *Operations Research* 50, no. 1 (2002): 192–96. For similar observations see also M. Shubik, "What Is an Application and When Is Theory a Waste of Time?," *Management Science* 33, no. 12 (1987): 1511–22; Martin Shubik, "Game Theory: Some Observations," *Yale School of Management Working Papers, Series B* 132 (2000).

19. See, e.g., the classic expositions of Daniel J. Kevles, *The Physicists: A History of a Scientific Community in Modern America* (New York: Knopf, 1978),

and Daniel Kleinman, *Politics on the Endless Frontier: Postwar Research Policy in the United States* (Durham, NC: Duke University Press, 1995).

20. On this point, see especially "World in a Matrix" in Erickson and others, *How Reason Almost Lost Its Mind: The Strange Career of Cold War Rationality* (Chicago: University of Chicago Press, 2013).

21. See Ian Hacking, *The Emergence of Probability: A Philosophical Study of Early Ideas about Probability, Induction, and Statistical Inference* (Cambridge: Cambridge University Press, 1984).

22. For an early statement of such distinctions, see especially C. H. Coombs, Howard Raiffa, and R. M. Thrall, "Mathematical Models and Measurement Theory," in *Decision Processes*, ed. R. M. Thrall, C. H. Coombs and R. L. Davis (New York: John Wiley and Sons, 1954), 19–38.

23. See Lorraine Daston, *Classical Probability in the Enlightenment* (Princeton, NJ: Princeton University Press, 1988), especially chap. 2, "Expectation and the Reasonable Man."

24. See especially "The Decline of the Classical Theory" in ibid., and Lorraine J. Daston, "Rational Individuals Versus Laws of Society: From Probability to Statistics," in *The Probabilistic Revolution*, vol. 1: *Ideas in History*, ed. Lorenz Krüger, Lorraine J. Daston, and Michael Heidelberger (Cambridge, MA: MIT Press, 1987), 295–304.

25. On the changing place of the rationality assumption in economics and political theory between the eighteenth century and the present day, see especially Morgan. "Economic Man as Model Man"; Amadae, *Rationalizing Capitalist Democracy*; Giocoli, *Modeling Rational Agents*; and the comments in Kenneth J. Arrow, "Rationality of Self and Others in an Economic System," *Journal of Business* 59, no. 4 (1986): S385–S399.

26. On rationality and the defense intellectuals, see David Halberstam, *The Best and the Brightest* (New York: Random House, 1972); Fred M. Kaplan, *The Wizards of Armageddon* (New York: Simon and Schuster, 1983); Carol Cohn, "Sex and Death in the Rational World of Defense Intellectuals," *Signs* 12, no. 4 (1987): 687–718; Alex Abella, *Soldiers of Reason: The Rand Corporation and the Rise of the American Empire* (Orlando, FL: Harcourt, 2008). On the 1960s "critique of rationality" see especially Theodore Roszak, *The Making of a Counter Culture: Reflections on the Technocratic Society and Its Youthful Opposition* (Garden City, NY: Doubleday, 1969); Maurice Isserman, *If I Had a Hammer: The Death of the Old Left and the Birth of the New Left* (Urbana: University of Illinois Press, 1993); David Steigerwald, *The Sixties and the End of Modern America* (New York: St. Martin's Press, 1995).

27. See especially "Enlightenment Reason, Cold War Rationality, and the Rule of Rules" in Erickson et al., *How Reason Almost Lost Its Mind*.

28. See especially Sharon Ghamari-Tabrizi, *The Worlds of Herman Kahn: The Intuitive Science of Thermonuclear War* (Cambridge, MA: Harvard Uni-

versity Press, 2005); Ghamari-Tabrizi, "Simulating the Unthinkable: Gaming Future War in the 1950s and 1960s," *Social Studies of Science* 30, no. 2 (2000): 163–223.

Chapter Two

1. "Books Published Today," *New York Times*, September 23, 1944, 17.

2. John von Neumann and Oskar Morgenstern, *Theory of Games and Economic Behavior* (Princeton, NJ: Princeton University Press, 1944), 31.

3. On the priority dispute, and on Borel's theory, see, e.g., J. von Neumann and M. Fréchet, "Communication on the Borel Notes," *Econometrica* 21, no. 1 (1953): 124–27; Émile Borel, "Sur la theorie des jeux," *Comptes Rendus de l'Academie des Sciences* 186, no. 25 (1928): 1689–91; Leonard, *Von Neumann, Morgenstern, and the Creation of Game Theory*, chap. 4; and L. Dell'Aglio, "Divergences in the History of Mathematics: Borel, von Neumann, and the Genesis of Game Theory," *Rivista di Storia della Scienza* 3 (1995): 1–46.

4. John von Neumann, "Zur Theorie der Gesellschaftsspiele," *Mathematische Annalen* 100 (1928): 295–320. This paper was subsequently published in English as John von Neumann, "On the Theory of Games of Strategy" in *Contributions to the Theory of Games*, vol. 4, ed. A. W. Tucker and R. D. Luce, trans. Sonya Bargmann (Princeton, NJ: Princeton University Press, 1959), 13–42.

5. Oskar Morgenstern, "The Collaboration between Oskar Morgenstern and John von Neumann on the Theory of Games," *Journal of Economic Literature* 14, no. 3 (1976): 805–16.

6. See especially Leonard, *Von Neumann, Morgenstern, and the Creation of Game Theory*; Leonard, "From Parlor Games to Social Science," 730–61; and the following, all in *Toward a History of Game Theory*, ed. E. Roy Weintraub (Durham, NC: Duke University Press, 1992): Robert Leonard, "Creating a Context for Game Theory," 29–76; Urs Rellstab, "New Insights Into the Collaboration between John von Neumann and Oskar Morgenstern on the *Theory of Games and Economic Behavior*," 77–94; and Andrew Schotter, "Oskar Morgenstern's Contribution to the Development of the Theory of Games," 95–112.

7. On the mathematization of economics between the 1930s to 1950s see especially E. Roy Weintraub and Philip Mirowski, "The Pure and the Applied: Bourbakism Comes to Mathematical Economics," *Science in Context* 7, no. 2 (1994): 245–72; E. Roy Weintraub, *How Economics Became a Mathematical Science* (Durham, NC: Duke University Press, 2002); and Giocoli, *Modeling Rational Agents*. On the exclusion of history that accompanied mathematization, see Hodgson, *How Economics Forgot History*.

8. Philip Mirowski, "What Were von Neumann and Morgenstern Trying to

Accomplish?," in *Toward a History of Game Theory*, ed. Weintraub, 113–47; and Leonard, *Von Neumann, Morgenstern, and the Creation of Game Theory*.

9. On the place of economics in the sweep of von Neumann's intellectual interests, see especially Mirowski, *Machine Dreams*; *John von Neumann and Modern Economics*, ed. M. H. I. Dore, Sukhamoy Chakravarty and Richard M. Goodwin (Oxford: Clarendon Press, 1989).

10. Martin Shubik, "Game Theory at Princeton, 1949–1955: A Personal Reminiscence," in *Toward a History of Game Theory*, ed. Weintraub, 153.

11. Von Neumann and Morgenstern, *Theory of Games and Economic Behavior*, 72–73. All quotations in this chapter are from the first edition unless otherwise specified.

12. See, e.g., the remarks in Ulrich Schwalbe and Paul Walker, "Zermelo and the Early History of Game Theory," *Games and Economic Behavior* 34 (2001): 123–37, or the remarks on Zermelo's theory of chess in Mirowski, "What Were von Neumann and Morgenstern Trying to Accomplish?," 118.

13. See, e.g., the following, all in *Contributions to the Theory of Games*, vol. 2, ed. H. W. Kuhn and A. W. Tucker (Princeton, NJ: Princeton University Press, 1953): H. W. Kuhn, "Extensive Games and the Problem of Information," 193–216; Norman Dalkey, "Equivalence of Information Patterns and Essentially Determinate Games," 217–44; and David Gale and F. M. Stewart, "Infinite Games with Perfect Information," 245–66.

14. Morgenstern, "Collaboration between Oskar Morgenstern and John von Neumann on the Theory of Games," 811; and Poundstone, *Prisoner's Dilemma*, 27.

15. S. Ulam et al., "John von Neumann, 1903–1957," in *The Intellectual Migration: Europe and America, 1930–1960*, ed. Donald Fleming and Bernard Bailyn (Cambridge, MA: Harvard University Press, 1968), 239.

16. W. V. Quine: "A Theorem on Parametric Boolean Functions," RAND RM-196, July 27, 1949; "Commutative Boolean Functions," RAND RM-199, August 10, 1949; "Notes on Information Patterns in Game Theory," RAND RM-216, August 17, 1949; and "On Functions of Relations with Especial Reference to Social Welfare" RAND RM-218, August 17, 1949.

17. Von Neumann to Frank Aydelotte, November 10, 1941, in Box 29, Folder 5: "Book File: Theory of Games and Economic Behavior: General Correspondence," John von Neumann Papers (hereafter von Neumann Papers), Manuscript Division, Library of Congress, Washington, DC.

18. E. Zermelo, "Untersuchungen Über Die Grundlagen Der Mengenlehre. I," *Mathematische Annalen* 65 (1908): 261–81, lays out the rationale and agenda for the axiomatic approach to set theory. See also Herbert Mehrtens, *Moderne, Sprache, Mathematik: Eine Geschichte des Streits um die Grundlagen der Disziplin und des Subjekts formaler Systeme* (Frankfurt am Main: Suhrkamp, 1990); and Paolo Mancosu, *From Hilbert to Brouwer: The Debate*

on the Foundations of Mathematics in the 1920s (New York: Oxford University Press, 1988).

19. On Brouwer and intuitionism, see Dirk van Dalen, *Mystic, Geometer, and Intuitionist: The Life of L.E.J. Brouwer* (New York: Oxford University Press, 1999).

20. Most accounts of von Neumann's early life draw upon a relatively small pool of sources; see Steve J. Heims, *John von Neumann and Norbert Wiener: From Mathematics to the Technologies of Life and Death* (Cambridge, MA: MIT Press, 1980), and Norman Macrae, *John von Neumann* (New York: Pantheon Books, 1992), for accounts based on these. For recollections of von Neumann's family and friends see Nicholas A. Vonneumann, *John von Neumann—As Seen by His Brother* (Meadowbrook, PA: Private Printing, 1987); Stanislaw Ulam, "John von Neumann, 1903–1957," *Annals of the History of Computing* 4, no. 2 (1982): 1–49; Ulam, *Adventures of a Mathematician* (New York: Scribner, 1976); Eugene Paul Wigner, *Symmetries and Reflections: Scientific Essays of Eugene P. Wigner* (Bloomington: Indiana University Press, 1967). For appreciations by fellow mathematicians, see Salomon Bochner, "John von Neumann," *Biographical Memoirs of the National Academy of Sciences* 32 (1958): 456–61, and J. Dieudonne, "John von Neumann," in *Dictionary of Scientific Biography*, ed. Charles C. Gillespie (New York: Scribner and Sons, 1982), 88–92.

21. On von Neumann's love of word games and paradoxes, see Ulam, *Adventures of a Mathematician*, 131, 193.

22. Vonneuman, *John von Neumann*, 32.

23. John von Neumann, "Eine Axiomatizierung der Mengenlehre," *Journal für die Reine und Angewandte Mathematick* 154 (1925): 219–40. On the place of von Neumann's work within the foundational debates in mathematics, see, e.g., Jean Van Heijenoort, *From Frege to Gödel: A Source Book in Mathematical Logic, 1879–1931* (Cambridge, MA: Harvard University Press, 1967).

24. Ernst Zermelo, "Über eine Anwendung der Mengenlehre auf die Theorie des Schachspiels," in *Proceedings of the 5th International Congress of Mathematicians, Cambridge, 22–28 August 1912*, vol. 2, ed. E. W. Hobson and A. E. H. Love (Cambridge: Cambridge University Press, 1913), 501–4.

25. Ibid., 501, translation by the author. On the relationship between logic and psychology in the later nineteenth century, see Martin Kusch, *Psychologism: A Case Study in the Sociology of Philosophical Knowledge* (London: Routledge, 1995).

26. See especially Gregory H. Moore, *Zermelo's Axiom of Choice: Its Origins, Development, and Influence* (New York: Springer-Verlag, 1982), on the reception of Zermelo's axioms.

27. Zermelo, "Über eine Anwendung," 501.

28. Dénes König, "Über eine Schlußweise aus dem Endlichen ins Unendliche," *Acta Scientiarum Mathematicarum, Universitatis Szegediensis* 3 (1927):

121–30, quoted in Schwalbe and Walker, "Zermelo and the Early History of Game Theory," 126.

29. Ibid. Results from this article were later incorporated into Dénes König, *Theorie der Endlichen und Unendlichen Graphen* (Leipzig: Teubner, 1936); see esp. chap. 6, sec. 2.

30. Schwalbe and Walker, "Zermelo and the Early History of Game Theory," 126.

31. Von Neumann and Morgenstern, *Theory of Games and Economic Behavior*, 60.

32. Von Neumann, "On the Theory of Games of Strategy," 14.

33. John von Neumann, "On the Theory of Games of Strategy," 13. All translations are from this article, unless explicitly stated otherwise.

34. Von Neumann and Morgenstern, *Theory of Games and Economic Behavior*, 84.

35. See, e.g., Heims, *John von Neumann and Norbert Wiener*, chap. 4; Dimand and Dimand, *A History of Game Theory*, chap. 1.

36. Von Neumann and Morgenstern, *Theory of Games and Economic Behavior*, 85.

37. Ibid., 176–77.

38. Oskar Morgenstern, "Vollkommene Voraussicht und wirtschaftliches Gleichgewicht," *Zeitschrift für Nationalökonomie* 6, no. 3 (1935): 337–57.

39. On the German historicists, see Erik Grimmer-Solem, *The Rise of Historical Economics and Social Reform in Germany, 1864–1894* (Oxford: Oxford University Press, 2003); Hodgson, *How Economics Forgot History*; Yuichi Shionoya, ed., *The German Historical School: The Historical and Ethical Approach to Economics* (London: Routledge, 2001); Keith Tribe, "Die Vernunft des List: National Economy and the Critique of Cosmopolitical Economy," in *Strategies of Economic Order: German Economic Discourse, 1750–1950* (Cambridge: Cambridge University Press, 1995); Erich Streissler and Karl Milford, "Theoretical and Methodological Positions of German Economists in the Middle of the Nineteenth Century," *History of Economic Ideas* 1, nos. 3/2 (1993): 66–71; Keith Tribe, *Governing Economy: The Reformation of German Economic Discourse, 1750–1840* (New York: Cambridge University Press, 1988); H. K. Betz, "How Does the German Historical School Fit," *History of Political Economy* 20, no. 3 (1988): 409–30.

40. The literature on the Austrian School is vast as well; see, e.g., Bruce Caldwell, *Hayek's Challenge: An Intellectual Biography of F. A. Hayek* (Chicago: University of Chicago Press, 2004).

41. The classic work on the socialist calculation debate is Friedrich A. von Hayek, N. G. Pierson, Ludwig von Mises, Georg Halm, and Enrico Barone, *Collectivist Economic Planning: Critical Studies on the Possibilities of Socialism* (London: G. Routledge, 1935). On the history of the debate in Austria and Ger-

many see Günther K. Chaloupek, "The Austrian Debate on Economic Calculation in a Socialist Economy," *History of Political Economy* 22, no.
4 (1990): 659–75; Günther Chaloupek and Werner Teufelsbauer, *Gesamtwirtschaftliche Planung in Westeuropa: Theoretische Entwicklungen und praktische Erfahrungen seit 1970*, Campus Forschung, Bd. 523 (Frankfurt/Main: Campus, 1987).

42. See Hayek, "Foreword," in Ludwig von Mises, *Socialism: An Economic and Sociological Analysis* (Indianapolis, IN: Liberty Fund, 1981).

43. See especially Leonard, "Between Worlds"; Leonard, *Von Neumann, Morgenstern, and the Creation of Game Theory*, chap. 6.

44. On Spann, see Anthony Carty, "Alfred Verdross and Othmar Spann: German Romantic Nationalism, National Socialism, and International Law," *European Journal of International Law* 6, no. 1 (1995): 1–21.

45. Leonard, "Between Worlds," 291.

46. See also Malcolm Rutherford, *Wesley Clair Mitchell* (London: Pickering and Chatto, 1997); Arthur F. Burns, *Wesley Clair Mitchell: The Economic Scientist* (New York: National Bureau of Economic Research, 1952); Philip Mirowski, "Problems in the Paternity of Econometrics: Henry Ludwell Moore," *History of Political Economy* 22, no. 4 (1990): 587–609.

47. In both Austria and the United States, the existence of postwar business cycles provided a justification for government intervention in the economy. See especially Hayek, "The Nature and History of the Problem" in Hayek et al., *Collectivist Economic Planning*, 1–40, and similar comments in "History and Politics" in Friedrich Hayek, *Capitalism and the Historians* (Chicago: University of Chicago Press, 1954).

48. See Oskar Morgenstern, *Wirtschaftsprognose: Eine Untersuchung ihrer Voraussetzungen und Moglichkeiten* (Wien: Julius Springer, 1928), especially "Wirkung und Anwendung der Prognose," 98.

49. Oskar Morgenstern, *Die Grenzen Der Wirtschaftspolitik* (Wien: Julius Springer, 1934), later published as Oskar Morgenstern, *The Limits of Economics*, trans. Vera Smith (London: W. Hodge, 1937); Oskar Morgenstern, "Das Zeitmoment in Der Wertlehre," *Zeitschrift für Nationalökonomie* 5, no. 4 (1934): 433–58; Oskar Morgenstern, "Logistik und Sozialwissenschaften," *Zeitschrift für Nationalökonomie* 7 (1936): 1–24; and Morgenstern, "Vollkommene Voraussicht und Wirtschaftliches Gleichgewicht," 337–57.

50. "The Contribution of Oskar Morgenstern" in *Essays in Mathematical Economics in Honor of Oskar Morgenstern*, ed. Martin Shubik (Princeton, NJ: Princeton University Press, 1967), vii–viii.

51. Von Neumann, "On the Theory of Games of Strategy," 16.

52. Ibid., 21 (emphasis added).

53. Ibid., 23.

54. Ibid.

55. Ibid., 26.

56. John von Neumann, D. Hilbert, and L. Nordheim, "Über die Grundlagen der Quantenmechanic," *Mathematische Annalen* 98 (1927): 1–30. Von Neumann's work on quantum mechanics was eventually synthesized in the course of his collaboration with Eugene Wigner and published together in John von Neumann, *Mathematische Grundlagen Der Quantenmechanik* (Berlin: Springer, 1932).

57. Building on these ideas, during the 1930s, von Neumann collaborated with the Harvard mathematician Garrett Birkhoff to develop probabilistic logical systems inspired by quantum mechanics; see Garrett Birkhoff and John von Neumann, "The Logic of Quantum Mechanics," *Annals of Mathematics* 37, no. 4 (1936): 823–43. On the hidden variables theorem see especially Louis Caruana, "John von Neumann's 'Impossibility Proof' in Historical Perspective," *Physis: Rivista Internazionale di Storia della Scienza* 32 (1995): 109–24; and Miklós Rédei and Michael Stöltzner, *John von Neumann and the Foundations of Quantum Physics* (Boston: Kluwer Academic, 2001).

58. Merrill M. Flood, "The Objectives of TIMS," *Management Science* 2, no. 2 (1956): 178.

59. Oskar Morgenstern diary, November 23, 1942; quoted in Mirowski, "What Were von Neumann and Morgenstern Trying to Accomplish?," 134.

60. Von Neumann and Morgenstern, *Theory of Games and Economic Behavior*, 178n.

61. Von Neumann, "On the Theory of Games of Strategy," 13.

62. Léon Walras, *Éléments d'économie politique pure: Ou, théorie de la richesse sociale* (Paris: F. Pichon, 1900); Gustav Cassel and Joseph McCabe, *The Theory of Social Economy* (New York: Harcourt, 1924); Gustav Cassel, *Theoretische Sozialökonomie*, Lehrbuch der allgemeinen volkswirtschaftslehre, 2. abt (Leipzig: C. F. Winter, 1921).

63. Mohammed Dore, "The Legacy of John von Neumann," in *John von Neumann and Modern Economics*, ed. Dore, Chakravarty and Goodwin, 82.

64. Heinz Kurz and Neri Salvadori, "Von Neumann's Growth Model and the 'Classical' Tradition," in *Understanding "Classical" Economics: Studies in Long-Period Theory* (New York: Routledge, 1998); Kurz and Salvadori, "Sraffa and von Neumann," *Review of Political Economy* 13, no. 2 (2001): 161–80.

65. Jacob Marschak, "Wirtschaftsrechnung und Gemeinwirtschaft," *Archiv für Sozialwissenschaft und Sozialpolitik* 51 (1924).

66. Quoted in E. Roy Weintraub, *General Equilibrium Analysis: Studies in Appraisal* (Cambridge: Cambridge University Press, 1985), 74n.

67. Ulam et al., "John von Neumann, 1903–1957," 237.

68. John von Neumann, "Über ein Ökonomisches Gleichungssystem und eine Verallgemeinerung des Brouwerschen Fixpunktsatzes," *Ergebnisse eines mathematischen Kolloquiums* 8 (1937): 73–83. An English translation was published as John von Neumann, "A Model of General Economic Equilibrium," *Re-*

view of Economic Studies 13, no. 33 (1945): 1–9. All quotes are taken from this translation unless otherwise indicated.

69. Ibid., 5.

70. Ibid.

71. Specifically, the mapping must be continuous and defined on a compact and convex set in R^n. Brouwer's original proof is for the case of the three-dimensional simplex; see L. E. J. Brouwer, "Beweis Der Invarianz Der Dimensionzahl," *Mathematische Annalen* 70 (1911): 161–65; but by the later 1920s it had been generalized to any number of dimensions; see, e.g., B. Knaster, C. Kuratowski, and S. Mazurkiewicz, "Ein Beweis des Fixpunktsatzes für N-dimensionale Simplexe," *Fundamenta Mathematicae* 14 (1929): 132–37.

72. Tinne Hoff Kjeldsen, "John von Neumann's Conception of the Minimax Theorem: A Journey through Different Mathematical Contexts." *Archive for the History of Exact Sciences* 56 (2001): 39–68.

73. von Neumann, "A Model of General Economic Equilibrium," 1 (emphasis in the original).

74. M. H. A. Newman, "Topology by S. Lefschetz: Analysis Situs by O. Veblen," *Mathematical Gazette* 16, no. 221 (1932): 352–53.

75. See, e.g., Ioan James, *History of Topology* (Amsterdam: Elsevier Science, 1999), on the historical development of topology as a distinct mathematical subfield.

76. See especially Solomon Lefschetz, "Reminiscences of a Mathematical Immigrant in the United States," *American Mathematical Monthly* 77, no. 4 (1970): 344–50, and Albert W. Tucker, "Solomon Lefschetz: A Reminiscence," *Two-Year College Mathematics Journal* 14, no. 3 (1983): 225–27, on working with Lefschetz.

77. S. Kakutani, "A Generalization of Brouwer's Fixed Point Theorem," *Duke Mathematical Journal* 8, no. 3 (1941): 457–59.

78. These appear in the proceedings of the colloquium for March 19, 1934; see K. Schlesinger, "Über die Produktionsgleichungen der Ökonomischen Wertlehre," *Ergebnisse eines mathematischen Kolloquiums* 6 (1933): 10–11; A. Wald, "Über die Eindeutige Positive Lösbarkeit der neuen Produktionsgleichungen," *Ergebnisse eines mathematischen Kolloquiums* 6 (1933): 12–20.

79. Von Neumann and Morgenstern, *Theory of Games and Economic Behavior*, 8 (emphasis added).

80. Ibid., 32.

81. Ibid., 31.

82. Ibid., 220.

83. Ibid., 221.

84. Ibid., 223.

85. See especially their discussion in chap. 12, sec. 66, on extensions of the utility concept, cited on p. 224.

86. Ibid., 224.

87. Ibid., 34.

88. See especially section 21.1 (p. 222) for an explanation of the game.

89. More precisely, the value obtained by coalitions is greater than the value obtained by individuals in a class of games that von Neumann and Morgenstern called "essential," as opposed to inessential games where players could not gain by joining coalitions.

90. Norbert Wiener, *Cybernetics: Or, Control and Communication in the Animal and the Machine* (New York: Wiley and Sons, 1948), 159–60.

91. Von Neumann and Morgenstern, *Theory of Games and Economic Behavior*, 40. See also sec. 30, "The Exact Form of the General Definitions," for a more precise discussion, on 264.

92. Von Neumann and Morgenstern, *Theory of Games and Economic Behavior*, 289.

93. Ibid., 41.

94. Ibid., 290.

95. Ibid., 45.

96. Von Neumann, "On the Theory of Games of Strategy," 33.

97. Von Neumann and Morgenstern, *Theory of Games and Economic Behavior*, 252. The axioms can be found on 241.

98. Ibid., 252.

99. L. S. Shapley, "Additive and Non-Additive Set Functions" (PhD diss., Princeton University, 1953).

100. Von Neumann and Morgenstern, *Theory of Games and Economic Behavior*, 291.

101. Ibid., 339.

102. Karl Menger, "The Formative Years of Abraham Wald and His Work in Geometry," *Annals of Mathematical Statistics* 23, no. 1 (1952): 18.

103. On the history of the Circle, see Edmund Runggaldier, *Carnap's Early Conventionalism: An Inquiry Into the Historical Background of the Vienna Circle* (Amsterdam: Rodopi, 1984); Klemens Szaniawski, *The Vienna Circle and the Lvov-Warsaw School* (Boston: Kluwer Academic Publishers, 1989); and Viktor Kraft, *The Vienna Circle, the Origin of Neo-Positivism: A Chapter in the History of Recent Philosophy* (New York: Philosophical Library, 1953).

104. Robert Leonard has discussed this episode of Morgenstern's career in Leonard, "Ethics and the Excluded Middle," 1–26, esp. 25. See also Karl Menger and Louise Golland, *Reminiscences of the Vienna Circle and the Mathematical Colloquium* (Dordrecht: Kluwer, 1994), and Thomas E. Uebel, *Rediscovering the Forgotten Vienna Circle: Austrian Studies on Otto Neurath and the Vienna Circle* (Boston: Kluwer Academic Publishers, 1991).

105. Karl Menger, *Moral, Wille und Weltgestaltung: Grundlegung zur Logik der Sitten* (Wien: J. Springer, 1934), translated as Karl Menger, *Morality, Deci-*

sion, and Social Organization: Toward a Logic of Ethics (Boston: D. Reidel Pub. Co., 1974). Menger presents the major results of his work in Karl Menger, "An Exact Theory of Social Groups and Relations," *American Journal of Sociology* 43, no. 5 (1938): 790–98.

106. Menger, *Morality, Decision, and Social Organization*, 31.

107. For more on this intellectual atmosphere, see Leonard, "Ethics and the Excluded Middle." On the Circle's ethical work (or relative lack thereof), see Menger's chapter, "The Circle on Ethics" in Menger and Golland, *Reminiscences of the Vienna Circle and the Mathematical Colloquium*, and his commentary, "Postscript to the English Edition," in Menger, *Morality, Decision, and Social Organization*.

108. See Schlesinger. "Über die Produktionsgleichungen der Ökonomischen Wertlehre," 10–11; Karl Menger, "Ein Satz über endliche Mengen mit Anwendungen auf die formale Ethik," *Ergebnisse eines mathematischen Kolloquiums* 6 (1933); Menger, "Bernoullische Wertlehre und Petersburger Spiel," *Ergebnisse eines mathematischen Kolloquiums* 6 (1933). Menger's latter paper would renew interest in the St. Petersburg Paradox among economists after it was translated and published in Morgenstern's *festschrift* in 1967. See Karl Menger, "The Role of Uncertainty in Economics," in *Essays in Mathematical Economics in Honor of Oskar Morgenstern*, ed. Shubik, 211–31; and Paul A. Samuelson, "St. Petersburg Paradoxes: Defanged, Dissected, and Historically Described," *Journal of Economic Literature* 15, no. 1 (1977): 24–55; Lloyd S. Shapley, "The St. Petersburg Paradox: A Con Game?," *Journal of Economic Theory* 14 (1977): 439–42; Robert J. Aumann, "The St. Petersburg Paradox: A Discussion of Some Recent Comments," *Journal of Economic Theory* 14 (1977): 443–45.

109. Menger, "Postscript to the English Edition," 114.

110. Morgenstern, *Limits of Economics*, 49.

111. The paper was later excerpted in the *Ergebnisse*, Heft 6.

112. Morgenstern, "Vollkommene Voraussicht und Wirtschaftliches Gleichgewicht," 344; translation by the author (emphasis added).

113. Ibid., 345.

114. Morgenstern, "Logistik und Sozialwissenschaften."

115. Morgenstern, "Collaboration between Oskar Morgenstern and John von Neumann on the Theory of Games," 806–7.

116. Ibid., 807.

117. Kenneth J. Arrow, "Jacob Marschak, July 23, 1898–July 27, 1977," *National Academy of Science Biographical Memoirs* 60 (1991): 129–46.

118. Ibid., 138.

119. On Wald's involvement with the Commission, see especially *Cowles Commission for Research in Economics Report for 1938* (Colorado Springs, CO: Cowles Commission). On the history of the Commission in this early period, see Carl F. Christ, "History of the Cowles Commission, 1932–1952," in *Economic*

Theory and Measurement: A Twenty Year Research Report, 1932–1952 (Baltimore, MD: Waverly Press, 1952).

120. H. W. Kuhn and A. W. Tucker, "John von Neumann's Work in the Theory of Games and Mathematical Economics," *Bulletin of the American Mathematical Society* 64 (1958): 100–122, cited in Kjeldsen, "John von Neumann's Conception of the Minimax Theorem," 58.

121. Von Neumann and Morgenstern, *Theory of Games and Economic Behavior*, 504n, 504.

122. Ibid., 507.

123. Ibid., 508.

124. Ibid., 513, 513n.

125. Ibid., 582.

126. Morgenstern, "Collaboration between Oskar Morgenstern and John von Neumann on the Theory of Games," 809.

127. For reviews of this literature see, e.g., Jacob Viner, "The Utility Concept in Value Theory and Its Critics, I," *Journal of Political Economy* 33, no. 4 (August 1925): 369–87; Viner, "The Utility Concept in Value Theory and Its Critics, II," *Journal of Political Economy* 33, no. 6 (December 1925): 638–59; J. R. Hicks and R. G. D. Allen. "A Reconsideration of the Theory of Value. Part I," *Economica* 1, no. 1 (1934): 52–76; Hicks and Allen, "A Reconsideration of the Theory of Value. Part II. A Mathematical Theory of Individual Demand Functions" *Economica* 1, no. 2 (May 1934): 196–219; George J. Stigler, "The Development of Utility Theory, I," *Journal of Political Economy* 58, no. 4 (August 1950): 307–27; Stigler, "The Development of Utility Theory, II," *Journal of Political Economy* 58, no. 5 (October 1950): 373–96.

128. Oskar Morgenstern, "Professor Hicks on Value and Capital," *Journal of Political Economy* 49, no. 3 (1941): 366.

129. See "The Axiomatic Treatment of Utility," in von Neumann and Morgenstern, *Theory of Games and Economic Behavior*, 2nd ed. (1947), 617–34.

130. Von Neumann and Morgenstern, *Theory of Games and Economic Behavior*, 16.

131. Ibid., 17.

132. J. Marschak, "Neumann's and Morgenstern's New Approach to Static Economics," *Journal of Political Economy* 54, no. 2 (1946): 97 (emphasis added).

133. Carl Kaysen, "A Revolution in Economic Theory?," *Review of Economic Studies* 14, no. 1 (1946): 13.

134. See, e.g., Morgenstern's caustic commentary in Maurice Allais and Ole Hagen, *Expected Utility Hypotheses and the Allais Paradox: Contemporary Discussions of Decisions under Uncertainty with Allais' Rejoinder* (Boston: D. Reidel, 1979).

135. See especially Milton Friedman and L. J. Savage, "The Utility Analy-

sis of Choices Involving Risk," *Journal of Political Economy* 56, no. 4 (1948): 279–304; Jacob Marschak, "Rational Behavior, Uncertain Prospects, and Measurable Utility," *Econometrica* 18, no. 2 (1950): 111–41; William J. Baumol, "The Neumann-Morgenstern Utility Index—An Ordinalist View," *Journal of Political Economy* 59, no. 1 (1951): 61–66; J. Marschak, "Why 'Should' Statisticians and Businessmen Maximize 'Moral Expectation'?," in *Proceedings of the Second Berkeley Symposium on Mathematical Statistics and Probability, July 31–August 12, 1950* (Berkeley: University of California Press, 1951); Milton Friedman and L. J. Savage, "The Expected-Utility Hypothesis and the Measurability of Utility," *Journal of Political Economy* 60, no. 6 (1952): 463–74; William J. Baumol, "Discussion," *American Economic Review* 43, no. 2 (1953): 415–16; Robert H. Strotz, "Cardinal Utility," *American Economic Review* 43, no. 2 (1953): 384–97; A. Alchian, "The Meaning of Utility Measurement," *American Economic Review* 43 (1953): 26–50; William J. Baumol, "The Cardinal Utility Which Is Ordinal," *Economic Journal* 68, no. 272 (1958): 665–72. The Neumann-Morgenstern theory even entered popular expositions of economics by the early 1950s, although it was presented in a critical light; see Dennis Holme Robertson, *Utility and All That: And Other Essays* (London: Allen and Unwin, 1952), especially 28–29. The literature dissecting the significance of the new utility theory was indeed vast, and we return to it in chapter 4 in broader context of interdisciplinary "behavioral science" in the 1950s.

136. D. Ellsberg, "Classic and Current Notions of 'Measurable Utility,'" *Economic Journal* 64, no. 255 (1954): 528–56.

137. Von Neumann and Morgenstern, *Theory of Games and Economic Behavior*, 588.

138. Ibid., 604.

139. Ibid., 608–9.

140. Will Lissner, "Mathematical Theory of Poker Is Applied to Business Problems," *New York Times*, March 10, 1946, 1, 36; John McDonald, *Strategy in Poker, Business and War* (New York: Norton, 1950).

141. Leonid Hurwicz, "*The Theory of Games and Economic Behavior*, by John von Neumann; Oskar Morgenstern," *Annals of Mathematical Statistics* 19, no. 3 (1948): 436–37; W. Edwards Deming, "Review of *Theory of Games and Economic Behaviour* by John von Neumann and Oskar Morgenstern," *Journal of the American Statistical Association* 40, no. 230 (1945): 263–65; M. G. K., "Review of *Theory of Games and Economic Behaviour* by John von Neumann and Oskar Morgenstern," *Journal of the Royal Statistical Society* 107, nos. 3/4 (1944): 293; C. A. B. Smith, "Review of *Theory of Games and Economic Behaviour* by John von Neumann and Oskar Morgenstern," *Mathematical Gazette* 29, no. 285 (1945): 131–33. For Wald's game-theoretic approach to problems of statistical inference see Abraham Wald, "Statistical Decision Functions which Minimize the Maximum Risk," *Annals of Mathematics* 46, no. 2 (1945): 265–80; Wald, "Gen-

eralization of a Theorem by V. Neumann Concerning Zero Sum Two Person Games," *Annals of Mathematics* 46, no. 2 (1945): 281–86.

142. See, e.g., T. Barna, "Review of *Theory of Games and Economic Behaviour* by John von Neumann and Oskar Morgenstern," *Economica* 13, no. 50 (1946): 136–38; Robert W. Harrison, "Review of *Theory of Games and Economic Behaviour* by John von Neumann and Oskar Morgenstern," *Journal of Farm Economics* 27, no. 3 (1945): 725–26.

143. Richard Stone, "The Theory of Games," *Economic Journal* 58, no. 230 (1948): 185–201.

144. Kaysen, "A Revolution in Economic Theory?," 14.

145. E. J. Gumbel, "Review of *Theory of Games and Economic Behaviour* by John von Neumann and Oskar Morgenstern," *Annals of the American Academy of Political and Social Science* 239 (1945): 209–10.

146. Louis Weisner, "Review of *Theory of Games and Economic Behavior* by John von Neumann and Oskar Morgenstern," *Science and Society* 9 (1945): 366–69.

147. George Chacko, "Economic Behaviour—A New Theory," *Indian Journal of Economics* 30 (1950): 349–65.

148. Marschak, "Neumann's and Morgenstern's New Approach to Static Economics," 114.

149. Herbert A. Simon, "Review of *Theory of Games and Economic Behaviour* by John von Neumann and Oskar Morgenstern," *American Journal of Sociology* 50, no. 6 (1945): 558–60.

150. Shubik, "Game Theory at Princeton," 151.

Chapter Three

1. Curtis LeMay and Bill Yenne, *Superfortress: The Boeing B-29 and American Air Power in World War II* (Yardley, PA: Westholme Publishing, 1988), 60.

2. OSRD Technical Report Number AMP 504-1-M14 contains the minutes of the committee meetings of Service Project AC-92: "Gunnery and Bombing Tactics of B-29 Planes (Bulletin Nos. 1–13)," AMG-Princeton, August 8, 1944, to February 13, 1945, Applied Mathematics Panel Technical Reports, National Archives II, College Park, MD (hereafter Applied Mathematics Panel Technical Reports).

3. Mirowski, *Machine Dreams*; and Mirowski, "When Games Grow Deadly Serious."

4. Leonard, *Von Neumann, Morgenstern, and the Creation of Game Theory*, 343.

5. See especially Robin E. Rider, "Operations Research and Game Theory: Early Connections," in *Toward a History of Game Theory*, ed. Weintraub, 225–39.

6. Nasar, *A Beautiful Mind*, especially chapters 16 and 20.

7. See John Isbell to Martin Shubik, July 15, 1976, Box 41, Folder: "Game Theory, 1962–1976," Morgenstern Papers.

8. The dean of American "industrial mathematics" in those days was Thornton C. Fry of Bell Telephone, whose period assessments of the state of his field are essential reading on the subject. It was only after the war that mathematicians working in industry would find professional representation in the form of the Society for Industrial and Applied Mathematics (1951). See Thornton C. Fry: "Mathematicians in Industry—the First 75 Years," *Science* 143, no. 3609 (1964): 934–38; "Mathematics as a Profession Today in Industry," *American Mathematical Monthly* 63, no. 2 (1956): 71–80; and "Industrial Mathematics," *American Mathematical Monthly* 48, no. 6 (1941): 1–38.

9. Although much attention has been paid to the extraordinary careers of Jerzy Neyman, Harold Hotelling, and Samuel S. Wilks, the 1930s were more broadly a time of increased interest and growing demand for statisticians in academia; see, e.g., Churchill Eisenhart, "S. S. Wilks's Princeton Appointment, and Statistics at Princeton before Wilks," in *A Century of Mathematics in America*, vol. 3 (Providence, RI: American Mathematical Society, 1989).

10. See, e.g., Nathan Reingold, "Refugee Mathematicians in the United States of America, 1933–1941: Reception and Reflection," *Annals of Science* 38 (1981): 313–38. Robert Kohler suggests some reasons for mathematicians' embrace of "purity" in "The Ph.D. Machine: Building on the Collegiate Base," *Isis* 81, no. 4 (1990): 638–62. But as Isbell's story suggests, "purity" was not only an artifact of institutional configurations, but was also essential to bounding mathematics and politics; see especially Kevles, *The Physicists*, 4, "Pure Science and Practical Politics," on the disciplinary significance of purity among American physicists.

11. See especially John W. Servos, "Mathematics and the Physical Sciences in America, 1880–1930," *Isis* 77, no. 4 (1986): 611–29.

12. On the composition and role of mathematicians in OR units see Charles W. McArthur, *Operations Analysis in the U.S. Army Eighth Air Force in World War II* (Providence, RI: American Mathematical Society, 1990), esp. chap. 1. On the origins and transatlantic migration of OR, see especially James Phinney Baxter, *Scientists Against Time* (Boston: Little, Brown, 1946); Mirowski, *Machine Dreams*; Maurice W. Kirby, *Operational Research in War and Peace: The British Experience from the 1930s to 1970s* (London: Imperial College Press, 2003); William Thomas, *Rational Action: The Sciences of Policy in Britain and America, 1940–1960* (Cambridge, MA: MIT Press, 2015).

13. Quoted in Larry Owens, "Mathematicians at War: Warren Weaver and the Applied Mathematics Panel, 1942–1945," in *History of Modern Mathematics*, ed. David Rowe and John McCleary (Boston: Academic Press, 1989), 296.

14. Phyllis Fox, "Mina Rees," in *Women of Mathematics: A Bio-Bibliographic*

Sourcebook, ed. Louise Grinstein and Paul Campbell (Westport, CT: Greenwood Press, 1987), 178.

15. Flood's experiences as consultant led to his only direct contribution to the economics literature in Merrill M. Flood, "Recursive Methods in Business-Cycle Analysis," *Econometrica* 8, no. 4 (1940): 333–53. On the mathematical clinic and Flood's company, see, e.g., Merrill Flood to Harold Dodds, December 12, 1942, in Box 1, Folder: "Correspondence: Classes 1942–46," Merrill M. Flood Papers (hereafter Flood Papers), Bentley Historical Library, University of Michigan, Ann Arbor.

16. Weaver to Rear Admiral J. A. Furer, July 23, 1942, in Entry 83: "Office Files of Warren Weaver," Box 4, Folder: "Navy Department," RG227, Office of Scientific Research and Development Records (hereafter OSRD Records), National Archives II, College Park, MD.

17. On the wartime computing programs and their prewar predecessors, see David Alan Grier, *When Computers Were Human* (Princeton, NJ: Princeton University Press, 2005).

18. "Foreword," in Office of Scientific Research and Development, National Defense Research Committee, *Mathematical Studies Relating to Military Physical Research*, vol. 1 of *Summary Technical Report of the Applied Mathematics Panel* (Washington, DC, 1946), vii–viii.

19. Philip M. Morse and George E. Kimball, *Methods of Operations Research* (New York: Wiley, 1951).

20. For an account of the conference, see "Diary of CE and MRH" in "Conference on B-29 Airplanes," AMG Report No. 192, AMG-Columbia, in Applied Mathematics Panel Technical Reports AMP-504.1-M12. The "Diary" account of the conference refers to a "J. Workman," but subsequent memoranda connected with the eventual project AC-92 refer to "E. J. Workman."

21. Ibid., 5.

22. For an outline of the directive of the contract, see "Memorandum for First Meeting, Steering Committee, AC-92," July 26, 1944, Bulletin no. 1, Project AC-92, in Applied Mathematics Panel Technical Reports, AMP-504.1-M14.

23. Merrill M. Flood, "Bulletin No. 1, Project AC-92," in Applied Mathematics Panel Technical Reports, AMP-504.1-M14.

24. This study is described in Charles W. Bray, "Training the B-29 Gunner," in *Human Factors in Military Efficiency: Training and Equipment*, vol. 2 of *Summary Technical Report of the Applied Psychology Panel, NDRC* (Washington DC, 1946). See also Bray, *Psychology and Military Proficiency: A History of the Applied Psychology Panel of the National Defense Research Committee* (Princeton, NJ: Princeton University Press, 1948), esp. chaps. 6–7.

25. An early delegation of tasks is laid out in "Memorandum for First Meeting, Steering Committee, AC-92."

26. See Owens, "Mathematicians at War," 293.

27. According to Flood, the original memo appeared as "Memorandum #8, Fire Control Research Office, Princeton University," and was later published in revised form after the war as M. M. Flood, "Aerial Bombing Tactics: General Considerations (A World War II Study)," RAND RM-913, September 2, 1952.

28. Flood, "Aerial Bombing Tactics," cited on pp. 1, 22, v.

29. Ibid., 1.

30. Ibid., iv.

31. "Analytical Studies in Aerial Warfare," ed. Mina Rees, in vol. 2 of Office of Scientific Research and Development, National Defense Research Committee, *Summary Technical Report of the Applied Mathematics Panel* (Washington DC, 1946), cited on 198 and 202.

32. Ibid., 202.

33. Ibid., 200.

34. Ibid., 203.

35. Warren Weaver, circular letter of December 14, 1944, Entry 154: "Applied Mathematics Panel: Records of Consultants and Aides, 1942–1946," Box 8, Folder: "Records of Certain Consultants and Aides to the Division," RG 227, OSRD Records.

36. Veblen to Weaver, December 17, 1945, in Box 8, Folder: "Records of Certain Consultants and Aides to the Division," RG 227, OSRD Records.

37. Paxson to Weaver, December 19, 1945, in Box 8, Folder: "Records of Certain Consultants and Aides to the Division," RG 227, OSRD Records.

38. Martin J. Collins, *Cold War Laboratory: RAND, the Air Force, and the American State, 1945–1950* (Washington, DC: Smithsonian Institution Press, 2002), 116-199; see also Weaver to Bowles, February 28, 1945, both in Box 3, Folder: "AMP—WW Correspondence," RG 227, OSRD Records.

39. The text of the original contract can be found in "Letter Contract No. W 33-038 ac-14105" (March 2, 1946) in "Conference of Social Scientists," Box 2, RAND Archives, RAND Corporation, Santa Monica, CA (hereafter RAND Archives). The literature on the development of RAND is by now extensive; see *inter alia* Abella, *Soldiers of Reason*; Martin J. Collins, *Cold War Laboratory: RAND, the Air Force, and the American State, 1945–1950*, Smithsonian History of Aviation and Spaceflight (Washington, DC: Smithsonian Institution Press, 2002); David Hounshell, "The Cold War, RAND, and the Generation of Knowledge, 1946–1962," *Historical Studies in the Physical and Biological Sciences* 27 (1997): 237–67; Paul Dickson, *Think Tanks* (New York: Athenaeum, 1971); Bruce L. R. Smith, *The Rand Corporation: Case Study of a Nonprofit Advisory Corporation* (Cambridge, MA: Harvard University Press, 1966).

40. Vaughn D. Bornet, "John Williams: A Personal Reminiscence," RAND D-19036, August 12, 1969, 1, in Box 1, John Williams Papers, RAND.

41. Thanks to Dr. Helmer (interview with the author, April 28, 2006) for this story.

42. Collins, *Cold War Laboratory*, 124–25.

43. Detailed notes on the conference, as well as transcripts of key presentations, are preserved in "Project RAND Conference of Social Scientists, September 14 to 19, 1947—New York," RAND R-106, June 9, 1948.

44. Ibid., 6. The theme of "organized complexity" was a significant one in Weaver's writings throughout the postwar period; see, for instance, Weaver's comments on operations research and interdisciplinary research teams in Warren Weaver, "Science and Complexity," *American Scientist* 36, no. 4 (1948): 536–44.

45. "Project RAND Conference of Social Scientists, September 14 to 19, 1947—New York," 6–7.

46. Ibid., 3.

47. Ibid., 8.

48. Williams to von Neumann, December 16, 1947, in Box 17, Folder: "Rand Corp 1947–55," von Neumann Papers.

49. Von Neumann to Williams, January 9, 1948, in ibid.

50. On von Neumann's visits to RAND, see Leonard, *Von Neumann, Morgenstern, and the Creation of Game Theory*, chap. 13.

51. See "Articles of Incorporation, 10 May 1948," in Box 4, RAND Archives.

52. R. D. Specht, "RAND: A Personal View of Its History," RAND P-1601 (October 23, 1958), 11.

53. "The RAND Corporation: Organization, 15 May 1948," Box 4, RAND Archives.

54. On publications, see R. L. Belzer, "Silent Duels, Specified Accuracies, One Bullet," RM-58 (October 18, 1948); L. Shapley, "Note on a Duel Game," RM-82 (December 28, 1948); R. Bellman and M. Girshick, "An Extension of Results on Duels with Two Opponents, One Bullet Each, Silent Guns, Equal Accuracy," RM-108 (February 22, 1949); L. S. Shapley, "Note on Duels with Continuous Firing," RM-118 (March 11, 1949); D. Blackwell, "Noisy Duel, One Bullet Each, Arbitrary Non-Monitone Accuracy," RM-131 (March 30, 1949); M. A. Girshick, "A Generalization of the Silent Duel, Two Opponents, One Bullet Each, Arbitrary Accuracy," RM-206 (August 1, 1949); and M. Girshick and D. Blackwell, "A Loud Duel with Equal Accuracy Where Each Duelist Has Only a Probability of Possessing a Bullet," RM-219 (August 19, 1949). On reconnaissance missions, see R. L. Belzer, "Solutions of a Special Reconnaissance Game," RM-202 (August 10, 1949); Lloyd Shapley, "A Tactical Reconnaissance Model," RM-205 (August 1949); H. Bohnenblust, L. Shapley, and S. Sherman, "Reconnaissance in Game Theory," RM-208 (August 12, 1949); S. Sherman, "Total Reconnaissance with Total Countermeasures; Simplified Model" P-106 (August 5, 1949).

55. See, e.g., J. Barkeley Rosser to Stephen C. Kleene, August 19, 1969, copy in possession of the author. For reviews of this literature see, e.g., Mel-

vin Dresher, "Games of Strategy," *Mathematics Magazine* 25 (1951): 93–99, and T. E. Caywood and C. J. Thomas, "Applications of Game Theory in Fighter Versus Bomber Combat," *Journal of the Operations Research Society of America* 3, no. 4 (1955): 402–11.

56. See H. Bohnenblust, M. Dresher, M. A. Girshick, T. Harris, O. Helmer, J. C. C. McKinsey, L. Shapley, and P. Snow, "Contributions to the Theory of Games," parts 1–5 (RAND RM-29, February 18, 1948), and Helmer, "Recent Developments in the Mathematical Theory of Games" (RAOP-16, April 30, 1948) (Lecture to be delivered at the meeting of the Institute for Mathematical Statistics, June 1948). For an overview of RAND's unclassified work in mathematics, see Mathematics Division, "List of Unclassified Mathematics Division Publications, Including Related Reports from Other Divisions," RAND RM-950 (September 16, 1952).

57. Helmer, "Recent Developments in the Mathematical Theory of Games," 16–18.

58. See especially Aaron L. Friedberg, *In the Shadow of the Garrison State: America's Anti-Statism and Its Cold War Grand Strategy*, Princeton Studies in International History and Politics (Princeton, NJ: Princeton University Press, 2000), on the budgetary politics of postwar demobilization.

59. E. W. Rawlings to General Spaatz, March 6, 1947, US Air Force Deputy Chief of Staff, Comptroller, Director of Management Analysis, Records, Entry 106, Box 697, Cost Control Division Subject File, 1944–50, Folder: "Administrative," RG 341, National Archives II, College Park MD. My understanding of the significance of figure 3.2 and its relationship with the programming activities at the Pentagon owes much to Judy L. Klein, "The Carnegie Institute of Technology's Cold War Hot House for Management Science, Rational Expectations, and Bounded Rationality," Max-Planck-Institut für Wissenschaftsgeschichte, Berlin (March 15–16, 2010); and Klein, "Cold War Dynamic Programming and the Science of Economizing: Bellman Strikes Gold in Policy Space," History of Science Society, Crystal City, VA, November 3, 2007.

60. George B. Dantzig, "Reminiscences about the Origins of Linear Programming," *Operations Research Letters* 1 (1982): 43–48.

61. Ibid., 45. This story is related in the many extant interviews with Dantzig; see the introduction to George B. Dantzig, *Linear Programming and Extensions* (Princeton, NJ: Princeton University Press, 1963); Dantzig, "Linear Programming," in *History of Mathematical Programming: A Collection of Personal Reminiscences*, ed. Jan Lenstra, Alexander Rinnooy Kan, and Alexander Schrijver (Amsterdam: North-Holland, 1991), 19–31; Robert Dorfman, "The Discovery of Linear Programming," *Annals of the History of Computing* 6 (1984): 283–95. An alternative history, founded in the history of mathematical economics, is sketched in T. C. Koopmans, "Introduction," in Koopmans, *Activity Analysis of Production and Allocation* (New York: Wiley, 1951).

62. Recall that in this paper von Neumann pointed out that his proof the existence of solutions to the systems of linear inequalities that defined economic equilibrium was similar to his 1928 proof of the existence of an optimal strategy to the two-person zero-sum game.

63. See, e.g., Dantzig, "Reminiscences about the Origins of Linear Programming," 45.

64. "Addresses of Those in Attendance at the Theory of Planning Colloquium, July 1948, Project RAND, Santa Monica, California," in Box 4, Folder: "Project RAND Seminar on Game Theory Summer 1948," Edwin Paxson Papers, RAND Corporation, Santa Monica, CA.

65. Morgenstern, "Note on the Formulation of the Study of Logistics," RM-614 (May 28, 1951); and "Prolegomena to a Theory of Organization," RM-734 (December 10, 1951).

66. See "The Econometric Society: Announcement of American Summer Session, Madison, Wisconsin, September 7–10, 1948," in Box 24, File: "Economics and the Theory of Games, 1948," Morgenstern Papers.

67. See Christ, "History of the Cowles Commission, 1932–1952," 46; Cowles Commission, "Report for Period January 1, 1948–June 30, 1949," Cowles Commission, Chicago, 1949. On Cowles's work for RAND, see Mirowski, *Machine Dreams,* chap. 5.

68. A significant exception is Tinne Hoff Kjeldsen, "The Emergence of Nonlinear Programming: Interactions between Practical Mathematics and Mathematics Proper," *Mathematical Intelligencer* 22, no. 3 (2000): 50–54; Kjeldsen, "A Contextualized Historical Analysis of the Kuhn-Tucker Theorem in Nonlinear Programming: The Impact of World War II," *Historia Mathematica* 27 (2000): 331–61. The standard reference on the history of the ONR is Harvey Sapolsky, *Science and the Navy: The History of the Office of Naval Research* (Princeton, NJ: Princeton University Press, 1990).

69. Mina Rees, Letter to Code 102 (Director of Science Division), November 1950, Entry 10: "Office of the Chief of Naval Research, 1947–56," Box 2, Folder: "Research Centers," RG 298, Office of Naval Research Records (hereafter ONR Records), National Archives II, College Park, MD.

70. I have not found any records detailing the precise agenda of the meeting; see Fox, "Mina Rees," 178, on its timing.

71. The original document is referenced as G. B. Document no. 425, Serial no. 315-X.

72. Barkley Rosser to General Board, USN, April 26, 1948, Box 63, Folder: "US Department of the Navy General Board," Morgenstern Papers.

73. Ibid., appendix I.

74. Ibid., appendix II.

75. See Morgenstern's "Budget ONR, April 1948," Box 63, Folder: "U.S. Department of the Navy General Board," Morgenstern Papers. For reminiscences

about Morgenstern's relationship with the ONR, see Martin Shubik, "Oskar Morgenstern: Mentor and Friend," *International Journal of Game Theory* 7, nos. 3/4 (1978): 131–35.

76. Morgenstern to Goldstein, February 15, 1949, Box 51, Folder: "RAND Corporation, 1948–1951," Morgenstern Papers.

77. Morgenstern to Hitch, November 19, 1948, and Hitch to Morgenstern, November 24, 1949 (L-535), in ibid.

78. Morgenstern to King, June 27, 1949, in ibid.

79. See especially "Note on the Formulation of the Study of Logistics," RM-614 (May 28, 1951), and "Prolegomena to a Theory of Organization."

80. See Rees-Lefschetz correspondence regarding grant N6ori-105 in Entry no. 1: "Mathematics Division Project Case Files (Contract Correspondence) 1945–1952," Box 1, Folder: "Princeton University: Correspondence through Dec 1949," RG 298, ONR Records.

81. Rees to Tucker, May 21, 1948, Entry no. 1: "Mathematics Division Project Case Files (Contract Correspondence) 1945–1952," Box 1, Folder: "Princeton University: Correspondence through Dec 1949," RG 298, ONR Records.

82. Rees to Lefschetz, May 18, 1948, in ibid.

83. Tucker to Dr. Fred Rigby, Head, Logistics Branch, Office of Naval Research, May 6, 1949, in ibid.

84. In his letter to Rees of October 29, 1948, Tucker asked for permission to give the paper, although he suspected that there would be no classification issues. See also Tucker to Fred Rigby, December 24, 1948, in ibid., on the paper.

85. See Tucker to Rigby, December 24, 1948; Rigby to Tucker, January 13, 1949, in ibid.

86. Tucker to Rees, January 7, 1949; see also Tucker to Rigby, December 24, 1948, in ibid: "It looks as though Mr. Gale will have a nice doctor's thesis in this theory of games."

87. Tucker to Dantzig, March 16, 1949, in ibid.

88. Mina Rees, "The Computing Program of the Office of Naval Research, 1946–1953," *Annals of the History of Computing* 4, no. 2 (1982): 102–20.

89. Harold Kuhn, *Lectures in the Theory of Games*, Annals of Mathematics Studies (Princeton, NJ: Princeton University Press, 2003).

90. *Contributions to the Theory of Games*, vol. 1, ed. Harold W. Kuhn and Albert W. Tucker, Annals of Mathematics Studies 24 (Princeton, NJ: Princeton University Press, 1950), x. A draft of the preface can also be found in Box 5, Folder: "Kuhn, Harold, 1949–1953," von Neumann Papers.

91. See Robert Dorfman to Oskar Morgenstern, May 22, 1950 (and attachments), Box 42, Folder: "Game Theory and Linear Programming, 1950–1951," Morgenstern Papers.

92. See, e.g., Oliver A. Gross and R. A. Wagner, "A Continuous Blotto Game," RM-424 (June 1950); Oliver A. Gross, "The Symmetric Blotto Game,"

RM-424 (July 1950); Oliver A. Gross, "An Infinite-Dimensional Extension of a Symmetric Blotto Game," RM-718 (November 8, 1951); M. P. Peisakoff, "Continuous Blotto," RM-736 (December 4, 1951).

93. *Contributions to the Theory of Games*, 1:x.

94. John von Neumann, "A Certain Zero-Sum Two-Person Game Equivalent to the Optimal Assignment Problem," in *Contributions to the Theory of Games*, vol. 2, ed. Kuhn and Tucker, 5–12. See also George W. Brown, "Some Notes on Computation of Games Solutions," RAND P-78 (April 25, 1949); George W. Brown and John von Neumann, "Solutions of Games by Differential Equations" P-142 (April 19, 1950); Kenneth Arrow and L. Hurwicz, "A Gradient Method for Approximating Saddle-Points and Constrained Maxima" P-223 (June 13, 1951). See also George W. Brown, "Iterative Solutions of Games by Fictitious Play," in *Activity Analysis of Production and Allocation*, ed. Tjalling C. Koopmans (New York: Wiley, 1951), 374–76; and John von Neumann, "A Numerical Method to Determine Optimum Strategy," *Naval Research Logistics Quarterly* 1 (1954): 109–15.

95. Richard Bellman, *Dynamic Programming* (Princeton, NJ: Princeton University Press, 1957); Richard Bellman and Stuart E. Dreyfus, *Applied Dynamic Programming* (Princeton NJ: Princeton University Press, 1962). On the significant problems associated with adapting simplified programming models to the practical needs of the Air Force, especially given the limited computational techniques available in the later 1940s, see especially "The Bounded Rationality of Cold War Operations Research," in Erickson et al., *How Reason Almost Lost Its Mind*, 51–80.

96. A. W. Tucker and R. D. Luce, "Introduction," in *Contributions to the Theory of Games*, vol. 4, edited by A. W. Tucker and R. D. Luce (Princeton, NJ: Princeton University Press, 1959), 2.

97. Ibid., 3. A counterexample demonstrating that not all games have stable set solutions is found in W. F. Lucas, "The Proof that a Game May Not Have a Solution," RAND RM-5543-PR (January 1968).

98. L. S. Shapley, "Notes on the *n*-Person Game—II: The Value of an *n*-Person Game," RM-670 (August 21, 1951), 1.

99. See Lloyd S. Shapley, "A Value for *n*-Person Games," in *Contributions to the Theory of Games*, vol. 2, ed. Kuhn and Tucker, 307–17.

100. John F. Nash, "Equilibrium Points in *n*-Person Games," *Proceedings of the National Academy of Sciences of the United States of America* 36, no. 1 (1950): 48–49.

101. Nasar, *A Beautiful Mind*, 94. Martin Shubik remembers asking von Neumann about the Nash solution while returning to Princeton by train sometime either in 1952. Von Neumann expressed his disinterest. See Shubik, "Game Theory at Princeton," 155.

102. The chronology of events leading up to Nash's dissertation is given in Nasar, *A Beautiful Mind*, chap. 10.

103. Apparently, he did this at Tucker's insistence. See ibid., 96.

104. Ibid. Nash published a slightly modified version of his dissertation as John Nash, "Non-Cooperative Games," *Annals of Mathematics* 54, no. 2 (1951): 286–95.

105. John Nash, "Two-Person Cooperative Games," *Econometrica* 21, no. 1 (1953): 128–40.

106. Ibid.

107. See Nasar, *A Beautiful Mind,* especially chap. 10, "Nash's Rival Idea"; Giocoli, *Modeling Rational Agents*; and Schmidt, *Game Theory and Economic Analysis.*

108. Howard Raiffa, *Memoir: Analytical Roots of a Decision Scientist* (CreateSpace Independent Publishing Platform, 2011), 26.

109. T. S. Motzkin *and others,* "The Double Description Method," in *Contributions to the Theory of Games, vol. 2,* ed. Kuhn and Tucker, 51–74.

110. In general, there are seventy-eight "qualitatively different" two-person non-zero-sum games, i.e., games in which the players hold distinct patterns of preferences over the different outcomes afforded to them.

111. See Howard Raiffa, "Game Theory at the University of Michigan," in *Toward a History of Game Theory,* ed. Weintraub, 173.

112. In this instance at least, the solutions suggested by equilibrium analysis and the maximization treatment coincide.

113. Raiffa, "Game Theory at the University of Michigan," 174.

114. This section was eventually published as H. Raiffa, H. W. Kuhn, and A. W. Tucker, "Arbitration Schemes for Generalizing Two-Person Games," in *Contributions to the Theory of Games, vol. 2,* ed. Kuhn and Tucker, 361–88.

115. Nasar, *A Beautiful Mind,* 366. See also the discussion of the psychology of the Nash Equilibrium in Mirowski, *Machine Dreams,* 341–49.

116. Nasar, *A Beautiful Mind,* 94.

117. See, e.g., Martin Shubik, *Game Theory in the Social Sciences: Concepts and Solutions* (Cambridge, MA: MIT Press, 1982), for a review of this literature.

118. RAND Corporation, Annual Report for 1949, RAND Archives.

119. Poundstone, *Prisoner's Dilemma,* 168.

120. Mirowski, *Machine Dreams,* 319–30, provides a number of such quotes.

121. See Richard Bellman, *Eye of the Hurricane* (Singapore: World Scientific, 1984), especially chap. 15, "Guilt by Association—1954." Kahn's trials and travails are detailed in Ghamari-Tabrizi, *Worlds of Herman Kahn.* On the impact of anticommunism on American science more broadly, see Jessica Wang, *American Science in an Age of Anxiety: Scientists, Anticommunism, and the Cold War* (Chapel Hill: University of North Carolina Press, 1999).

122. These techniques have attracted significant historical attention in their own right; see, e.g., Ghamari-Tabrizi, "Simulating the Unthinkable," 163–223,

on war gaming; Kaya Tolon, "Futures Studies: A New Social Science Rooted in Cold War Strategic Thinking," in *Cold War Social Science: Knowledge Production, Liberal Democracy, and Human Nature*, ed. Mark Solovey and Hamilton Cravens (New York: Palgrave Macmillan, 2012), 45–62, on futures studies.

123. E. S. Quade and Wayne I. Boucher, eds., *Systems Analysis and Policy Planning: Applications in Defense* (New York: Elsevier, 1968), 2; the original quote was italicized. See also E. S. Quade, *Analysis for Military Decisions* (Chicago: Rand McNally, 1964) for a primer on systems analysis as practiced at RAND.

124. Melvin Dresher, "Mathematical Models of Conflict," in *Systems Analysis and Policy Planning*, ed. Quade and Boucher, 228.

125. See Collins, *Cold War Laboratory*, chap. 5.

126. See especially Rider, "Operations Research and Game Theory," 225–39, on the way that interest in programming rapidly crowded out game theory in the OR literature.

127. See Mirowski, *Machine Dreams*, 260–61; similar connections have been noted in literature on the history of computing; see especially Edwards, *Closed World*. The connection between computers, linear programming, and a culture of "closed worlds" is well borne out by the later history of programming, for instance, in the McNamara Pentagon with the Programming Planning and Budgeting System (PPBS) described in Gregory Palmer, *The McNamara Strategy and the Vietnam War: Program Budgeting in the Pentagon, 1960–1968*, Contributions in Political Science 13 (Westport, CT: Greenwood Press, 1978), and George Dantzig's later plans for scientifically managed cities in George B. Dantzig and Thomas L. Saaty, *Compact City: A Plan for a Livable Urban Environment* (San Francisco: W. H. Freeman, 1973).

128. See, e.g., "Pushbutton Planning: In Sight for Businessmen," *Business Week* 15 (December 1951): 95–103; "Industry Bristles at Robot Planning," *Business Week* (January 5, 1952): 23; and Kurt Singer, "Robot Economics," *Economic Record* 25 (July 1949): 48–73.

129. See especially William Baumol, *Economic Theory and Operations Analysis* (Englewood Cliffs, NJ: Prentice-Hall, 1961); S. Vajda, *An Introduction to Linear Programming and the Theory of Games* (New York: Wiley, 1960), and R. G. D. Allen, *Mathematical Economics* (London: Macmillan, 1956). The history of the relationship between programming and economic theory is far beyond the scope of this chapter, although for fascinating speculation on the future of the role of linear programming theory in economics from the perspective of the 1940s, see Paul A. Samuelson, "Market Mechanisms and Maximization," RAND P-69 (March 28, 1949).

130. R. Duncan Luce and Howard Raiffa, *Games and Decisions* (New York: Wiley, 1957), 57.

Chapter Four

1. For a survey of these developments, see especially Angela M. O'Rand, "Mathematizing Social Science in the 1950s: The Early Development and Diffusion of Game Theory," in *Toward a History of Game Theory*, ed. Weintraub, 177–204.

2. D. Wade Hands, "On Operationalisms and Economics," *Journal of Economic Issues* 38, no. 4 (2004): 953–68; Paul A. Samuelson, "Consumption Theory in Terms of Revealed Preference," *Economica* 15, no. 60 (1948): 243–53; Samuelson, "The Empirical Implications of Utility Analysis," *Econometrica* 6, no. 4 (1938): 344–56; Samuelson, "A Note on the Pure Theory of Consumer's Behaviour," *Economica* 5, no. 17 (1938): 61–71.

3. Isaac, *Working Knowledge*, chap. 3; S. S. Stevens, "On the Theory of Scales of Measurement," *Science, n.s.* 103, no. 2684 (1946): 677–80; Stevens, "The Operational Definition of Psychological Concepts," *Psychological Review* 42, no. 6 (1935): 517; Stevens, "The Operational Basis of Psychology," *American Journal of Psychology* 47, no. 2 (1935): 323–30.

4. See, e.g., Rebecca Lemov, "'Hypothetical Machines': The Science Fiction Dreams of Cold War Social Science," *Isis* 101, no. 2 (2010): 401–11, on sociologist Robert K. Merton's imagined "introspectometer," which might give interviewers direct access to a subject's inner world.

5. See, e.g., Sarah Elizabeth Igo, *The Averaged American: Surveys, Citizens, and the Making of a Mass Public* (Cambridge, MA: Harvard University Press, 2007), on opinion polling in mid-twentieth-century American culture.

6. Cohen-Cole, *The Open Mind*, 142. See also Gerd Gigerenzer, "Where Do New Ideas Come From? A Heuristic of Discovery in the Cognitive Sciences," in *Observation and Experiment in the Natural and Social Sciences*, ed. M. C. Galavotti (Dordrecht: Kluwer, 2003), 99–139; Gigerenzer, "From Tools to Theories: A Heuristic of Discovery in Cognitive Psychology," *Psychological Review* 98, no. 2 (1991): 254–67.

7. See, e.g., Joel Isaac, "Epistemic Design: Theory and Data in Harvard's Department of Social Relations," in *Cold War Social Science, ed.* Solovey and Cravens, 79–95; and Isaac, *Working Knowledge*.

8. See, e.g., von Neumann, "On the Theory of Games of Strategy," 42.

9. See, e.g., John von Neumann to A. F. Carpenter, November 29, 1939, and March 29, 1940, Box 3, Folder 2: "C Miscellany," von Neumann Papers.

10. On the history of Hex, see Martin Gardner, *The Scientific American Book of Mathematical Puzzles and Diversions* (New York: Simon and Schuster, 1959). The game was originally invented by Danish engineer Piet Hein in 1942 and was produced commercially by Parker Brothers in 1952. As it turns out, the proof that the first player can always win is equivalent to the proof of the Brouwer fixed-point theorem; see David Gale, "The Game of Hex and the Brou-

wer Fixed-Point Theorem," *American Mathematical Monthly* 86, no. 10 (1979): 818–27.

11. The descriptions of the tournaments can be found in the internal news-letter of the mathematics division, *RANDom Notes*, RAND Archives. For par-ticipant recollections of kriegsspiel at RAND, see Richard Bellman, "RAND, Summer of 1948," in Richard Bellman, *Eye of the Hurricane*.

12. Flood to Mendel, January 25, 1946, Box 1, Folder: "Correspondence—Mendel, Clifford," Flood Papers.

13. See Flood, "The Score at Bridge," in ibid.

14. Bellman, "On a Simple Game Possessing Some of the Qualities of Poker and Blackjack, I," RM-139 (April 18, 1949); Bellman, "On a Simple Game Pos-sessing Some of the Qualities of Poker and Blackjack, II," RM-140 (April 20, 1949); and Bellman and David Blackwell, "On a Simple Game Possessing Some of the Qualities of Poker and Blackjack, III," RM-147 (April 25, 1949).

15. J. F. Nash, "Non-Cooperative Games" (Ph.D. diss., Princeton University, 1950), 17–20.

16. Lloyd Shapley, "'So Long Sucker'—A Four-Person Game," Box 42, Folder: "Games—New Material, 1950–1973," Morgenstern Papers.

17. Nasar, *A Beautiful Mind*, 102.

18. Merrill M. Flood, "Some Experimental Games" RAND RM-789-1 (re-vised June 20, 1952), 5.

19. Ibid., 6.

20. This story is retold in Poundstone, *Prisoner's Dilemma*, 101–3. The origi-nal source is Flood, "Some Experimental Games"; the quote is on 16.

21. See, e.g., Ariel Rubinstein, "Comments on the Interpretation of Game Theory," *Econometrica* 59, no. 4 (1991): 909–24, and S. Abu Turab Rizvi, "Game Theory to the Rescue?," *Contributions to Political Economy* 13 (1994): 1–28.

22. Flood, "Some Experimental Games," 2. Among game theorists, the behavior of children would become a frequent point of reference; see, e.g., Thomas C. Schelling, *The Strategy of Conflict* (Cambridge, MA: Harvard Uni-versity Press, 1960), 11–13, on deterrence in child-rearing.

23. See, e.g., Merrill Flood, "Illustrative Example of Application of Koop-mans' Transportation Theory to Scheduling Military Tanker Fleet," RM-267 (October 17, 1949).

24. See Tucker to Flood, "A Two-Person Dilemma," Box 1, Folder: "Notes, 1929–1967," Flood Papers.

25. Ibid.

26. The players in question were apparently UCLA economist Armen Al-chian and John Williams; see Poundstone, *Prisoner's Dilemma*, 106.

27. Flood, "Some Experimental Games," 24.

28. Ibid., 24n.

29. The author has seen at least one RAND draft memo on the topic by Olaf Helmer and Norman Dalkey from June 1950.

30. M. M. Flood, "A Preference Experiment," P-258 (December 5, 1951).

31. See, e.g., Edwards, *Closed World*; Mirowski, *Machine Dreams;* Galison, "Ontology of the Enemy."

32. Wiener, *Cybernetics.*

33. Arturo Rosenblueth, Norbert Wiener, and Julian Bigelow, "Behavior, Purpose and Teleology," *Philosophy of Science* 10, no. 1 (1943): 18–24.

34. Pamela McCorduck, *Machines Who Think: A Personal Inquiry Into the History and Prospects of Artificial Intelligence* (San Francisco: W. H. Freeman, 1979), 66 (quoted in Edwards, *Closed World*, 188)

35. On McCulloch and Pitts's analogies between neurons and circuits, see Walter Pitts, "Some Observations on the Simple Neuron Circuit," *Bulletin of Mathematical Biophysics* 4, no. 3 (1942): 121–29; Warren McCulloch and Walter Pitts, "A Logical Calculus of Ideas Immanent in Nervous Activity," *Bulletin of Mathematical Biophysics* 5 (1943): 115–33.

36. C. E. Shannon, "A Mathematical Theory of Communication," *Bell System Technical Journal* 27 (1948): 379–423, 623–56; for the classic work on the history of the Macy Conferences and their aftermath, see Steve J. Heims, *The Cybernetics Group* (Cambridge, MA: MIT Press, 1991).

37. On the history of the Psycho-Acoustic Laboratory, see Edwards, *Closed World*, chap. 4, "The Machine in the Middle." On Harvard's funding for fields in this area, see Harvard University, *The Behavioral Sciences at Harvard: Report by a Faculty Committee* (Cambridge, MA: Harvard University, 1954); on ONR spending on the behavioral sciences, see Sapolsky, *Science and the Navy*, and Ron Robin, *The Making of the Cold War Enemy: Culture and Politics in the Military-Intellectual Complex* (Princeton, NJ: Princeton University Press, 2001), 51.

38. On von Neumann's involvement with the Macy Conferences, see Heims, *John von Neumann and Norbert Wiener*, 203.

39. Wiener, *Cybernetics*, 159–60.

40. John von Neumann, *The Computer and the Brain* (New Haven, CT: Yale University Press, 1958). On the differences between Wiener's and von Neumann's perspectives on game theory, see Heims, *John von Neumann and Norbert Wiener*, 153, 307–15.

41. "Project RAND Conference of Social Scientists, September 14 to 19, 1947—New York," iii.

42. For a discussion of the application of such social-scientific techniques to national security issues see "Panel II: Attitudes and Opinions" in ibid.

43. Ruth Benedict, *The Chrysanthemum and the Sword: Patterns of Japanese Culture* (Boston: Houghton Mifflin, 1946).

44. Robin, *The Making of the Cold War Enemy*, 10.

45. Andrew D. Grossman, *Neither Dead nor Red: Civil Defense and Amer-*

ican Political Development during the Early Cold War (New York: Routledge, 2001), 60.

46. On the history of the foundation, see Francis X. Sutton, "The Ford Foundation: The Early Years," *Daedalus* 116, no. 1 (1987): 41–91; Kathleen D. McCarthy, "From Cold War to Cultural Development: The International Cultural Activities of the Ford Foundation, 1950–1980," *Daedalus* 116, no. 1 (1987): 93–117; John Caldwell and Pat Caldwell, *Limiting Population Growth and the Ford Foundation Contribution* (London: F. Pinter, 1986); Richard Magat, *The Ford Foundation at Work: Philanthropic Choices, Methods, and Styles* (New York: Plenum Press, 1979); Francis X. Sutton, "The Ford Foundation's Development Program in Africa," *African Studies Bulletin* 3, no. 4 (1960): 1–7; Dwight Macdonald, *The Ford Foundation: The Men and the Millions* (New York: Reynal, 1956).

47. Ford Foundation, *Report of the Study for the Ford Foundation on Policy and Program* (Detroit, November 1949), 17–18.

48. Ibid., 90.

49. This dimension of Cold War psychology has become the focus of literature on civil defense programs in the 1950s; see, e.g., Robert Earnest Miller, *"The War that Never Came: Civilian Defense, Mobilization, and Morale during World War II"* (PhD diss., University of Cincinnati, 1991); and Grossman, *Neither Dead nor Red.*

50. On the conversion of the loan into a grant and the history of the foundation's involvement with RAND, see "Excerpt from Docket, July 15–16, 1953—Board of Trustees Meeting," Grant File 52-140, Ford Foundation Archives (hereafter FFA), New York.

51. See, e.g., Frank Collbohm to Henry T. Heald, November 21, 1957, Grant File 52-140, FFA.

52. On the origins of the term and its meaning in the context of the Ford Foundation, see especially Solovey, *Shaky Foundations*, chap. 3; and Pooley and Solovey, "Marginal to the Revolution."

53. "The Ford Foundation Behavioral Sciences Program Final Report, 1951–1957," Report #002074, FFA, 2.

54. The reports issued by this study include University of North Carolina, *Survey of Behavioral Science* (Chapel Hill: University of North Carolina, 1954); Stanford University Executive Committee on the Survey of the Behavioral Sciences, *The Stanford Survey of the Behavioral Sciences, 1953–1954: Report of the Executive Committee and Staff* (Stanford, CA: Stanford University, 1954); Harvard University, *The Behavioral Sciences at Harvard*; University of Michigan Committee on the Survey of the Behavioral Sciences, *Survey of the Behavioral Sciences: Report of the Faculty Committee and Report of the Visiting Committee* (Ann Arbor: University of Michigan, 1954); University of Chicago Behavioral Sciences Self-Study Committee, *A Report on the Behavioral Sciences at the University of Chicago* (Chicago: University of Chicago, 1954).

55. R. Duncan Luce, "R. Duncan Luce," in *A History of Psychology in Autobiography, vol. 8*, ed. Gardner Lindzey (Stanford, CA: Stanford University Press, 1989), 247.

56. On the founding of the Center, see Kurt Lewin, "The Research Center for Group Dynamics at Massachusetts Institute of Technology," *Sociometry* 8, no. 2 (1945): 126–36; for a more critical view on this history see J. L. Moreno, "How Kurt Lewin's 'Research Center for Group Dynamics' Started," *Sociometry* 16, no. 1 (1953): 101–4.

57. For a critical review of the literature in small group studies to 1953 see Fred L. Strodtbeck and A. Paul Hare, "Bibliography of Small Group Research (From 1900 Through 1953)," *Sociometry* 17, no. 2 (1954): 107–78, J. L. Moreno, "Old and New Trends in Sociometry: Turning Points in Small Group Research," *Sociometry* 17, no. 2 (1954): 179–93.

58. Kurt Lewin, Ronald Lippitt, and Ralph K. White, "Patterns of Aggressive Behavior in Experimentally Created 'Social Climates,'" *Journal of Social Psychology* 10 (1939): 271–99; Kurt Lewin and Ronald Lippitt, "An Experimental Approach to the Study of Autocracy and Democracy: A Preliminary Note," *Sociometry* 1, nos. 3/4 (1938): 292–300.

59. On Lewin's life and career more generally, see Alfred J. Marrow, *The Practical Theorist: The Life and Work of Kurt Lewin* (New York: Basic Books, 1969); Kurt Lewin and Martin Gold, *The Complete Social Scientist: A Kurt Lewin Reader* (Washington, DC: American Psychological Association, 1999); and Marvin Ross Weisbord, *Productive Workplaces Revisited: Dignity, Meaning, and Community in the 21st Century* (San Francisco: Jossey-Bass, 2004), esp. chaps. 4 and 5. For a snapshot to 1950 on the close connections that developed between Lewin's students and the ONR, see Harold S. Guetzkow, *Groups, Leadership and Men: Research in Human Relations: Reports on Research Sponsored by the Human Relations and Morale Branch of the Office of Naval Research, 1945–1950* (Pittsburgh, PA: Carnegie Press, 1951).

60. Kurt Lewin, *Field Theory in Social Science* (New York: Harper and Brothers, 1951), xi.

61. See, e.g., Kurt Lewin, Fritz Heider, and Grace M. Heider, *Principles of Topological Psychology* (New York: McGraw-Hill, 1936).

62. J. L. Moreno, *Who Shall Survive? A New Approach to the Problem of Human Interrelations* (Washington, DC: Nervous and Mental Disease Pub. Co., 1934).

63. This work later appeared as R. Duncan Luce and Albert D. Perry, "A Method of Matrix Analysis of Group Structure," *Psychometrika* 14, no. 1 (1949): 95–116; and R. Duncan Luce, "Connectivity and Generalized Cliques in Sociometric Group Structure," *Psychometrika* 15 (1950): 169–90.

64. Luce, "R. Duncan Luce," 248.

65. For highly sketchy accounts of Bavelas's work and its funding sources dur-

ing these years, see Stewart Brand, *The Media Lab: Inventing the Future at MIT* (New York: Viking, 1987), 135, and Christopher Simpson, *Universities and Empire: Money and Politics in the Social Sciences during the Cold War* (New York: New Press, 1998), 12, 23; specific sponsors for Luce's research during these years are mentioned in Josiah Macy Jr., Lee S. Christie, and R. Duncan Luce, "Coding Noise in Task-Oriented Groups," *Journal of Abnormal and Social Psychology* 48, no. 3 (1953): 401–9.

66. Luce, "R. Duncan Luce," 248. Bavelas's PhD dissertation (Alex Bavelas, "Some Mathematical Properties of Psychological Space" [MIT, Cambridge, MA, 1948]), sheds some light on the relationship between Lewin's vision for theory in social psychology and Bavelas's research in the later 1940s following Lewin's death.

67. Macy, Christie, and Luce, "Coding Noise in Task-Oriented Groups," 401.

68. Luce, "R. Duncan Luce," 267.

69. R. Duncan Luce, "Two Decomposition Theorems for a Class of Finite Oriented Graphs," *American Journal of Mathematics* 74, no. 3 (1952): 701–22; and R. Duncan Luce, "Networks Satisfying Minimality Conditions," *American Journal of Mathematics* 75, no. 4 (1953): 825–38.

70. Bavelas, "Some Mathematical Properties of Psychological Space"; Alex Bavelas, "Communication Patterns in Task-Oriented Groups," *Journal of the Acoustical Society of America* 22, no. 6 (1950): 725–30.

71. R. Duncan Luce, "Ψ-Stability: A New Equilibrium Concept for *n*-Person Game Theory," in *Mathematical Models of Human Behavior: Proceedings of a Symposium* (New York: Dunlap and Associates, 1955), 32.

72. See R. Duncan Luce, "A Definition of Stability for *n*-Person Games," *Annals of Mathematics* 59, no. 3 (1954): 357–66; see also Luce, "k-Stability of Symmetric and of Quota Games," *Annals of Mathematics* 62, no. 3 (1955): 517–27; and Luce, "Ψ-Stability," 32–44.

73. The results of this study were published as Paul F. Lazarsfeld, *Radio and the Printed Page: An Introduction to the Study of Radio and Its Role in the Communication of Ideas* (New York: Duell, Sloan and Pearce, 1940). On Lazarsfeld in America, see especially Paul Lazarsfeld, "An Episode in the History of Social Research: A Memoir," in *Intellectual Migration*, ed. Fleming and Bailyn, 270–337; Marjorie Fiske and Paul F. Lazarsfeld, "The Columbia Office of Radio Research," *Hollywood Quarterly* 1, no. 1 (1945): 51–59; David L. Sills, "Paul F. Lazarsfeld," in *Biographical Memoirs, vol. 56* (Washington DC: National Academy Press, 1987).

74. See Paul F. Lazarsfeld, "Recent Developments in Latent Structure Analysis," *Sociometry* 18, no. 4 (1955): 391–403. On the quantitative methodology developed at the BASR, see Lazarsfeld, "The Logical and Mathematical Foundation of Latent Structure Analysis," in Stouffer et al., *The American Soldier, vol. 4, Measurement and Prediction* (Princeton, NJ: Princeton University Press,

1949). On the distinctions between Lazarsfeld's "latent structure analysis" and more generalized factor analysis earlier promoted by the English psychologist Charles Spearman, see Bert F. Green Jr., "Latent Structure Analysis and Its Relation to Factor Analysis," *Journal of the American Statistical Association* 47, no. 257 (1952): 71–76.

75. Luce, "R. Duncan Luce," 268. See also *Developments in Mathematical Psychology: Information, Learning, and Tracking: A Study of the Behavioral Models Project, Bureau of Applied Social Research, Columbia University*, ed. R. Duncan Luce (Glencoe, IL: Free Press, 1960).

76. See, e.g., R. Duncan Luce, *Individual Choice Behavior: A Theoretical Analysis* (New York: Wiley, 1959); R. Duncan Luce and Ward Edwards, "The Derivation of Subjective Scales from Just Noticeable Differences," *Psychological Review* 65, no. 4 (1958): 222–37; R. Duncan Luce, "Semiorders and a Theory of Utility Discrimination," *Econometrica* 24, no. 2 (1956): 178–91; and R. Duncan Luce, D. H. Krantz, Patrick Suppes, and Amos Tversky, *Foundations of Measurement, vols. 1–3* (New York: Academic Press, 1971–89).

77. Luce and Raiffa, *Games and Decisions*, x.

78. Howard Raiffa and Stephen E. Fienberg, "The Early Statistical Years, 1947–1967: A Conversation with Howard Raiffa," *Statistical Science* 23, no. 1 (2008): 136–49; see also Raiffa, *A Memoir*, 39.

79. Letter to Alexander Ruthven, September 29, 1950, Grant File 50-255, FFA.

80. The planning process is described in a letter from James P. Adams to Paul G. Hoffman, August 9, 1951, Grant File 50-255, FFA.

81. "Interdisciplinary Program in the Application of Mathematics to the Behavioral Sciences; Report of the First Two and One-Half Years," November 23, 1953, Grant File 50-255, FFA.

82. Raiffa and Fienberg, "Early Statistical Years," 140.

83. "Research Seminar in Quantitative Economics Report, Academic Year 1951–1952," p. 1, Grant File 50-255, FFA.

84. C. H. Coombs, L. R. Klein, and R. M. Thrall, "Proposal for a Research Program in the Measurement of Values," p. 2, Grant File no. 52-98, FFA.

85. On Coombs, see Amos Tversky, "Clyde Hamilton Coombs 1912–1988," *National Academic of Sciences Biographical Memoirs* 61 (1992): 59–77.

86. Thurstone's first essay on the measurement of social values came in L. L. Thurstone, "The Method of Paired Comparisons for Social Values," *Journal of Abnormal and Social Psychology* 21 (1927): 384–400; the food studies are related in L. L. Thurstone, "An Experiment in the Prediction of Choice," in *Proceedings of the Fourth Research Conference* (Chicago: Council on Research, American Meat Institute, 1952); L. L. Thurstone, "Methods of Food-Tasting Experiments," in *Proceedings of the Second Conference on Research* (Chicago: American Meat Institute, March, 1950).

87. Thurstone's first foray into economics is found in L. L. Thurstone, "The

Indifference Function," *Journal of Social Psychology* 3 (1931): 139–67. For a retrospective assessment of Thurstone's pioneering application of psychophysical methods in social psychology, see C. H. Coombs, "Thurstone's Measurement of Social Values Revisited Forty Years Later," *Journal of Personality and Social Psychology* 6 (1967): 85–91.

88. Coombs explores the connections between traditional psychophysical analysis and the more general problem of psychological measurement in C. H. Coombs, "Mathematical Models in Psychological Scaling," *Journal of the Statistical Association* 46 (1951): 480–89; C. H. Coombs, "Psychological Scaling Without a Unit of Measurement," *Psychological Review* 57 (1950): 145–58.

89. Coombs, Klein, and Thrall, "Proposal for a Research Program in the Measurement of Values," 2.

90. Friedman and Savage, "Utility Analysis of Choices Involving Risk"; Marschak, "Rational Behavior, Uncertain Prospects, and Measurable Utility," 111–41; Marschak, "Why 'Should' Statisticians and Businessmen Maximize 'Moral Expectation'?"

91. Coombs, Klein, and Thrall, "Proposal for a Research Program in the Measurement of Values," 2.

92. Amadae, *Rationalizing Capitalist Democracy*, 85.

93. Kenneth J. Arrow, *Social Choice and Individual Values* (New Haven, CT: Yale University Press, 1963), 3–6.

94. Frederick Mosteller and Philip Nogee, "An Experimental Measurement of Utility," *Journal of Political Economy* 59, no. 5 (1951): 371–404. For recollections of the experiments see Frederick Mosteller, *The Pleasures of Statistics: The Autobiography of Frederick Mosteller* (New York: Springer Verlag, 2010), 197–98.

95. Coombs, Klein, and Thrall, "Proposal for a Research Program in the Measurement of Values," 2.

96. Ibid., 4.

97. The project is described in Robert L. Chapman *and others*, "The Systems Research Laboratory's Air Defense Experiments," *Management Science* 5, no. 3 (1959): 250–69; Robert L. Chapman and John L. Kennedy, "The Background and Implications of the Systems Research Laboratory Studies," in *Symposium on Air Force Human Engineering, Personnel, and Training Research*, National Academy of Sciences—National Research Council Publication (Washington, DC: National Academy of Sciences—National Research Council, 1956); John L. Kennedy, "The Uses and Limitations of Mathematical Models, Game Theory, and Systems Analysis in Planning and Problem Solution," in *Current Trends: Psychology in the World Emergency* (Pittsburgh, PA: University of Pittsburgh Press, 1952).

98. "Agenda Material, Seventh Semi-Annual Meeting, Board of Trustees, 6, 7, 8 October 1955," in Box 11, Folder: "RAND Corporation," Philip M. Morse Papers, Special Collections, Massachusetts Institute of Technology.

99. "Summer Seminar on Design of Experiments on Decision Processes, Autumn '52," p. 1, Grant File 52-98, FFA.

100. Ibid.

101. List of participants in *Decision Processes*, ed. Thrall, Coombs, and Davis, 329–30.

102. R. L. Davis, "Introduction," in ibid., 4.

103. Kalisch et al., "Some Experimental *n*-Person Games," in ibid., 301–27.

104. Leo A. Goodman, "On Methods of Amalgamation," in ibid., 39–48.

105. Paul J. Hoffman, Leon Festinger, and Douglas H. Lawrence, "Tendencies toward Group Comparability in Competitive Bargaining," in ibid., cited on 232.

106. See the following, all in ibid.: Flood, "Game-Learning Theory and Some Decision-Making Experiments," 139–58; Robert R. Bush, Frederick Mosteller, and G. L. Thompson, "A Formal Structure for Multiple-Choice Situations," 99–126; and Flood, "Environmental Non-Stationarity in a Sequential Decision-Making Experiment," 287–99. Flood collaborated with a number of statisticians and mathematical psychologists on learning models and the two-armed bandit in the early 1950s. See especially Flood to Mosteller, March 29, 1955, Box 1, Folder: "Correspondence—General, 1947–1987," Flood Papers, for a description of how the problem was invented and conveyed to RAND in the summer of 1951.

107. Jacob Marschak, "Towards an Economic Theory of Organization and Information," in *Decision Processes*, ed. Thrall, Coombs, and Davis, 187–220.

108. See Marschak and Roy Radner, "Structural and Operational Communication Processes in Teams, I," *Cowles Commission Discussion Papers*, Economics 2076 (abstracted in *Econometrica* [July 1953]: 485; Hurwicz, "Theory of Economic Organization," *Econometrica* (July 1951): 54; M. Beckmann, "On Marschak's Model of an Arbitrage Firm," *Econometrica* (April 1953): 347.

109. Alan Newell, "Session on Experimental Techniques: Experimental Study of Organizations—3," June 26, 1952, Grant File 52-98, FFA.

110. Ward Edwards, review of *Decision Processes,* in *American Journal of Psychology* 68, no. 3 (September 1955): 505–7.

111. Bernard Berelson to C. H. Coombs, December 24, 1952, Grant File 52-98, FFA.

112. See C. H. Coombs, H. Raiffa, and R. M. Thrall, "Mathematical Models and Measurement Theory" in *Decision Processes*, ed. Thrall, Coombs, and Davis, 19–37.

113. See J. Marschak, "On Mathematics for Economists," *Review of Economics and Statistics* 29, no. 4 (1947): 269–73, for a review of the new theories—especially game theory—that seemed such promising additions to the social sciences after the war.

114. See William Madow to Bernard Berelson, "Program for Furthering the Mathematical Training of Social Scientists," n.d., p. 3, Grant File 53-1, FFA.

115. See "Docket Excerpt, Executive Committee Meeting, Behavioral Sciences," June 18, 1956, Grant File 53-1, FFA.

116. William Madow, "Report of the Committee on the Mathematical Training of Social Scientists (December, 1952 to June 15, 1953)," p. 4, Grant File no. 53-0001, FFA.

117. Ibid., 3.

118. Berelson to Coombs, December 24, 1952.

119. See Raiffa, *A Memoir*; and Raiffa and Fienberg, "Early Statistical Years," 142.

120. Luce and Raiffa, *Games and Decisions*, 6.

121. Ibid., 104.

122. Herbert A. Simon, "Review of *Games and Decisions: Introduction and Critical Survey*," *American Sociological Review* 23, no. 3 (1958): 342–43. For a critical assessment of Luce and Raiffa's judgement on the value of game theory in the social sciences, see M. H. Peston, "Review of *Games and Decisions: Introduction and Critical Survey*," *Economica* 27, no. 106 (1960): 185–87; on Simon's thinking in the wake of the conference, see Herbert A. Simon, "A Behavioral Model of Rational Choice," *Quarterly Journal of Economics* 69, no. 1 (1955): 99–118.

123. Luce and Raiffa, *Games and Decisions*, xi.

124. Hunter Heyck, "Producing Reason," in *Cold War Social Science*, ed. Solovey and Cravens, 99–116.

Chapter Five

1. McDonald, *Strategy in Poker, Business and War.*

2. Bertrand Russell, *Common Sense and Nuclear Warfare* (New York: Simon and Schuster, 1959). On the mathematization of Chicken, Glenn H. Snyder, "'Prisoner's Dilemma' and 'Chicken' Models in International Politics," *International Studies Quarterly* 15, no. 1 (1971): 66–103; Anatol Rapoport and Albert M. Chammah, "The Game of Chicken," *American Behavioral Scientist* 3 (1966): 10–28; Anatol Rapoport, "Chicken à la Kahn: A Critique of Herman Kahn, *on Escalation: Metaphors and Scenarios* (New York: Praeger, 1965)," *Virginia Quarterly Review* 41, no. 3 (1965): 370–89; Herman Kahn, *Thinking about the Unthinkable* (New York: Horizon Press, 1962).

3. See especially Thomas C. Schelling, *Arms and Influence* (New Haven, CT: Yale University Press, 1966); Schelling, *Strategy of Conflict*; Schelling, "An Essay on Bargaining," *American Economic Review* 46, no. 3 (1956): 281–306.

4. See, e.g., Graham Allison's classic study of the Cuban missile crisis in Graham T. Allison, *Essence of Decision: Explaining the Cuban Missile Crisis* (Boston: Little, Brown, 1971).

5. The literature that touches on the relationship between game theory and the evolution of US nuclear doctrine vast; see *inter alia* Lawrence Freedman, *The Evolution of Nuclear Strategy* (New York: St. Martin's Press, 1981); Kaplan, *The Wizards of Armageddon*; Marc Trachtenberg, *History and Strategy* (Princeton, NJ: Princeton University Press, 1991); Andrew David May, "The RAND Corporation and the Dynamics of American Strategic Thought, 1946–1962" (PhD diss., Emory University, 1998); Robert Ayson, *Thomas Schelling and the Nuclear Age* (New York: Frank Cass, 2004); Ghamari-Tabrizi, *Worlds of Herman Kahn*; Abella, *Soldiers of Reason*.

6. See James G. Miller, "Toward a General Theory for the Behavioral Sciences," in *The State of the Social Sciences*, ed. Leonard D. White (Chicago: University of Chicago Press, 1956), 29.

7. Debora Hammond, *The Science of Synthesis* (Boulder: University Press of Colorado, 2003), relates Miller's recollections of Marquis' involvement in guiding the committee (see especially 168).

8. Grant File 50-123, FFA.

9. Grant File 53-90, FFA.

10. A somewhat patchy collection of the committee's meeting minutes can be found in Box 3, Folder: "Committee on Behavioral Sciences," Nicholas Rashevsky Papers (hereafter Rashevsky Papers), Regenstein Library, University of Chicago. The faculty roster from 1953 can be found in the same folder.

11. N. Rashevsky, *Mathematical Theory of Human Relations* (Bloomington, IN: Principia Press, 1947). On Rashevsky, see Tara Abraham, "Nicholas Rashevsky's Mathematical Biophysics," *Journal of the History of Biology* 37 (2004): 333–85.

12. Donald Marquis, "Scientific Methodology in Human Relations," in *Experiments in Social Process*, ed. J. G. Miller (New York: McGraw-Hill, 1950), 3.

13. Ibid., 12.

14. See Rapoport, "Cycle Distributions in Random Nets," *Bulletin of Mathematical Biophysics* 10 (1948): 145–57; "Outline of a Probabilistic Approach to Animal Sociology," pts. I–III, *Bulletin of Mathematical Biophysics* 11 (1949): 183–96, 273–81; "Nets with Distance Bias," *Bulletin of Mathematical Biophysics* 13 (1951): 85–91; "Spread of Information Through a Population with Socio-Structural Bias, I–III," *Bulletin of Mathematical Biophysics* 15 (1953): 523–33, 535–46, and 16 (1954): 75–81; and Anatol Rapoport and Lionel I. Rebhun, "On the Mathematical Theory of Rumor Spread," *Bulletin of Mathematical Biophysics* 14 (1952): 375–83.

15. Christ, "History of the Cowles Commission, 1932–1952," 64–65.

16. The subject matter overlapped substantially with the content of Marschak's 1950 publication on the topic; see Marschak, "Rational Behavior, Uncertain Prospects, and Measurable Utility."

17. Marschak, "Towards an Economic Theory of Organization and Information."

18. On the bargaining experiments, see Hoffman, Festinger, and Lawrence, "Tendencies Toward Group Comparability in Competitive Bargaining," 231–53. Extensive notes on Marschak's presentation can be found in the minutes of the meeting of the Committee on the Behavioral Sciences theory group, February 10, 1953 (Box 3, Folder: "Committee on Behavioral Sciences," Rashevsky Papers).

19. Rashevsky to Marschak, April 18, 1956, in Box 6, Folder: "Cowles Commission," Rashevsky Papers.

20. See especially Hammond, *Science of Synthesis*; and Gregg Mitman, *The State of Nature: Ecology, Community, and American Social Thought, 1900–1950* (Chicago: University of Chicago Press, 1992).

21. Miller, "Toward a General Theory for the Behavioral Sciences," 30. See also Committee on Behavioral Sciences, University of Chicago, *Symposium: Profits and Problems of Homeostatic Models in the Behavioral Sciences* (Chicago: Committee on Behavioral Sciences, 1954) for the proceedings of the committee's symposium on systems, and Roy R. Grinker, *Toward a Unified Theory of Human Behavior* (New York: Basic Books, 1956)

22. Committee on Behavioral Sciences Theory Group Minutes, January 13, 1953, Box 3, Folder: "Committee on Behavioral Sciences," Rashevsky Papers.

23. Aside from James Grier Miller, *Living Systems* (New York: McGraw-Hill, 1978), the best statements of the systems philosophy that guided the group at Chicago and Michigan can be found in James Grier Miller: "Living Systems: Basic Concepts," *Behavioral Science* 10 (1965): 193–237; "Living Systems: Structure and Process," *Behavioral Science* 10 (1965): 337–79; and "Living Systems: Cross-Level Hypotheses," *Behavioral Science* 10 (1965): 380–411. Miller discusses the most significant levels of organization (cell, organ, organism, group) in a series of articles from the early 1970s. See Miller: "Living Systems: The Cell," *Currents in Modern Biology* 4 (1971): 78–206; "Living Systems: The Organ," *Currents in Modern Biology* 4 (1971): 78–206; "Living Systems: The Organism," *Quarterly Review of Biology* 48, no. 2 (1971): 92–276; Miller, "Living Systems: The Group," *Behavioral Science* 16 (1971): 302–98; "Living Systems: The Society," *Behavioral Science* 20 (1975): 366–416.

24. See Hammond, *Science of Synthesis*, 169.

25. Anatol Rapoport, *Certainties and Doubts: A Philosophy of Life* (Toronto: Black Rose Books, 2000), 106.

26. Box 3, Folder: "Fundraising, 1949–1952," Rashevsky Papers.

27. Ibid., 5.

28. The transition is documented in the Cowles Commission Annual Reports for 1954 and 1955.

29. The events surrounding the establishment of the institute are outlined in a number of places. See, e.g., Box 1, Folder: "History—Background: First Ten Years," Mental Health Research Institute Records (hereafter MHRI Records), Bentley Library, University of Michigan–Ann Arbor; and "The First Ten Years: A Summary of Activities of the Mental Health Research Institute from Its Founding Through July, 1965," in *Mental Health Research Institute Tenth Annual Report* (Ann Arbor: University of Michigan, 1965).

30. On Boulding's longstanding involvement with behavioral science at Michigan, see Fontaine, "Stabilizing American Society."

31. Early faculty appointments are reported in the Mental Health Research Institute annual reports 1–3 (1957–59). Rejected offers and announcements of visitors and new arrivals can be found in the biweekly faculty meeting minutes in the MHRI records. See Box 2, Folder: "Academic Staff Meetings, 1956–1968," MHRI Records.

32. The rationale for the building's design is described in *Mental Health Research Institute Second Annual Report* (Ann Arbor: University of Michigan, 1958). The original architectural plans of the building can be found in Box 1, Folder: "Admin Files—Building: Architectural and Financial, 1956–1960," MHRI Records.

33. The scheduled times for these can be found in Box 2, Folder: "Academic Staff Meetings, 1956–1968," MHRI Records.

34. This was exclusive of support for buildings and equipment. See especially *Mental Health Research Institute Third Annual Report* (Ann Arbor: University of Michigan, 1959), 5.

35. Ibid., 17.

36. See Nicholas Rashevsky, "A Problem in Mathematical Biophysics of Interaction of Two or More Individuals Which May Be of Interest in Mathematical Sociology," *Bulletin of Mathematical Biophysics* 8 (1947): 9–15; Anatol Rapoport, "Forms of Output Distribution between Two Individuals Motivated by a Satisfaction Function," *Bulletin of Mathematical Biophysics* 9 (1947): 109–22; Rapoport, "Mathematical Theory of Motivation Interactions of Two Individuals: I," *Bulletin of Mathematical Biophysics* 9 (1947): 17–28; Rapoport, "Mathematical Theory of Motivation Interactions of Two Individuals: II," *Bulletin of Mathematical Biophysics* 9 (1947): 41–61.

37. Rapoport, *Certainties and Doubts*, 118.

38. Anatol Rapoport, "Some Game-Theoretical Aspects of Parasitism and Symbiosis," *Bulletin of Mathematical Biophysics* 18 (1956): 15–30.

39. Rapoport had envisioned experimental studies of the interaction model from an early date, although did not immediately pursue them; see Anatol Rapoport and Alfonso Shimbel, "Suggested Experimental Procedure for Determining the Satisfaction Function of Animals," *Bulletin of Mathematical Biophysics* 9 (1947): 169–77.

40. Anatol Rapoport and C. C. Foster, "Parasitism and Symbiosis in an *n*-Person Non-Constant Sum Game," *Bulletin of Mathematical Biophysics* 18 (1956): 231.

41. A description of Rapoport's work on rumors can be found in Box 2, Folder: "History of the Committee," Rashevsky Papers. See also Rapoport, *Certainties and Doubts*, 94–95, and Rapoport and Rebhun, "On the Mathematical Theory of Rumor Spread," 375–83, for references to the project.

42. Rapoport, *Certainties and Doubts*, 119. The date of the start of funding is not mentioned in the first year report (1957) but can be surmised from the MHRI's academic staff meeting of June 13, 1957, where Miller mentioned that the contract had been renewed for a second year.

43. See "Mental Health Research Staff meeting, June 13, 1957," in Box 2, Folder: "Academic Staff Meetings 1956–1968," MHRI Records.

44. A seminal work in this literature was Roy R. Grinker, *Men under Stress*, ed. John P. Spiegel (Philadelphia: Blakiston, 1945).

45. Rapoport describes the experiments and quantitative results in a number of articles from the late 1950s; see Anatol Rapoport, "Quantification of Performance on a Logical Task with Uncertainty," in *Symposium on Information Theory in Biology, Gatlinburg, Tennessee, October 29–31, 1956,* ed. Hubert Palmer Yockey (New York: Pergamon Press Symposium Publications Division, 1958), 230–38; Anatol Rapoport, William Hays, and J. David Birch, "The Formation of Hypotheses and Styles in the Process of Solving a Logical Task," in *Decisions, Values, and Groups, vol. 1,* ed. Dorothy Willner (New York: Pergamon Press, 1958), 50–69; and Anatol Rapoport, "A Study of Disjunctive Reaction Times," *Behavioral Science* 4 (1959): 299–315.

46. Rapoport describes the game in a number of places; see, e.g., "Mental Health Research Advisory Committee, Minutes of Meeting, January 9, 1956," p. 2, in Box 2, Folder: "Committees: Advisory Committee Minutes, January 1956–June 1964," MHRI Records.

47. Rapoport, *Certainties and Doubts*, 118.

48. Rapoport describes these experiments in ibid., 120–21. See also Anatol Rapoport, "Prisoner's Dilemma: Reflections and Recollections," *Simulation and Gaming* 26, no. 4 (1995): 489–502; and Anatol Rapoport, "Prisoner's Dilemma—Recollections and Observations," in *Game Theory as a Theory of Conflict Resolution,* ed. Anatol Rapoport (Dordrecht: Reidel, 1974), 17–34.

49. See Mental Health Research Institute Seventh Annual Report (Ann Arbor: University of Michigan, 1963), 8. The last entry for "Conflict and Cooperation in Small Groups" appears in the eleventh annual report (1967–68).

50. Anatol Rapoport and Albert M. Chammah, *Prisoner's Dilemma: A Study in Conflict and Cooperation* (Ann Arbor: University of Michigan Press, 1965).

51. Ibid., 11.

52. Ibid., vi.

53. Anatol Rapoport, Melvin Guyer, and David G. Gordon, *The 2×2 Game* (Ann Arbor: University of Michigan Press, 1976).

54. Erica Frydenberg, *Morton Deutsch: A Life and Legacy of Mediation and Conflict Resolution* (Brisbane: Australian Academic Press, 2005), 67–68.

55. Morton Deutsch, "Trust and Suspicion," *Journal of Conflict Resolution* 2, no. 4 (1958): 265–79.

56. Morton Deutsch and Robert M. Krauss, "The Effect of Threat upon Interpersonal Bargaining," *Journal of Abnormal and Social Psychology* 61, no. 2 (1960): 181–89.

57. Robert Axelrod, *The Evolution of Cooperation* (New York: Basic Books, 1984), 28.

58. See, e.g., Luce's critical comments on Rapoport and Chammah's book in R. Duncan Luce, "Dilemma of Research Strategy," *Science* 151, no. 3708 (1966): 318–19, and the literature review in Barry R. Schlenker and Thomas V. Bonoma, "Fun and Games: The Validity of Games for the Study of Conflict," *Journal of Conflict Resolution* 22, no. 1 (1978): 7–38.

59. Matt Price, "Roots of Dissent: The Chicago Met Lab and the Origins of the Franck Report," *Isis* 86, no. 2 (1995): 222–44; Wang, *American Science in an Age of Anxiety*; Kelly Moore, *Disrupting Science: Social Movements, American Scientists, and the Politics of the Military, 1945–1975* (Princeton, NJ: Princeton University Press, 2008).

60. Miller, ed., *Experiments in Social Process*, vii.

61. Quincy Wright, "Criteria for Judging the Relevance of Researches on the Problems of Peace: To What Extent It Is Possible to Establish Criteria for the Delimitation of Research of Direct Relevance to the Problem of Peaceful Adjustment in International Relations," in *Research for Peace: Essays*, ed. Ingemund Gullvåg (Amsterdam: North-Holland, 1954), 25.

62. Theodore F. Lentz, *Towards a Science of Peace: Turning Point in Human Destiny* (New York: Bookman Associates, 1955), 119.

63. Dwight D. Eisenhower, "State of the Union Address, 1958."

64. Raymond A. Bauer, "National Support for Behavioral Science," *Behavioral Science* 3, no. 3 (1958): 217–27.

65. Cynthia Earl Kerman, *Creative Tension: The Life and Thought of Kenneth Boulding* (Ann Arbor: University of Michigan Press, 1974), 66. See also 48 and 68 on reading Richardson at Stanford.

66. Paul N. Edwards, *A Vast Machine: Computer Models, Climate Data, and the Politics of Global Warming* (Cambridge MA: MIT Press, 2010), 93–96.

67. For Richardson's comparisons between the study of conflict and other sciences see especially the comments in L. F. Richardson: "Hints from Physics and Meteorology as to Mental Periodicities," *British Journal of Psychology* 28 (1937): 212–15; "Mathematical Psychology of War," *Nature* 135 (1935): 830–31; "Thresholds When Sensation Is Regarded as Quantitative," *British Journal of*

Psychology 19 (1928): 158–66; and *Mathematical Psychology of War* (Oxford: Hunt, 1919). On Richardson more generally, see Oliver M. Ashford, *Prophet—or Professor? The Life and Work of Lewis Fry Richardson* (Bristol: Hilger, 1985); E. Gold, "Lewis Fry Richardson. 1881–1953," *Obituary Notices of Fellows of the Royal Society* 9, no. 1 (1954): 216–35; Stephen A. Richardson, "Lewis Fry Richardson (1881–1953): A Personal Biography," *Conflict Resolution* 1, no. 3 (1957): 300–304; and "A Bibliography of Lewis Fry Richardson's Studies on the Causation of Wars with a View to their Avoidance" *Conflict Resolution* 1, no. 3 (September 1957): 205–7.

68. See Richardson, *Mathematical Psychology of War.*

69. L. F. Richardson: "Could an Arms-Race End without Fighting?," *Nature* 168 (1951): 567–68; "War Moods: I and II," *Psychometrika* 13 (1948): 147–74, 197–232; "Frequency of Occurrence of Wars and Other Fatal Quarrels," *Nature* 148 (1941): 784; and *Generalized Foreign Politics: A Study in Group Psychology* (Cambridge: Cambridge University Press, 1939).

70. Rapoport, *Certainties and Doubts,* 112.

71. Kenneth Ewart Boulding, *The Economics of Peace* (New York: Prentice-Hall, 1945).

72. See, e.g., Anatol Rapoport, Gwen Goodrich Rapoport, and Alfonso Shimbel, "Sanity and the Cold War," *Measure: A Critical Journal* 2 (Spring 1951): 159–74.

73. Kerman, *Creative Tension,* 69.

74. "An Editorial," *Conflict Resolution* 1, no. 1 (March 1957): 1–2.

75. Anatol Rapoport, "Lewis F. Richardson's Mathematical Theory of War," *Journal of Conflict Resolution* 1 (1957): 249–99.

76. See Taylor Drysdale (Chief, Research and Analysis Division, Department of the Air Force) to William Barth (executive editor, Journal of Conflict Resolution), dated January 14, 1959, praising the contents of a recent volume of the *Journal* (Box 3, Folder, "Correspondence Pre 1961, Alpha-D," Center for Research on Conflict Resolution [hereafter CRCR]), Bentley Library, University of Michigan–Ann Arbor). For data on the center's military support, see "Sources of Support" in Box 7, Folder: "Review of Center, 1959–1971."

77. Listings of new institutions devoted to peace research worldwide can be found in the *International Newsletter on Peace Research,* edited by Elise Boulding (Ann Arbor: International Consultative Committee on Peace Research, 1963–64).

78. See especially Kenneth Ewart Boulding, *Conflict and Defense: A General Theory* (New York: Harper, 1962), which laid out what Boulding hoped would prove a conceptual framework for conflict resolution; and Emile Benoit and Kenneth Ewart Boulding, *Disarmament and the Economy* (New York: Harper and Row, 1963) on the economics of disarmament. On Boulding's central place in the pedagogy of the program, see "Center for Research on Conflict Resolu-

tion: Review of Program, July 1, 1959–June 30, 1965," Box 7, Folder: "Review of Center, 1959–1971," CRCR.

79. See, e.g., J. David Singer: *Human Behavior and International Politics: Contributions from the Social-Psychological Sciences*, Rand McNally Political Science (Chicago: Rand McNally, 1965); "International Politics and the Behavioral Sciences," *International Review of Political Sciences* 1 (1964): 103–14; "Soviet and American Foreign Policy Attitudes: A Content Analysis of Elite Articulations," *Journal of Conflict Resolution* 8, no. 4 (1964): 424–85; "Cosmopolitan Attitudes and International Relations Courses: Some Tentative Conclusions," *Journal of Politics* (1964); and "Inter-Nation Influence: A Formal Model," *American Political Science Review* 2 (1963): 420–30, for early research stemming from these projects.

80. Marc Pilisuk et al., "War Hawks and Peace Doves: Alternative Resolutions of Experimental Conflicts," *Journal of Conflict Resolution* 9, no. 4 (1965): 491–508; Pilisuk and Rapoport, "Stepwise Disarmament and Sudden Destruction in a Two-Person Game: A Research Tool," *Journal of Conflict Resolution* 8, no. 1 (1964): 36–49.

81. Lists of major research projects and associated bibliographies can be found in the document, "Center for Research on Conflict Resolution: Review of Program, July 1, 1959–June 30, 1965." The annual reports of the Mental Health Research Institute provide synopses of work on "Conflict Processes in Social Systems," "International Political Systems," and other projects related to the Center's staff. These are the best records for tracking progress on these projects after 1965, although see also "Center for Research on Conflict Resolution Projected Program, July 1, 1965–June 30, 1970," Box 7, Folder: "Review of Center, 1959–1971," CRCR.

82. Richardson, "Generalized Foreign Politics," 87.

83. Robert Dodge, *The Strategist* (Hollis, NH: Hollis Publishing, 2006), 47–48.

84. Schelling, "An Essay on Bargaining," 281–306.

85. Ibid., 282.

86. Dodge, *The Strategist*, 60–61.

87. Schelling, personal communication with the author, July 28, 2005.

88. Thomas C. Schelling, "Bargaining, Communication, and Limited War," *Conflict Resolution* 1, no. 1 (1957): 19–36.

89. Ibid., 19.

90. Thomas C. Schelling, "The Strategy of Conflict: Prospectus for a Reorientation of Game Theory," *Journal of Conflict Resolution* 2, no. 3 (1958): 203–64.

91. Ibid., 203.

92. Herman Kahn and Irwin Mann: "Techniques of Systems Analysis," *RAND* RM-1829-1-PR (1957); "War Gaming," *RAND* P-1167 (1957); "Game Theory," *RAND* P-1166 (1957); "Ten Common Pitfalls," RAND RM-1937 (1957).

93. Schelling, *Strategy of Conflict*. In addition to the papers he published through *Journal of Conflict Resolution*, Schelling also produced numerous internal RAND publications in this period. See T. C. Schelling: "Toward a Theory of Strategy for International Conflict" (RAND P-1648, May 8, 1958); "The Reciprocal Fear of Surprise Attack" (RAND P-1342, April 16, 1958); "Re-Interpretation of the Solution Concept for 'Non-Cooperative' Games" (RAND P-1385, June 2, 1958); "Surprise Attack and Disarmament" (RAND P-1574, December 10, 1958); "Nuclear Weapons and Limited War" (RAND P-1620, February 20, 1959); "The Role of Theory in the Study of Conflict" (RAND RM-2515, January 1, 1960); "The Retarded Science of International Strategy," *Midwest Journal of Political Science* 4, no. 2 (1960): 107–37; and "For the Abandonment of Symmetry in Game Theory," *Review of Economics and Statistics* 41, no. 3 (1959): 213–24.

94. Rapoport's paper was circulated internally within the MHRI during the fall of 1958; in the MHRI's academic staff meeting of October 24, 1958, James Miller specifically drew attention to preprint no. 23, "Critiques of Game Theory" by Rapoport. See Box 2, Folder: "Academic Staff Meetings 1956–1968," MHRI Records. The paper was subsequently published as Anatol Rapoport, "Critiques of Game Theory," *Behavioral Science* 4 (1959): 49–66.

95. Rapoport, "Critiques of Game Theory," 52.

96. Ibid., 56 and 58.

97. Anatol Rapoport, *Fights, Games, and Debates* (Ann Arbor: University of Michigan Press, 1960), 242. For positive joint reviews of Rapoport and Schelling's books, see Oskar Morgenstern, "Review of Rapoport, *Fights, Games, and Debates,*" *Southern Economic Journal* 28, no. 1 (1961): 103–5; Martin Shubik, "Review of Schelling, *The Strategy of Conflict*, and Rapoport, *Fights, Games and Debates,*" *Journal of Political Economy* 69, no. 5 (1961): 501–3.

98. Albert Wohlstetter, "The Delicate Balance of Terror," *Foreign Affairs* 37 (1959): 211–34.

99. Herman Kahn, *On Thermonuclear War* (Princeton, NJ: Princeton University Press, 1960), esp. 291–95.

100. For biographical detail on Ellsberg see Tom Wells, *Wild Man: The Life and Times of Daniel Ellsberg* (New York: Palgrave, 2001).

101. On Kahn's departure from RAND, see Ghamari-Tabrizi, *Worlds of Herman Kahn*, 37.

102. Anatol Rapoport, "New Logic for the Test Ban," *Nation* 192 (1961): 284.

103. Rapoport, "Prisoner's Dilemma: Reflections and Recollections," 497.

104. Anatol Rapoport and J. David Singer, "Memorandum to the Peace Movement: An Alternative to Slogans," *Nation* 194 (1962): 250; see also Rapoport and Singer, "The Armers and the Disarmers," *Nation* 196 (1963): 174–77.

105. Anatol Rapoport, "Formal Games as Probing Tools for Investigating Behavior Motivated by Trust and Suspicion," *Journal of Conflict Resolution* 7, no. 3 (1963): 574.

106. Anatol Rapoport, *Strategy and Conscience* (New York: Harper and Row, 1964), xix–xxi.

107. Ibid., xxiii.

108. On "chicken," see ibid., 116.

109. Ibid., 172–73.

110. A. J. Muste, "Strategy and Conscience: An Exchange," *Liberation* (1965): 17–18.

111. *New York Times Book Review,* July 19, 1964, 6.

112. D. G. Brennan, "Review of *Strategy and Conscience*," *Bulletin of the Atomic Scientists* 21, no. 10 (1965): 28.

113. Ibid., 26.

114. Anatol Rapoport, "The Sources of Anguish: A Reply to D. G. Brennan's Review of *Strategy and Conscience, Liberation,* August 1965," *Bulletin of the Atomic Scientists* 21, no. 10 (1965): 31.

115. Thomas C. Schelling, "Review of *Strategy and Conscience*, by Anatol Rapoport," *American Economic Review* 54, no. 6 (1964): 1083.

116. See especially Solly Zuckerman, "Judgment and Control in Modern Warfare," *Foreign Affairs* 40, no. 2 (1962): 196–212; P. M. S. Blackett, "Operational Research and Nuclear Weapons," *Journal of the Royal United Services Institution* 106, no. 623 (1961): 342–54; Blackett, "Critique of Some Contemporary Defense Thinking," *Encounter* 16, no. 4 (1961): 9–17.

117. Irving Louis Horowitz, *The War Game: Studies of the New Civilian Militarists* (New York: Ballantine Books, 1963). See also Irving Louis Horowitz, *The Rise and Fall of Project Camelot: Studies in the Relationship Between Social Science and Practical Politics* (Cambridge, MA: MIT Press, 1967); Anatol Rapoport, "The Scientific Relevance of C. Wright Mills," in *The New Sociology: Essays in Social Science and Social Theory, in Honor of C. Wright Mills*, ed. Irving Louis Horowitz and C. Wright Mills (New York: Oxford University Press, 1963), 94–107.

118. Philip Green, *Deadly Logic: The Theory of Nuclear Deterrence* (Columbus: Ohio State University Press, 1966); Norman Moss, *Men Who Play God: The Story of the Hydrogen Bomb* (London: Gollancz, 1968).

119. See, e.g., Kathleen Archibald, *Strategic Interaction and Conflict: Original Papers and Discussion* (Berkeley: Institute of International Studies, University of California, Berkeley, 1966), and the papers and commentary in A. Mensch, *Theory of Games: Techniques and Applications* (New York: American Elsevier Publishing, 1966).

120. The Arms Control and Disarmament Agency funded a number of such studies at this time, including one overseen by Russell Ackoff of the University of Pennsylvania Management Science Center (resulting in the monograph, *A Model Study of the Escalation and De-Escalation of Conflict*, March 1, 1967), and another by Mathematica, which produced a multivolume study of *Models of*

Gradual Reduction of Arms (September 1967). The latter study helped to support the research of a number of game theorists who subsequently won great fame, including Robert Aumann, John Harsanyi, John Mayberry, Herbert Scarf, and Reinhard Selten.

121. On these models, see John C. Harsanyi: "Subjective Probability and the Theory of Games: Comments on Kadane and Larkey's Paper," *Management Science* 28, no. 2 (1982): 120–24; "Rational-Choice Models of Political Behavior vs. Functionalist and Conformist Theories," *World Politics* 21, no. 4 (1969): 513–38; "Games with Incomplete Information Played by 'Bayesian' Players, I–III. Part II. Bayesian Equilibrium Points," *Management Science* 14, no. 5 (1968): 320–34; and "Games with Incomplete Information Played by 'Bayesian' Players, I–III. Part III. The Basic Probability Distribution of the Game," *Management Science* 14, no. 7 (1968): 486–502.

122. Rapoport, *Certainties and Doubts*, 136 and subsequent.

123. On the timing of Boulding's departure see Kerman, *Creative Tension*, 354.

124. For press coverage of the center's closing see Box 7, Folder: "Center Closing, 1971," CRCR. On the problems facing the center in the wake of Boulding's departure for the University of Colorado, see Box 7, Folder: "Meeting to Redefine Center for Research on Conflict Resolution, 1968," and on the impact of state fiscal tightening see Box 7, Folder: "Review of Center, 1959–1971," CRCR.

125. J. David Singer to Frank H. T. Rhodes, June 1, 1972, in Box 3, Folder: "JCR Move to Yale, 1971–1972," J. David Singer Papers, Bentley Library, University of Michigan, Ann Arbor. The "Camelot fiasco" is a reference to the uproar the followed revelations of the extent to which a number of social scientists had been assisting the war effort in Vietnam; see Horowitz, *The Rise and Fall of Project Camelot*; Ellen Herman, *The Romance of American Psychology: Political Culture in the Age of Experts* (Berkeley: University of California Press, 1995), especially chap. 6, "Project Camelot and Its Aftermath"; Jefferson P. Marquis, "The Other Warriors: American Social Science and Nation Building in Vietnam," *Diplomatic History* 24, no. 1 (2000): 79–105; and Mark Solovey, "Project Camelot and the 1960s Epistemological Revolution: Rethinking the Politics-Patronage-Social Science Nexus," *Social Studies of Science* 31, no. 2 (2001): 171–206.

Chapter Six

1. Richard Dawkins, *The Selfish Gene* (Oxford: Oxford University Press, 1976), 84. In the 1989 edition, Dawkins reflects that this comment may have been "a bit over the top," but that it nevertheless reflected his determination at the time to more effectively publicize the relatively new concept.

2. See John Maynard Smith, *Evolution and the Theory of Games* (New York:

Cambridge University Press, 1982); for a more recent survey of the field, see Lee Alan Dugatkin and Hudson Kern Reeve, *Game Theory and Animal Behavior* (New York: Oxford University Press, 1998), or Josef Hofbauer and Karl Sigmund, *The Theory of Evolution and Dynamical Systems: Mathematical Aspects of Selection* (Cambridge: Cambridge University Press, 1988), which illustrates the connections that have developed between evolutionary game theory and dynamical systems theory.

3. The broader history of the "problem" of altruism, from Darwin on down, forms the narrative framework of Oren Harman's biographical study of Price; see Oren S. Harman, *The Price of Altruism: George Price and the Search for the Origins of Kindness* (New York: W. W. Norton, 2010).

4. Robert H. MacArthur: "On the Relative Abundance of Bird Species," *Proceedings of the National Academy of Sciences of the United States of America* 43, no. 3 (1957): 293–95; and "Fluctuations of Animal Populations and a Measure of Community Stability," *Ecology* 36 (1955): 533–36.

5. Richard Levins, "Theory of Fitness in a Heterogeneous Environment" (PhD diss., Columbia University, 1965). See also Levins, *Evolution in Changing Environments* (Princeton, NJ: Princeton University Press, 1968).

6. This account is taken from Edward O. Wilson, *Naturalist* (Washington, DC: Island Press, 1994), 254. Richard Lewontin (interview with the author, December 22, 2004) credits similar inspirations for the group's work.

7. Richard C. Lewontin, "Evolution and the Theory of Games," *Journal of Theoretical Biology* 1 (1961): 382–403; Lawrence B. Slobodkin, "The Strategy of Evolution," *American Scientist* 52, no. 3 (1964): 342–57.

8. Another attempt at the use of game theory can be found in C. H. Waddington, *The Strategy of the Genes: A Discussion of Some Aspects of Theoretical Biology* (London: Allen and Unwin, 1957), which was loosely related to Lewontin's work on development in *drosophila*; I do not discuss it here. Oskar Morgenstern was aware of all three works by the late 1960s; see Box 42, Folder: "Game Theory and Philosophy, 1961–1969," Morgenstern Papers.

9. On the relationship between the Chicago school of population ecology and political and social commitments of its members, see Mitman, *The State of Nature*.

10. Alfred E. Emerson, "Dynamic Homeostasis: A Unifying Principle in Organic, Social, and Ethical Evolution," *Scientific Monthly* 78, no. 2 (1954): 67–85.

11. "The Population Bomb" (New York: Hugh Moore Fund), 3. For more on the intellectual and political history of the Cold War–era "population problem," see especially Faye D. Ginsburg and Rayna R. Reiter, *Conceiving the New World Order: The Global Politics of Reproduction* (Berkeley: University of California Press, 1995); John Sharpless, "Population Science, Private Foundations, and Development: The Transformation of Demographic Knowledge in the United States, 1945–1965," in *International Development and the Social Sci-*

ences: Essays in the History and Politics of Knowledge, ed. Frederick Cooper and Randall Packard (Berkeley: University of California Press), 176–200; Matthew James Connelly, *Fatal Misconception: The Struggle to Control World Population* (Cambridge, MA: Belknap Press of Harvard University Press, 2008); Thomas Robertson, *The Malthusian Moment: Global Population Growth and the Birth of American Environmentalism* (New Brunswick, NJ: Rutgers University Press, 2012).

12. These are often discussed as part of the so-called group selection controversy, e.g., in Mark E. Borrello: *Evolutionary Restraints: The Contentious History of Group Selection* (Chicago: University of Chicago Press, 2010); and "Synthesis and Selection," *Journal of the History of Biology* 36, no. 3 (2003): 531–66. Other literature treating group selection focuses primarily on the debate in the United States, especially George C. Williams's critique of the ideas of Warder Clyde Allee and Afred Emerson at the University of Chicago; see George C. Williams, *Adaptation and Natural Selection: A Critique of Some Current Evolutionary Thought* (Princeton, NJ: Princeton University Press, 1966); Elliott Sober and David Sloan Wilson, *Unto Others: The Evolution and Psychology of Unselfish Behavior* (Cambridge, MA: Harvard University Press, 1998); and Mitman, *The State of Nature.*

13. Vero Copner Wynne-Edwards, *Animal Dispersion in Relation to Social Behavior* (Edinburgh: Oliver and Boyd, 1962), 131.

14. Alexander Carr-Saunders, *The Population Problem: A Study in Human Evolution* (Oxford: Clarendon Press, 1922). On Carr-Saunders, see "Obituary: Sir Alexander Carr-Saunders, 14 January 1886–6 October 1966," *Population Studies* 20, no. 3 (March 1967): 365–69.

15. "The Animal 'Wisdom' Man Has Lost," Aberdeen *Press and Journal,* May 9, 1962, in V. C. Wynne-Edwards Papers, Box 6 loc. 5137.1, File: "Animal Dispersion 1962–64," V. C. Wynne-Edwards Papers (hereafter Wynne-Edwards Papers), Special Collections, Queens University, Kingston, ON.

16. J. Gordon Blower, "More Natural Selection," *The Guardian,* September 21, 1962, Box 6 loc. 5134, File: "Animal Dispersion 1962–64," Wynne-Edwards Papers.

17. See Wynne-Edwards's later comments in V. C. Wynne-Edwards, "Population, Affluence, and the Environment," *Environmental Education* 3 (1973): 10–18; and "Ecology and the Evolution of Social Ethics," in *Biology and the Human Sciences: The Herbert S. Spencer Lectures 1970,* ed. J. W. S. Pringle (Oxford: Clarendon Press, 1972), 49–69, the former of which comments on the work of the Club of Rome. An earlier version of some ideas is also found in V. C. Wynne-Edwards, "The Overfishing Principle Applied to Natural Populations and Their Food-Resources: And a Theory of Natural Conservation," in *Proceedings of the XIIth International Ornithological Congress, Helsinki 1958* (Helsinki: Tilgmannin Kirjapaino,1960), 790–94.

18. See MacArthur to Lack, October 22, 1962, correspondence file "M," David Lack Papers (hereafter Lack Papers), Alexander Library of Ornithology, University of Oxford.

19. This notion is related to Ernst Mayr's quasi-historical concept of "population thinking" (as opposed to typological thinking), which is identified as a major intellectual achievement of modern evolutionary biology in the historiography of the evolutionary synthesis. For commentary, see Elliott Sober, "Evolution, Population Thinking, and Essentialism," *Philosophy of Science* 47, no. 3 (1980): 350–83.

20. Edward O. Wilson, "The Social Biology of Ants," *Annual Review of Entomology* 8 (1963): 345–68.

21. Edward O. Wilson, "The Ergonomics of Caste in the Social Insects," *American Naturalist* 102, no. 923 (1968): 41.

22. See, e.g., Bert Hölldobler and Edward O. Wilson, *The Ants* (Cambridge, MA: Belknap Press of Harvard University Press, 1990); George F. Oster and Edward O. Wilson, *Caste and Ecology in the Social Insects* (Princeton, NJ: Princeton University Press, 1978).

23. This is despite the general rejection of group selection during the 1970s. See Richard Dawkins, *The Extended Phenotype: The Gene as the Unit of Selection* (Oxford: Freeman, 1982), 56, and Edward O. Wilson, *Sociobiology: The New Synthesis* (Cambridge, MA: Belknap Press of Harvard University Press, 1975), chap. 5.

24. Oster and Wilson, *Caste and Ecology in the Social Insects*, 302.

25. Lewontin, "Evolution and the Theory of Games," 382–403.

26. To borrow the language of Levins, the feasible fitness surface is simply the convex hull of the S_is.

27. See Luce and Raiffa, *Games and Decisions*, especially chapter 4 on the connection between game theory and decision theory in the 1950s.

28. Lewontin, "Evolution and the Theory of Games," 402.

29. On Levins's estimation of game theory see Levins, *Evolution in Changing Environments*, 99.

30. David Lambert Lack, *The Natural Regulation of Animal Numbers* (Oxford: Clarendon Press, 1954). Within months of the appearance of Lack's book, Wynne-Edwards had already proposed the title for *Animal Dispersion*; see Box 17, Folder: "Untitled Notebook #1 N.D.," Wynne-Edwards Papers.

31. Maynard Smith to Lack, October 27, 1955, Folder 121, Lack Papers.

32. John Maynard Smith, "In Haldane's Footsteps," in *Leaders in the Study of Animal Behavior*, ed. Donald A. Dewsbury (Lewisburg, PA: Bucknell University Press, 1985), 351. See also John Maynard Smith, "Fertility, Mating Behaviour and Sexual Selection in *Drosophila Subobscura*," *Journal of Genetics* 54 (1956): 261–79.

33. "Introductory Talk by Professor Michael Swann," Box 1, File: "Animal Dispersion," Wynne-Edwards Papers. The history of the *Third Programme* is related in Humphrey Carpenter and Jennifer R. Doctor, *The Envy of the World: Fifty Years of the BBC Third Programme and Radio 3, 1946–1996* (London: Weidenfeld and Nicolson, 1996).

34. "Studio 6C, 1930–2200, 17th October, 1963," Box 1, loc. 2287.9, File: "Animal Dispersion," Wynne-Edwards Papers; cited on 6, 9, 17.

35. John Maynard Smith, "Group Selection and Kin Selection," *Nature* 200 (1964): 1145–47; Christopher Perrins, "Survival of Young Swifts in Relation to Brood-Size," *Nature* 201, no. 4924 (1964): 1147–48. A year later, Maynard Smith disputed the necessity of group selection for the evolution of another phenomenon documented by Wynne-Edwards: the tendency of birds to raise a general alarm at some cost to themselves if a predator approached their flock; see John Maynard Smith, "The Evolution of Alarm Calls," *American Naturalist* 99, no. 904 (1965): 59–63.

36. V. C. Wynne-Edwards, "Survival of Young Swifts in Relation to Brood-Size," *Nature* 201, no. 4924 (1964): 1148–49. Wynne-Edwards's assessment here dovetails with that of his biographer, Mark Borrello, who suggests that group selection ultimately sprang from Wynne-Edwards's experience as a practicing field ecologist; see Borrello, *Evolutionary Restraints*, chap. 8.

37. G. Ainsworth Harrison, "The Genetic Structure of Human Populations," in *The Royal Society Population Study Group Abstracts of Proceedings 2* (London: Royal Society, October 21, 1966).

38. W. D. Hamilton, *The Narrow Roads of Gene Land, vol. 2* (Oxford: Oxford University Press, 2001), xxxiv.

39. Ibid., xlix.

40. This focus is apparent in many of Hamilton's writings from this period; see *ibid.*, xli, and W. D. Hamilton, "Review of K. Mather, *Human Diversity,*" *Population Studies* 19, no. 2 (1965): 203–5.

41. Hamilton, *Narrow Roads of Gene Land,* 2:xxxvii.

42. Ibid., xlvi.

43. The objectives for the MAPW are found on a pamphlet announcing a conference to be held at the Royal College of Surgeons, May 10–11, 1952, Box 41, Folder 3, Lionel Penrose Papers (hereafter Penrose Papers), University College London Archives. Penrose's antiwar involvement in this period also led to L. S. Penrose, *On the Objective Study of Crowd Behaviour* (London: H. K. Lewis, 1952), which proposed a mental health model of mass behavior.

44. John Maynard Smith, in *Conference on the Pathogenesis of War,* Medical Association for the Prevention of War, September 1961, cited in Paul Smoker, "On Mathematical Models in Arms Races," *Journal of Peace Research* 2, no. 1 (1965): 94–95.

45. Patrick Deighan and Paul Smoker to Penrose, September 20, 1962, Box 41, Folder 8: "Papers, Notice of Meeting, and Correspondence in Connexion with the Peace Research Group," Penrose Papers.

46. Cedric A. B. Smith to "Peace Research Group," August 13, 1962, Box 41, Folder 8, Penrose Papers.

47. Ted and Enola Lentz, "Summer Safari—1964," Box 6, Folder "Theodore Lentz, 1961–1968," CRCR. On the Tavistock Institute's history as a center for social psychological research, see Eric Trist and Hugh Murray, *The Social Engagement of Social Science: A Tavistock Anthology* (Philadelphia: University of Pennsylvania Press, 1990).

48. Anatol Rapoport: "Research for Peace, I," *The Listener, BBC-TV Review, London* I (1966): 455–56, 475; "Research for Peace, II," *The Listener, BBC-TV, London* (1966): 504–5, 508.

49. "Annual Report of the Council," *Journal of the Royal Statistical Society, Series A (General)* 126, no. 4 (1963): 572–91, 575.

50. "Report of Meeting Held at the Russell Hotel on Tuesday 11th September at 8:00 p.m.," Box 41, Folder 8, Penrose Papers.

51. Hamilton thanks Cedric Smith for this help in his 1964 paper, and likewise Smith credits Hamilton's work on the biology of altruism in his 1966 inaugural lecture as Weldon Professor of Biometry; see Cedric A. B. Smith, *Life, Form, and Number: An Inaugural Lecture Delivered at University College London 17 May 1965* (London: H. K. Lewis, 1966).

52. W. D. Hamilton: "The Genetical Evolution of Social Behavior, I," *Journal of Theoretical Biology* 7 (1964): 1–26; "The Genetical Evolution of Social Behavior, II," *Journal of Theoretical Biology* 7 (1964): 27–52; "The Evolution of Altruistic Behavior," *American Naturalist* 97 (1963): 354–56.

53. Smith's review can be found in *The Mathematical Gazette,* July 1945, 131–35.

54. W. D. Hamilton, "Extraordinary Sex Ratios," *Science* 156 (1967): 477–88.

55. Hamilton to Price, March 21, 1968, Item KPX1_5.5, George Price Archive, W. D. Hamilton Papers (hereafter Hamilton Papers), British Library, London.

56. Anatol Rapoport, "Escape from Paradox," *Scientific American* 217 (1967): 50–56.

57. Hamilton to Price, March 21, 1968, Item KPX1_5.5, Price Archive, Hamilton Papers.

58. S. Dillon Ripley, "Preface," in *Man and Beast: Comparative Social Behavior,* ed. J. F. Eisenberg and Wilton S. Dillon (Washington, DC: Smithsonian Institution Press, 1971), 6, 11.

59. W. D. Hamilton, "Selection of Selfish and Altruistic Behavior in Some Extreme Models," in *Man and Beast,* ed. Eisenberg and Dillon, 57–91.

60. Ibid., 82.

61. Hamilton to Wilton S. Dillon, Smithsonian Institution, January 30, 1970, no number yet assigned, Hamilton Papers.

62. Robert Trivers, "The Evolution of Reciprocal Altruism," *Quarterly Review of Biology* 46 (1971): 35–57.

63. Pnina Abir-Am, "The Molecular Transformation of 20th Century Biology," in *Science in the Twentieth Century*, ed. John Krige and Dominique Pestre (Stanford, CA: Stanford University Press, 1997), 495–524; Kay, *Who Wrote the Book of Life?*; Evelyn Fox Keller, *The Century of the Gene* (Cambridge, MA: Harvard University Press, 2000).

64. John Maynard Smith and D. Michie, "Machines That Play Games," *New Scientist* 12 (1964): 367–69. See also John Maynard Smith, "Natural Selection and the Concept of Protein Space," *Nature* 225 (1970): 563–64; Maynard Smith, "The Spatial and Temporal Organization of Cells," in *Mathematics and Computer Science in Biology and Medicine* (London: Medical Research Council, 1965), and Maynard Smith's reflections in his article, "The Concept of Information in Biology," *Philosophy of Science* 67, no. 2 (2000): 177–94.

65. Price's transition into biology is discussed in Steve A. Frank, "George Price's Contributions to Evolutionary Genetics," *Journal of Theoretical Biology* 175 (1995): 373–88; James Schwartz, "Death of an Altruist," *Lingua Franca* 10, no. 5 (2000): 51–61; and Harman, *Price of Altruism*.

66. No copy of Price's first letter is extant, but a reply letter from Hamilton to Price of March 21, 1968, Item KPX1_5.5, Price Archive, Hamilton Papers, suggests the content of his first query. Price's work during his first year in London and his reading of *The Naked Ape* is discussed in two letters to his family dated December 10, 1967, and n.d., items KPX1_4.1 and KPX1_4.2, Price Archive, Hamilton Papers.

67. Frank, "George Price's Contributions to Evolutionary Genetics"; Schwartz, "Death of an Altruist"; and Hamilton, "Spite and Price," in W. D. Hamilton, *Narrow Roads of Gene Land: The Collected Papers of W. D. Hamilton* (New York: W. H. Freeman, 1996).

68. Price, "Supplementary Details of Intended Research," May 1969, Item KPX1_5.4 (emphasis added), Price Archive, Hamilton Papers.

69. Ibid.

70. Cedric Smith, "SRC Research Grant Application," March 31, 1969, Folder: "MRC Application," Item KXP1_10.1, Price Archive, Hamilton Papers.

71. Hamilton notes sending off the paper to *Nature* in a letter to Price of July 21, 1970, Item KXP1_4.8, Price Archive, Hamilton Papers. The papers were published as George R. Price, "Selection and Covariance," *Nature* 227 (1970): 520–21, and W. D. Hamilton, "Selfish and Spiteful Behavior in an Evolutionary Model," *Nature* 228 (1970): 1218–20.

72. Many colleagues of Price—not to mention historians—have looked for a copy of this paper; see, e.g., Steve Frank to Hamilton, n.d., Item Z1x102_1.1,

Price Archive, Hamilton Papers, and Frank, "George Price's Contributions to Evolutionary Genetics."

73. This account was first laid out in a short paper published by Maynard Smith in 1976 in which he sought to put his personal stamp on the history of game theory in biology; see John Maynard Smith, "Evolution and the Theory of Games," *American Scientist* 64 (1976): 41–45.

74. The closest thing we have to a reconstruction of this problem is presented in Maynard Smith's "Game Theory and the Evolution of Fighting," in John Maynard Smith, *On Evolution* (Edinburgh: Edinburgh University Press, 1972).

75. Ibid., 21 (emphasis added).

76. Maynard Smith would develop this insight in greater detail in John Maynard Smith, "Optimization Theory in Evolution," *Annual Review of Ecology and Systematics* 9 (1978): 31–56.

77. Price to Hamilton, September 21, 1970, Item Z1X102_1.1.2.4, Price Archive, Hamilton Papers. Price describes the UCL computer facilities in his letter to Miss Sara Smith, (CPL Recruitment Division, 14 Old Park Lane, London), August 28, 1974, no number assigned, Price Archive, Hamilton Papers.

78. See, e.g., W. D. Hamilton, "Geometry for the Selfish Herd," *Journal of Theoretical Biology* 31 (1971): 295–311.

79. These events are described in Hamilton, "Friends, Romans, Groups . . ." in Hamilton, *Narrow Roads of Gene Land: The Collected Papers of W. D. Hamilton*, and Schwartz, "Death of an Altruist."

80. For their part, Price and Hamilton resented what they saw as Maynard Smith's appropriation of Price's ideas; see Ullica Segerstråle, *Defenders of the Truth: The Battle for Science in the Sociobiology Debate and Beyond* (Oxford: Oxford University Press, 2000), chap. 4.

81. According to Maynard Smith, Price insisted that the strategy "mouse" be used instead of the more familiar "dove" on grounds that war-gaming with the latter would constitute a sacrilege against the Holy Ghost. Maynard Smith reinstated "dove" in his later papers on game theory; see, e.g., John Maynard Smith, "Game Theory and the Evolution of Behaviour," *Proceedings of the Royal Society of London, Series B, Biological Sciences* 205 (1979): 475–88.

82. John Maynard Smith and George R. Price, "The Logic of Animal Conflict," *Nature* 246 (1973): 15–18.

83. Hamilton, "Friends, Romans, Groups . . . ," 325.

84. The rhetorical features of this process have been examined by Sergio Sismondo in "Modeling Strategies: Creating Autonomy for Biology's Theory of Games," *History and Philosophy of the Life Sciences* 19 (1997): 147–61. See further John Maynard Smith, "The Theory of Games and the Evolution of Animal Conflicts," *Journal of Theoretical Biology* 47 (1974): 209–21, John Maynard Smith and George R. Parker, "The Logic of Asymmetric Contests," *Animal Behavior* 24 (1976): 159–75; Maynard Smith, *Evolution and the Theory of Games*.

85. See, e.g., Hamilton to Price, March 21, 1968, Item KPX1_5.5, Price Archive, Hamilton Papers.

86. Anatol Rapoport, "Game Theory without Rationality," *Behavioral and Brain Sciences* 7, no. 1 (1984): 114–15; and John Maynard Smith, "Game Theory without Rationality," *Behavioral and Brain Sciences* 7, no. 1 (1984): 117–22.

87. Axelrod, *The Evolution of Cooperation*; Robert Axelrod and W. D. Hamilton, "The Evolution of Cooperation," *Science* 211 (1981): 1390–96.

Chapter Seven

1. John Maynard Smith, interview with the author, January 10, 2004.

2. See Reinhard Selten, "A Note on Evolutionary Stable Strategies in Asymmetric Animal Conflicts," *Journal of Theoretical Biology* 84 (1980): 91–101, and Peter Hammerstein and Reinhard Selten, "Game Theory and Evolutionary Biology," in *Handbook of Game Theory with Economic Applications, vol. 2*, ed. Robert J. Aumann and S. Hart (Amsterdam: Elsevier, 1994), 929–93.

3. R. J. Aumann, "What Is Game Theory Trying to Accomplish?," in *Frontiers of Economics*, ed. Kenneth Joseph Arrow and S. Honkapohja (Oxford: Basil Blackwell, 1985), 43.

4. Franklin M. Fisher, "Games Economists Play: A Noncooperative View," *RAND Journal of Economics* 20, no. 1 (1989): 113.

5. Kreps, *Game Theory and Economic Modelling*, 1.

6. On the movement of ideas between defense strategy and social policy see Robert Ayson, *Thomas Schelling and the Nuclear Age;* David Jardini, "Out of the Blue Yonder: The RAND Corporation's Diversification into Social Welfare Research, 1946–1968" (PhD diss., Carnegie Mellon University, 1996); David Jardini, "Out of the Blue Yonder: The Transfer of Systems Thinking from the Pentagon to the Great Society, 1961–1965," in *Systems, Experts, and Computers: The Systems Approach in Management and Engineering, World War II and After,* ed. Agatha C. Hughes and Thomas P. Hughes (Cambridge, MA: MIT Press, 2000); and Jennifer S. Light, *From Warfare to Welfare: Defense Intellectuals and Urban Problems in Cold War America* (Baltimore, MD: Johns Hopkins University Press, 2003).

7. Schmidt, *Game Theory and Economic Analysis*, 4.

8. Martin Shubik, "The Role of Game Theory in Economics," *Kyklos* 7, no. 2 (1953): 21.

9. On the relationship between the new game theoretic models and Cournot's analysis, see especially Leonard, "Reading Cournot, Reading Nash," 492–511.

10. See, e.g., ibid., 507.

11. Martin Shubik, "Information, Theories of Competition, and the Theory of Games," *Journal of Political Economy* 60, no. 2 (1952): 146.

12. Shubik, "The Role of Game Theory in Economics," 22.

13. Ibid., 23.

14. Martin Shubik, *Strategy and Market Structure: Competition, Oligopoly, and the Theory of Games* (New York: Wiley, 1959).

15. G. C. Archibald, review of Shubik, *Strategy and Market Structure*, *Biometrika* 48, nos. 3/4 (1961): 479.

16. E. M. Fels, review of Martin Shubik, *Strategy and Market Structure*, *Weltwirtschaftliches Archiv* 87 (1961): 12.

17. Archibald, review of Shubik, *Strategy and Market Structure*, 479. See also similar sentiments in Sidney Weintraub, "Oligopoly and Game Theory: A Review of Martin Shubik's *Strategy and Market Structure*," *Kyklos* 13, no. 3 (1960): 400–406.

18. John C. Harsanyi, review of Martin Shubik, *Strategy and Market Structure*, *Econometrica* 29, no. 2 (1961): 268.

19. John C. Harsanyi, "A Bargaining Model for the Cooperative *n*-Person Game," in *Contributions to the Theory of Games, vol. 4*, ed. Kuhn and Tucker, 325.

20. Ibid., 347.

21. See especially Harsanyi, "Games with Incomplete Information Played by 'Bayesian' Players, I–III: Part I. The Basic Model," *Management Science* 14, no. 3 (1967): 159–82; "Part II. Bayesian Equilibrium Points," 320–34; and "Part III. The Basic Probability Distribution of the Game," 486–502. Also Harsanyi, "Bargaining in Ignorance of the Opponent's Utility Function," *Journal of Conflict Resolution* 6, no. 1 (1962): 29–38.

22. See chapter 1, "Overview of the Notion of 'Threat' and Its Relation to Bargaining Theories," 13, in *Models of Gradual Reduction of Arms* (Report Prepared for the US Arms Control and Disarmament Agency, Mathematica, Princeton, NJ, September 1967).

23. Ibid., 57.

24. J. Harsanyi, "Individualistic and Functionalistic Explanations in the Light of Game Theory: The Example of Social Status," in *Problems in the Philosophy of Science*, ed. I. Lakatos and A. Musgrave (Amsterdam: North-Holland, 1968), 305.

25. Ibid., 309.

26. See, e.g., John C. Harsanyi, "Bargaining and Conflict Situations in the Light of a New Approach to Game Theory," *American Economic Review* 55, nos. 1/2 (1965): 447–57.

27. Harsanyi, "Individualistic and Functionalistic Explanations in the Light of Game Theory," 309.

28. G. Debreu, "A Social Equilibrium Existence Theorem," *Proceedings of the National Academy of Sciences* 38, no. 10 (1952): 886–93; Kenneth J. Arrow and Gerard Debreu, "Existence of an Equilibrium for a Competitive Economy,"

Econometrica 22, no. 3 (1954): 265–90; Gerard Debreu, *Theory of Value: An Axiomatic Analysis of Economic Equilibrium* (New York: Wiley, 1959). For clarification of the precise relationship between postwar general equilibrium models and the earlier work of von Neumann and Wald, see the historical comments in Arrow and Debreu, "Existence of an Equilibrium for a Competitive Economy," 288, and the discussion in *John von Neumann and Modern Economics*, ed. Dore, Chakravarty and M. Goodwin.

29. On the relationship between the equilibrium models of the 1950s and the earlier work of von Neumann and Wald, see, e.g., Arrow and Debreu, "Existence of an Equilibrium for a Competitive Economy," 288. For computational approaches to general equilibrium see Herbert Scarf, "An Example of an Algorithm for Calculating General Equilibrium Prices," *American Economic Review* 59, no. 4 (1969): 669–77.

30. See, e.g., Arrow and Debreu, "Existence of an Equilibrium for a Competitive Economy," 265, on hopes for the new theory; on the history and significance of equilibrium in economic thought, see especially Bruna Ingrao and Giorgio Israel, *The Invisible Hand: Economic Equilibrium in the History of Science* (Cambridge, MA: MIT Press, 1990); Philip Mirowski, *More Heat Than Light: Economics as Social Physics, Physics as Nature's Economics* (Cambridge: Cambridge University Press, 1989); Philip Mirowski, R. Shaw, and Arnold Mandell, "The Rise and Fall of the Concept of Equilibrium in Economic Analysis," *Recherches économiques de Louvain/Louvain Economic Review* 55, no. 4 (1989): 447–68.

31. For historical perspectives on this mathematization of economics connected with general equilibrium models in the mid-twentieth century see especially Weintraub and Mirowski, "The Pure and the Applied," 245–72; Weintraub, *How Economics Became a Mathematical Science.*

32. See, e.g., the emerging applications of fixed-point approximation mentioned in Herbert E. Scarf, "Fixed-Point Theorems and Economic Analysis: Mathematical Theorems Can Be Used to Predict the Probable Effects of Changes in Economic Policy," *American Scientist* 71, no. 3 (1983): 289–96.

33. This comes across most strongly in Andrew Schotter, "Core Allocations and Competitive Equilibrium: A Survey," *Zeitschrift für Natinoalökonomie* 33 (1973): 281–313. The classic papers connecting cooperative game theory with general equilibrium theory are Martin Shubik, "Edgeworth Market Games," in *Contributions to the Theory of Games, vol. 4*, ed. Kuhn and Tucker, and Gerard Debreu and Herbert Scarf, "A Limit Theorem on the Core of an Economy," *International Economic Review* 4, no. 3 (1963): 235–46.

34. Andrew Schotter and Gerhard Schwödiauer, "Economics and the Theory of Games: A Survey," *Journal of Economic Literature* 18, no. 2 (1980): 480.

35. "The Prize in Economics 1994—Press Release," Nobelprize.org, Febru-

ary 29, 2012, http://www.nobelprize.org/nobel_prizes/economics/laureates/1994/
press.html.

36. This position is expressed *inter alia* in Schmidt, *Game Theory and Economic Analysis*; Christian Schmidt, "Game Theory and Economics: An Historical Survey," *Revue d'économie politique* 100, no. 5 (1990), cited 593, 589. For an excellent counterpoint to this argument see especially Leonard, "Reading Cournot, Reading Nash."

37. Eric van Damme, "On the Contributions of John C. Harsanyi, John F. Nash, and Reinhard Selten," *International Journal of Game Theory* 24 (1995): 9.

38. See S. A. T. Rizvi, "The Sonnenschein-Mantel-Debreu Results after Thirty Years," *History of Political Economy* 38, suppl 1 (2006): 228–45; S. Abu Turab Rizvi, "The Microfoundations Project in General Equilibrium Theory," *Cambridge Journal of Economics* 18 (1994): 357–77; Rizvi, "Game Theory to the Rescue?," 1–28. The seminal article undermining the general equilibrium framework is Hugo Sonnenschein, "Market Excess Demand Functions," *Econometrica* 40, no. 3 (1972): 549–63; more readable is Donald Saari, "Mathematical Complexity of Simple Economics," *Notices of the AMS* 42, no. 2 (1995): 222–30. For broader perspectives on the demise of general equilibrium theory, see Frank Ackerman, "Still Dead after All These Years: Interpreting the Failure of General Equilibrium Theory," *Journal of Economic Methodology* 9, no. 2 (2002): 119–39.

39. For an excellent statement of this explanation to game theory's growth in popularity during the 1980s see Giocoli, "Three Alternative (?) Stories on the Late 20th Century Rise of Game Theory," 194–200.

40. See especially Paul Milgrom and John Roberts, "Limit Pricing and Entry Under Incomplete Information: An Equilibrium Analysis," *Econometrica* 50, no. 2 (1982): 443–59; Paul Milgrom and Nancy Stokey, "Information, Trade and Common Knowledge," *Journal of Economic Theory* 26, no. 1 (1982): 17–27; Paul Milgrom and John Roberts, "Predation, Reputation, and Entry Deterrence," *Journal of Economic Theory* 27, no. 2 (1982): 280–312, for prominent examples of this literature; and Kofi O. Nti and Martin Shubik, "Entry in Oligopoly Theory: A Survey," *Eastern Economic Journal* 5, nos. 1/2 (1979): 271–89, for a broader survey of game theoretic models of market entry to 1979. Suggestively, many of the seminal works modeling market predation in terms of game theory draw analogies between the competing firms and superpower rivalry.

41. Giocoli, "Three Alternative (?) Stories," 201–5. See also the historical comments in Roger B. Myerson: "Perspectives on Mechanism Design in Economic Theory," *American Economic Review* 98, no. 3 (2008): 586–603; "Comments on 'Games with Incomplete Information Played by "Bayesian" Players,'" 1818–24; "Learning Game Theory from John Harsanyi," 20–25; "Nash Equilibrium and the History of Economic Theory," *Journal of Economic Literature* 37,

no. 3 (1999): 1067–82; and *Game Theory: Analysis of Conflict* (Cambridge, MA: Harvard University Press, 1997).

42. Myerson, "Nash Equilibrium and the History of Economic Theory," 1069.

43. Kenneth J. Arrow and Leonid Hurwicz, "Some Remarks on the Equilibria of Economic Systems," *Econometrica* 28, no. 3 (1960): 640–46; Kenneth J. Arrow, H. D. Block, and Leonid Hurwicz, "On the Stability of the Competitive Equilibrium, II," *Econometrica* 27, no. 1 (1959): 82–109; Leonid Hurwicz, "Input-Output Analysis and Economic Structure," *American Economic Review* 45, no. 5 (1955): 945.

44. Leonid Hurwicz, "The Design of Mechanisms for Resource Allocation," *American Economic Review* 63, no. 2 (1973): 23. See also Leonid Hurwicz, "Inventing New Institutions: The Design Perspective," *American Journal of Agricultural Economics* 69, no. 2 (1987): 395–402; Hurwicz, "On Informationally Decentralized Systems," in *Decision and Organization: A Volume in Honor of Jacob Marschak*, ed. C. B. McGuire and Roy Radner (Amsterdam: North-Holland, 1972), 297–336; Hurwicz, "On the Concept and Possibility of Informational Decentralization," *American Economic Review* 59, no. 2 (1969): 513–24.

45. John Harsanyi, *Rational Behavior and Bargaining Equilibrium in Games and Social Situations* (Cambridge: Cambridge University Press, 1977).

46. Horace W. Brock, "A Critical Discussion of the Work of John C. Harsanyi," *Theory and Decision* 9 (1978): 351.

47. Reinhard Selten: "Reexamination of the Perfectness Concept for Equilibrium Points in Extensive Games," *International Journal of Game Theory* 4, no. 1 (1975): 25–55; "Spieltheoretische Behandlung eines Oligopolmodells mit Nachfrageträgheit," *Journal of Institutional and Theoretical Economics* 131, no. 2 (1975): 374; "Spieltheoretische Behandlung eines Oligopolmodells mit Nachfrageträgheit: Teil I: Bestimmung Des Dynamischen Preisgleichgewichts," *Journal of Institutional and Theoretical Economics* 121, no. 2 (1965): 301–24; "Spieltheoretische Behandlung eines Oligopolmodells mit Nachfrageträgheit: Teil II: Eigenschaften Des Dynamischen Preisgleichgewichts," *Journal of Institutional and Theoretical Economics* 121, no. 4 (1965): 667–89; "A Simple Model of Imperfect Competition, Where 4 Are Few and 6 Are Many," *International Journal of Game Theory* 2 (1973): 141–201.

48. Selten, "Reexamination of the Perfectness Concept for Equilibrium Points in Extensive Games" *International Journal of Game Theory* 4, no. 1 (1975): 25–55; John C. Harsanyi and Reinhard Selten, "A Generalized Nash Solution for Two-Person Bargaining Games with Incomplete Information," *Management Science* 18, no. 5 (1972), cited P89.

49. This literature is by now enormous; see especially E. van Damme, *Stability and Perfection of Nash Equilibria* (Princeton, NJ: Princeton University Press, 1991); van Damme, *Refinements of the Nash Equilibrium Concept* (Berlin: Springer-Verlag, 1983); and Ken Binmore, Martin J. Osborne, and Ariel Rubenstein, "Noncooperative Models of Bargaining," in *Handbook of Game The-*

ory with Economic Applications, vol. 1, ed. Robert Aumann and Sergiu Hart (Amsterdam: Elsevier, 1992), 179–225, for snapshots of its development to the early 1990s.

50. Robert Sugden, "Review of *A General Theory of Equilibrium Selection in Games,*" *Economica* 56, no. 224 (1989): 530.

51. Ken Binmore, "Review of *A General Theory of Equilibrium Selection in Games,*" *Journal of Economic Literature* 27, no. 3 (1989): 1172.

52. George J. Stigler, *The Organization of Industry* (Homewood, IL: R. D. Irwin, 1968), 1.

53. Robert H. Porter, "A Review Essay on Handbook of Industrial Organization," *Journal of Economic Literature* 29, no. 2 (1991): 553–72.

54. D. Fudenberg and J. Tirole, "Noncooperative Game Theory for Industrial Organization: An Introduction and Overview," in *Handbook of Industrial Organization, vol. 1,* ed. R. Schmalensee and R. D. Willig (Amsterdam: North-Holland, 1989), 322.

55. Sam Peltzman, "The Handbook of Industrial Organization," *Journal of Political Economy* 99, no. 1 (1991): 206.

56. Porter, "A Review Essay on *Handbook of Industrial Organization,*" 554.

57. Rubinstein, "Comments on the Interpretation of Game Theory," 923.

58. Ibid.

59. Horst Todt, "Reinhard Selten and the Scientific Climate in Frankfurt during the Fifties," in *The Selten School of Behavioral Economics: A Collection of Essays in Honor of Reinhard Selten,* ed. Axel Ockenfels and Abdolkarim Sadrieh (New York: Springer Verlag, 2010), 31.

60. Ibid.

61. See especially Heinz Sauermann and Reinhard Selten, "Ein Oligopolexperiment," *Journal of Institutional and Theoretical Economics* 115, no. 3 (1959): 427–71.

62. Klaus Abbink and Jordi Brandts, "Drei Oligopolexperimente," in *Selten School of Behavioral Economics,* ed. Ockenfels and Sadrieh, 54. On the overlapping histories of experimental economics and behavioral economics, see especially Floris Heukelom, "What to Conclude from Psychological Experiments: The Contrasting Cases of Experimental and Behavioral Economics," *History of Political Economy* 43, no. 4 (2011): 649–81; Ivan Moscati, "Early Experiments in Consumer Demand Theory: 1930–1970," *History of Political Economy* 39, no. 3 (2007): 359–401; Esther-Mirjam Sent, "Behavioral Economics: How Psychology Made Its (Limited) Way Back Into Economics," *History of Political Economy* 36, no. 4 (2005): 735–60; Philippe Fontaine and Robert Leonard, *The Experiment in the History of Economics* (London: Routledge, 2005).

63. Reinhard Selten, "The Chain Store Paradox," *Theory and Decision* 9, no. 2 (1978): 128.

64. Ibid., 132–33.

65. David M. Kreps and Robert Wilson, "Reputation and Imperfect Information," *Journal of Economic Theory* 27, no. 2 (1982): 253–79. Similar work on supergames is also found in David M. Kreps et al., "Rational Cooperation in the Finitely Repeated Prisoners' Dilemma," *Journal of Economic Theory* 27, no. 2 (1982): 245–52.

66. Selten, "Chain Store Paradox," 152.

67. Reinhard Selten, "Bounded Rationality," *Journal of Institutional and Theoretical Economics* 146, no. 4 (1990): 651.

68. Selten's comments on this occasion came as a response to a paper by Kahneman; see Daniel Kahneman, "New Challenges to the Rationality Assumption," *Journal of Institutional and Theoretical Economics* 150, no. 1 (1994): 18–36.

69. Reinhard Selten, "New Challenges to the Rationality Assumption: Comment," *Journal of Institutional and Theoretical Economics* 150, no. 1 (1994): 42. Selten makes similar comments in a number of places, e.g., Gerd Gigerenzer and Reinhard Selten, "Rethinking Rationality," in *Bounded Rationality: The Adaptive Toolbox*, ed. Gerd Gigerenzer and Reinhard Selten (Cambridge, MA: MIT Press, 2001), 1–12.

70. Ibid.

71. Jacob K. Goeree and Charles A. Holt, "Stochastic Game Theory: For Playing Games, Not Just for Doing Theory," *Proceedings of the National Academy of Sciences of the United States of America* 96, no. 19 (1999): 10564.

72. Abdolkarim Sadrieh, "Reinhard Selten a Wanderer," in *Selten School of Behavioral Economics*, 5.

73. See, e.g., K. G. Binmore, *Calculus* (Cambridge: Cambridge University Press, 1983).

74. Ken Binmore, Joe Swierzbinski, and Chris Proulx, "Does Minimax Work? An Experimental Study," *Economic Journal* 111, no. 473 (2001): 445–64; Binmore, Osborne, and Rubenstein, "Noncooperative Models of Bargaining"; Ken Binmore, *Essays on the Foundations of Game Theory* (Oxford: Basil Blackwell, 1990); Ken Binmore, Avner Shaked, and John Sutton, "An Outside Option Experiment," *Quarterly Journal of Economics* 104, no. 4 (1989): 753–70; Binmore, Shaked, and Sutton, "A Further Test of Noncooperative Bargaining Theory: Reply," *American Economic Review* 78, no. 4 (1988): 837–39; Binmore, Shaked, and Sutton, "Testing Noncooperative Bargaining Theory: A Preliminary Study," *American Economic Review* 75, no. 5 (1985): 1178–80; K. G. Binmore and M. J. Herrero, "Matching and Bargaining in Dynamic Markets," *Review of Economic Studies* 55, no. 1 (1988): 17–31; K. G. Binmore and P. G. Bennett, "Why Game Theory 'Doesn't Work,'" in *Analysing Conflict and Its Resolution: Some Mathematical Contributions* (London: Clarendon, 1987); Ken Binmore and Partha Dasgupta, *The Economics of Bargaining* (Oxford: Basil Blackwell, 1987); Binmore and Dasgupta, *Economic Organizations as Games* (Oxford: Basil Blackwell, 1986); Ken Binmore, Ariel Rubinstein, and Asher Wo-

linsky, "The Nash Bargaining Solution in Economic Modelling," *RAND Journal of Economics* 17, no. 2 (1986): 176–88; K. G. Binmore, "Equilibria in Extensive Games," *Economic Journal* 95 (1985): 51–59.

75. Binmore, *Essays on the Foundations of Game Theory*, vii.

76. Ibid., vii, 5, 22.

77. Ibid., 5–6.

78. In other places, Binmore makes a similar argument that confusion over the *application* of game theory—not the theory itself—was the central reason that game theory did not immediately bear fruit in the social sciences. See, e.g., Binmore and Dasgupta, *Economics of Bargaining*, 1.

79. Binmore, *Essays on the Foundations of Game Theory*, 15. See p. 16 for Binmore's discussion of Dawkins.

80. Binmore and Dasgupta, *Economic Organizations as Games*, 12.

81. Binmore, "Equilibria in Extensive Games," 56; Binmore and Dasgupta, *Economic Organizations as Games*, 8.

82. Binmore, "Equilibria in Extensive Games," 52.

83. Binmore and Dasgupta, *Economic Organizations as Games*, 12.

84. See, e.g., K. G. Binmore, *Natural Justice* (New York: Oxford University Press, 2005); Binmore, *Just Playing* (Cambridge, MA: MIT Press, 1998); Binmore, *Game Theory and the Social Contract* (Cambridge, MA: MIT Press, 1994); Ken Binmore and Larry Samuelson, "An Economist's Perspective on the Evolution of Norms," *Journal of Institutional and Theoretical Economics* 150, no. 1 (1994): 45–63; Ken Binmore, "Social Contract I: Harsanyi and Rawls," *Economic Journal* 99, no. 395 (1989): 84–102; see especially Robert Sugden, "Ken Binmore's Evolutionary Social Theory," *Economic Journal* 111, no. 469 (2001): F213–F243, for a review of this work.

85. Binmore and Dasgupta, *Economic Organizations as Games*, 10.

86. On the theory of auctions, see, e.g., William Vickrey, "Counterspeculation, Auctions, and Competitive Sealed Tenders," *Journal of Finance* 16, no. 1 (1961): 8–37; Roger B. Myerson, "Optimal Auction Design," *Mathematics of Operations Research* 6, no. 1 (1981): 58–73.

87. Ken Binmore and Paul Klemperer, "The Biggest Auction Ever: The Sale of the British 3G Telecom Licences," *Economic Journal* 112, no. 478 (2002): C84.

88. Ibid., C85.

89. See, e.g., the remarks in Martin Shubik, "Game Theory and Operations Research: Some Musings 50 Years Later," *Operations Research* 50, no. 1 (2002): 192–96; Shubik, "Game Theory: Some Observations"; and Schmidt, *Game Theory and Economic Analysis*.

90. Myerson, "Nash Equilibrium and the History of Economic Theory," 1079.

91. Herbert Gintis, *The Bounds of Reason: Game Theory and the Unification of the Behavioral Sciences* (Princeton, NJ: Princeton University Press, 2009), 45.

Works Cited

Archival Holdings Consulted

Alexander Library of Ornithology, University of Oxford, Oxford
    David Lack Papers
British Library, London
    George Price Archive in the W. D. Hamilton Papers
    W. D. Hamilton Papers
Duke University, David M. Rubenstein Rare Book and Manuscript Library,
    Durham, NC
    Oskar Morgenstern Papers
Ford Foundation Archives, New York, NY
    Grant Files
Library of Congress, Manuscript Division, Washington, DC
    John von Neumann Papers
Massachusetts Institute of Technology, Special Collections, Cambridge, MA
    Philip M. Morse Papers
National Archives II, College Park, MD
    Applied Mathematics Panel Technical Reports
    Office of Naval Research, Records
    Office of Scientific Research and Development, Records
    US Air Force Deputy Chief of Staff, Comptroller, Records
Queens University Archives, Kingston, Ontario
    V. C. Wynne-Edwards Papers
RAND Corporation, Santa Monica, CA
    Edwin Paxson Papers
    Rand Archives
    John Williams Papers
University College London, Special Collections, London
    Lionel Penrose Papers

University of Chicago, Regenstein Library, Chicago IL
  Nicholas Rashevsky Papers
University of Michigan–Ann Arbor, Bentley Library, Ann Arbor, MI
  Center for Research on Conflict Resolution Records
  J. D. Singer Papers
  Mental Health Research Institute Records
  Merrill M. Flood Papers

Bibliography

Abbink, Klaus, and Jordi Brandts. "Drei Oligopolexperimente." In *The Selten School of Behavioral Economics: A Collection of Essays in Honor of Reinhard Selten*, edited by Axel Ockenfels and Abdolkarim Sadrieh, 53–72. New York: Springer Verlag, 2010.

Abella, Alex. *Soldiers of Reason: The Rand Corporation and the Rise of the American Empire*. Orlando, FL: Harcourt, 2008.

Abir-Am, Pnina. "The Molecular Transformation of 20th Century Biology." In *Science in the Twentieth Century*, edited by John Krige and Dominique Pestre, 495–524. Stanford, CA: Stanford University Press, 1997.

Abraham, Tara. "Nicholas Rashevsky's Mathematical Biophysics." *Journal of the History of Biology* 37 (2004): 333–85.

Ackerman, Frank. "Still Dead After All These Years: Interpreting the Failure of General Equilibrium Theory." *Journal of Economic Methodology* 9, no. 2 (2002): 119–39.

Alchian, A. "The Meaning of Utility Measurement." *American Economic Review* 43 (1953): 26–50.

Allais, Maurice, and Ole Hagen, eds. *Expected Utility Hypotheses and the Allais Paradox: Contemporary Discussions of Decisions under Uncertainty with Allais' Rejoinder*. Dordrecht: D. Reidel, 1979.

Allen, R.G.D. *Mathematical Economics*. London: Macmillan, 1956.

Allison, Graham T. *Essence of Decision: Explaining the Cuban Missile Crisis*. Boston: Little, Brown, 1971.

Amadae, S. M. *Rationalizing Capitalist Democracy: The Cold War Origins of Rational Choice Liberalism*. Chicago: University of Chicago Press, 2003.

Archibald, G. C. "Review of Shubik, Strategy and Market Structure." *Biometrika* 48, nos. 3/4 (1961): 479–80.

Archibald, Kathleen. *Strategic Interaction and Conflict: Original Papers and Discussion*. Berkeley, CA: Institute of International Studies, University of California, 1966.

Arrow, Kenneth J. "Jacob Marschak, July 23, 1898–July 27, 1977." *National Academy of Science Biographical Memoirs* 60 (1991): 129–46.

———. "Rationality of Self and Others in an Economic System." *Journal of Business* 59, no. 4 (1986): S385–S399.

———. *Social Choice and Individual Values*. 2nd ed. New Haven, CT: Yale University Press, 1963.

Arrow, Kenneth J., H. D. Block, and Leonid Hurwicz. "On the Stability of the Competitive Equilibrium, II." *Econometrica* 27, no. 1 (1959): 82–109.

Arrow, Kenneth J., and Gerard Debreu. "Existence of an Equilibrium for a Competitive Economy." *Econometrica* 22, no. 3 (1954): 265–90.

Arrow, Kenneth J., and Leonid Hurwicz. "A Gradient Method for Approximating Saddle-Points and Constrained Maxima." RAND P-223, June 13, 1951.

———. "Some Remarks on the Equilibria of Economic Systems." *Econometrica* 28, no. 3 (1960): 640–46.

Ashford, Oliver M. *Prophet—or Professor? The Life and Work of Lewis Fry Richardson*. Bristol: Hilger, 1985.

Aumann, Robert J. "The St. Petersburg Paradox: A Discussion of Some Recent Comments." *Journal of Economic Theory* 14 (1977): 443–45.

———. "What Is Game Theory Trying to Accomplish?" In *Frontiers of Economics*, edited by Kenneth J. Arrow and S. Honkapohja, 28–76. Oxford: Basil Blackwell, 1985.

Axelrod, Robert. *The Evolution of Cooperation*. New York: Basic Books, 1984.

Axelrod, Robert, and W. D. Hamilton. "The Evolution of Cooperation." *Science* 211 (1981): 1390–96.

Ayson, Robert. *Thomas Schelling and the Nuclear Age*. New York: Frank Cass, 2004.

Barna, T. "Review of *Theory of Games and Economic Behaviour* by John von Neumann and Oskar Morgenstern." *Economica* 13, no. 50 (1946): 136–38.

Bauer, Raymond A. "National Support for Behavioral Science." *Behavioral Science* 3, no. 3 (1958): 217–27.

Baumol, William J. "The Cardinal Utility Which Is Ordinal." *Economic Journal* 68, no. 272 (1958): 665–72.

———. "Discussion." *American Economic Review* 43, no. 2 (1953): 415–16

———. *Economic Theory and Operations Analysis*. Englewood Cliffs, NJ: Prentice-Hall, 1961.

———. "The Neumann-Morgenstern Utility Index—An Ordinalist View." *Journal of Political Economy* 59, no. 1 (1951): 61–66

Bavelas, Alex. "Communication Patterns in Task-Oriented Groups." *Journal of the Acoustical Society of America* 22, no. 6 (1950): 725–30.

———. "Some Mathematical Properties of Psychological Space." PhD diss., January 9, 1948.

Baxter, James Phinney. *Scientists Against Time*. Boston: Little, Brown, 1946.

Bellman, Richard. *Dynamic Programming*. Princeton, NJ: Princeton University Press, 1957.

———. *Eye of the Hurricane*. Singapore: World Scientific, 1984.

———. "On a Simple Game Possessing Some of the Qualities of Poker and Black-jack, I." RAND RM-139, April 18, 1949.

———. "On a Simple Game Possessing Some of the Qualities of Poker and Black-jack, II." RAND RM-140, April 20, 1949.

Bellman, Richard, and David Blackwell. "On a Simple Game Possessing Some of the Qualities of Poker and Blackjack, III." RAND RM-147, April 25, 1949.

Bellman, Richard, and Stuart E. Dreyfus. *Applied Dynamic Programming*. Princeton NJ: Princeton University Press, 1962.

Bellman, Richard, and M. Girshick, "An Extension of Results on Duels with Two Opponents, One Bullet Each, Silent Guns, Equal Accuracy" RAND RM-108, February 22, 1949.

Belzer, R. L. "Silent Duels, Specified Accuracies, One Bullet." RAND RM-58, October 18, 1948.

———. "Solutions of a Special Reconnaissance Game." RAND RM-202, August 10, 1949.

Benedict, Ruth. *The Chrysanthemum and the Sword: Patterns of Japanese Culture*. Boston: Houghton Mifflin, 1946.

Benoit, Emile, and Kenneth Ewart Boulding. *Disarmament and the Economy*. New York: Harper and Row, 1963.

Betz, H. K. "How Does the German Historical School Fit?" *History of Political Economy* 20, no. 3 (1988): 409–30.

Bierman, H. Scott, and Luis Fernandez. *Game Theory with Economic Applications*. Reading, MA: Addison-Wesley, 1993.

Binmore, Ken. *Calculus*. Cambridge: Cambridge University Press, 1983.

———. "Equilibria in Extensive Games." *Economic Journal* 95 (1985): 51–59.

———. *Essays on the Foundations of Game Theory*. Oxford: Basil Blackwell, 1990.

———. *Game Theory: A Very Short Introduction*. Oxford: Oxford University Press, 2007.

———. *Game Theory and the Social Contract*. Cambridge, MA: The MIT Press, 1994.

———. *Just Playing*. Cambridge, MA: MIT Press, 1998.

———. *Natural Justice*. New York: Oxford University Press, 2005.

———. "Review of 'A General Theory of Equilibrium Selection in Games.'" *Journal of Economic Literature* 27, no. 3 (1989): 1171–73.

———. "Social Contract I: Harsanyi and Rawls." *Economic Journal* 99, no. 395 (1989): 84–102.

———. "Why Game Theory 'Doesn't Work.'" In *Analysing Conflict and Its Resolution: Some Mathematical Contributions,* edited by P. G. Bennett, 23–42. London: Clarendon, 1987.

Binmore, Ken, and Partha Dasgupta. *Economic Organizations as Games*. Oxford: Basil Blackwell, 1986.

———. *The Economics of Bargaining*. Oxford: Basil Blackwell, 1987.

Binmore, K. G., and M. J. Herrero. "Matching and Bargaining in Dynamic Markets." *Review of Economic Studies* 55, no. 1 (1988): 17–31.

Binmore, Ken, and Paul Klemperer. "The Biggest Auction Ever: The Sale of the British 3G Telecom Licences." *Economic Journal* 112, no. 478 (2002): C74–C96.

Binmore, Ken, Martin J. Osborne, and Ariel Rubenstein. "Noncooperative Models of Bargaining." In *Handbook of Game Theory with Economic Applications*, vol. 1, edited by Robert Aumann and Sergiu Hart, 179–225. Amsterdam: Elsevier, 1992.

Binmore, Ken, Ariel Rubinstein, and Asher Wolinsky. "The Nash Bargaining Solution in Economic Modelling." *RAND Journal of Economics* 17, no. 2 (1986): 176–88.

Binmore, Ken, and Larry Samuelson. "An Economist's Perspective on the Evolution of Norms." *Journal of Institutional and Theoretical Economics* (JITE)/ *Zeitschrift für die gesamte Staatswissenschaft* 150, no. 1 (1994): 45–63.

Binmore, Ken, Avner Shaked, and John Sutton. "A Further Test of Noncooperative Bargaining Theory: Reply." *American Economic Review* 78, no. 4 (1988): 837–39

———. "An Outside Option Experiment." *Quarterly Journal of Economics* 104, no. 4 (1989): 753–70.

———. "Testing Noncooperative Bargaining Theory: A Preliminary Study." *American Economic Review* 75, no. 5 (1985): 1178–80.

Binmore, Ken, Joe Swierzbinski, and Chris Proulx. "Does Minimax Work? An Experimental Study." *Economic Journal* 111, no. 473 (2001): 445–64.

Birkhoff, Garrett, and John von Neumann. "The Logic of Quantum Mechanics." *Annals of Mathematics* 37, no. 4 (1936): 823–43.

Blackett, P.M.S. "Critique of Some Contemporary Defense Thinking." *Encounter* 16, no. 4 (1961): 9–17.

———. "Operational Research and Nuclear Weapons." *Journal of the Royal United Services Institution* 106, no. 623 (1961): 342–54.

Blackwell, D. "Noisy Duel, One Bullet Each, Arbitrary Non-Monotone Accuracy." RM-131, March 30, 1949.

Bochner, Salomon. "John von Neumann." *Biographical Memoirs of the National Academy of Sciences* 32 (1958): 456–61.

Bohnenblust, H., M. Dresher, M. A. Girshick, T. Harris, O. Helmer, J.C.C. McKinsey, L. Shapley, and P. Snow. "Contributions to the Theory of Games," Parts I–V. RAND RM-29, February 18, 1948.

Bohnenblust, H., L. S. Shapley, and S. Sherman. "Reconnaissance in Game Theory." RM-208, August 12, 1949.

Borel, Émile. "Sur la theorie des jeux." *Comptes Rendus de l'Academie des Sciences* 186, no. 25 (1928): 1689–91.

Borrello, Mark E. *Evolutionary Restraints: The Contentious History of Group Selection.* Chicago: University of Chicago Press, 2010.

———. "Synthesis and Selection." *Journal of the History of Biology* 36, no. 3 (2003): 531–66.

Boulding, Kenneth Ewart. *Conflict and Defense: A General Theory.* New York: Harper, 1962.

———. *The Economics of Peace.* New York: Prentice-Hall, 1945.

Brand, Stewart. *The Media Lab: Inventing the Future at MIT.* New York: Viking, 1987.

Bray, Charles W. *Psychology and Military Proficiency: A History of the Applied Psychology Panel of the National Defense Research Committee.* Princeton, NJ: Princeton University Press, 1948.

Brennan, D. G. "Review of *Strategy and Conscience.*" *Bulletin of the Atomic Scientists* 21, no. 10 (1965): 18–24.

Brock, Horace W. "A Critical Discussion of the Work of John C. Harsanyi." *Theory and Decision* 9 (1978): 349–67.

Brouwer, L. E. J. "Beweis der Invarianz der Dimensionzahl." *Mathematische Annalen* 70 (1911): 161–65.

Brown, George W. "Iterative Solutions of Games by Fictitious Play." In *Activity Analysis of Production and Allocation,* edited by Tjalling C. Koopmans, 374–76. New York: Wiley, 1951.

———. "Some Notes on Computation of Games Solutions" RAND P-78, April 25, 1949.

Brown, George W., and John von Neumann. "Solutions of Games by Differential Equations." P-142, April 19, 1950.

Burns, Arthur F. *Wesley Clair Mitchell: The Economic Scientist.* New York: National Bureau of Economic Research, 1952.

Caldwell, Bruce. *Hayek's Challenge: An Intellectual Biography of F. A. Hayek.* Chicago: University of Chicago Press, 2004.

Caldwell, John, and Pat Caldwell. *Limiting Population Growth and the Ford Foundation Contribution.* London: F. Pinter, 1986.

Carpenter, Humphrey, and Jennifer R. Doctor. *The Envy of the World: Fifty Years of the BBC Third Programme and Radio 3, 1946–1996.* London: Weidenfeld and Nicolson, 1996.

Carr-Saunders, Alexander. *The Population Problem: A Study in Human Evolution.* Oxford: Clarendon Press, 1922.

Carty, Anthony. "Alfred Verdross and Othmar Spann: German Romantic Nationalism, National Socialism, and International Law." *European Journal of International Law* 6, no. 1 (1995): 1–21.

Caruana, Louis. "John von Neumann's 'Impossibility Proof' in Historical Perspective." *Physis: Rivista Internazionale di Storia della Scienza* 32 (1995): 109–24.

Cassel, Gustav. *Theoretische Sozialökonomie. 2. verb. Aufl. ed. Lehrbuch der allgemeinen volkswirtschaftslehre . . . 2. abt.* Leipzig: C. F. Winter, 1921.

Cassel, Gustav, and Joseph McCabe. *The Theory of Social Economy.* New York: Harcourt, 1924.

Caywood, T. E., and C. J. Thomas, "Applications of Game Theory in Fighter versus Bomber Combat." *Journal of the Operations Research Society of America* 3, no. 4 (1955): 402–11.

Chacko, George. "Economic Behaviour—A New Theory." *Indian Journal of Economics* 30 (1950): 349–65.

Chaloupek, Günther K. "The Austrian Debate on Economic Calculation in a Socialist Economy." *History of Political Economy* 22, no. 4 (1990): 659–75.

Chaloupek, Günther, and Werner Teufelsbauer. *Gesamtwirtschaftliche Planung in Westeuropa: Theoretische Entwicklungen Und Praktische Erfahrungen Seit 1970.* Campus Forschung; Bd. 523. Frankfurt/Main: Campus, 1987.

Chapman, Robert L. "The Background and Implications of the Systems Research Laboratory Studies." In *Symposium on Air Force Human Engineering, Personnel, and Training Research.* National Academy of Sciences–National Research Council Publication. Washington, DC: National Academy of Sciences–National Research Council, 1956.

Chapman, Robert L., John L. Kennedy, Allen Newell, and William C. Biel. "The Systems Research Laboratory's Air Defense Experiments." *Management Science* 5, no. 3 (1959): 250–69.

Christ, Carl F. "History of the Cowles Commission, 1932–1952." In *Economic Theory and Measurement: A Twenty Year Research Report, 1932–1952.* Baltimore, MD: Waverly Press, 1952.

Cohen-Cole, Jamie. *The Open Mind: Cold War Politics and the Sciences of Human Nature.* Chicago: University of Chicago Press, 2014.

Cohn, Carol. "Sex and Death in the Rational World of Defense Intellectuals." *Signs* 12, no. 4 (1987): 687–718.

Collins, Martin J. *Cold War Laboratory: RAND, the Air Force, and the American State, 1945–1950.* Smithsonian History of Aviation and Spaceflight. Washington, DC: Smithsonian Institution Press, 2002.

Committee on Behavioral Sciences, University of Chicago. *Symposium: Profits and Problems of Homeostatic Models in the Behavioral Sciences.* Chicago: University of Chicago, 1954.

Connelly, Matthew James. *Fatal Misconception: The Struggle to Control World Population.* Cambridge, MA: Belknap Press of Harvard University Press, 2008.

Coombs, C. H. "Mathematical Models in Psychological Scaling." *Journal of the Statistical Association* 46 (1951): 480–89.

———. "Psychological Scaling Without a Unit of Measurement." *Psychological Review* 57 (1950): 145–58.

———. "Thurstone's Measurement of Social Values Revisited Forty Years Later." *Journal of Personality and Social Psychology* 6 (1967): 85–91.

Coombs, C. H., Howard Raiffa, and R. M. Thrall. "Mathematical Models and Measurement Theory." In *Decision Processes*, ed. Thrall, Coombs, and Davis, 19–37.

Craig, Cecil C. "Early Days in Statistics at Michigan." *Statistical Science* 1 (1986): 292–93.

Creager, Angela N. H., E. Lunbeck, and M. Norton Wise, eds. *Science without Laws: Model Systems, Cases, Exemplary Narratives*. Durham, NC: Duke University Press, 2007.

Crowther-Heyck, Hunter. *Herbert A. Simon: The Bounds of Reason in Modern America*. Baltimore, MD: Johns Hopkins University Press, 2005.

———. "Patrons of the Revolution: Ideals and Institutions in Postwar Behavioral Science." *Isis* 97, no. 3 (2006): 420–46.

Dalkey, Norman. "Equivalence of Information Patterns and Essentially Determinate Games." In *Contributions to the Theory of Games*, vol. 2, ed. Kuhn and Tucker, 217–44.

Dantzig, George B. "Linear Programming." In *History of Mathematical Programming: A Collection of Personal Reminiscences*, edited by Jan Lenstra, Alexander Rinnooy Kan and Alexander Schrijver, 19–31. Amsterdam: North-Holland, 1991.

———. *Linear Programming and Extensions*. Princeton, NJ: Princeton University Press, 1963.

———. "Reminiscences about the Origins of Linear Programming." *Operations Research Letters* 1 (1982): 43–48.

Dantzig, George B., and Thomas L. Saaty. *Compact City: A Plan for a Livable Urban Environment*. San Francisco: W. H. Freeman, 1973.

Daston, Lorraine. *Classical Probability in the Enlightenment*. Princeton, NJ: Princeton University Press, 1988.

———. "Rational Individuals versus Laws of Society: From Probability to Statistics." In *The Probabilistic Revolution*, vol. 1: *Ideas in History*, edited by Lorenz Krüger, Lorraine J. Daston, and Michael Heidelberger, 295–304. Cambridge, MA: MIT Press, 1987.

Dawkins, Richard. *The Extended Phenotype: The Gene as the Unit of Selection*. Oxford: Freeman, 1982.

———. *The Selfish Gene*. Oxford: Oxford University Press, 1976.

de Chadarevian, S., and N. Hopwood, eds. *Models: The Third Dimension of Science*. Stanford, CA: Stanford University Press, 2004.

Debreu, Gerard. "A Social Equilibrium Existence Theorem." *Proceedings of the National Academy of Sciences* 38, no. 10 (1952): 886–93.

———. *Theory of Value: An Axiomatic Analysis of Economic Equilibrium*. New York: Wiley, 1959.

Debreu, Gerard, and Herbert Scarf. "A Limit Theorem on the Core of an Economy." *International Economic Review* 4, no. 3 (1963): 235–46.

Dell'Aglio, L. "Divergences in the History of Mathematics: Borel, Von Neumann, and the Genesis of Game Theory." *Rivista di Storia della Scienza* 3 (1995): 1–46.

Deming, W. Edwards. "Review of *Theory of Games and Economic Behaviour* by John von Neumann and Oskar Morgenstern." *Journal of the American Statistical Association* 40, no. 230 (1945): 263–65.

Dennis, Michael Aaron. "Our First Line of Defense: Two University Laboratories in the Postwar American State." *Isis* 85 (1994): 427–55.

Deutsch, Morton. "Trust and Suspicion." *Journal of Conflict Resolution* 2, no. 4 (1958): 265–79.

Deutsch, Morton, and Robert M. Krauss. "The Effect of Threat upon Interpersonal Bargaining." *Journal of Abnormal and Social Psychology* 61, no. 2 (1960): 181–89.

Dickson, Paul. *Think Tanks*. New York: Athenaeum, 1971.

Dieudonne, J. "John von Neumann." In *Dictionary of Scientific Biography*, edited by Charles C. Gillespie, 88–92. New York: Scribner and Sons, 1982.

Dimand, Mary Ann, and Robert W. Dimand. *The Foundations of Game Theory*. An Elgar Reference Collection. Cheltenham: Edward Elgar Publishing, 1997.

———. *A History of Game Theory*. New York: Routledge, 1996.

Dodge, Robert. *The Strategist*. Hollis, NH: Hollis Publishing, 2006.

Donaldson, Peter J. "On the Origins of the United States Government's International Population Policy." *Population Studies* 44, no. 3 (1990): 385–99.

Dore, M. H. I., Sukhamoy Chakravarty, and Richard M. Goodwin, eds. *John von Neumann and Modern Economics*. Oxford: Clarendon Press, 1989.

Dorfman, Robert. "The Discovery of Linear Programming." *Annals of the History of Computing* 6 (1984): 283–95.

Dresher, Melvin. "Games of Strategy." *Mathematics Magazine* 93 (1951): 93–99.

———. "Mathematical Models of Conflict." In *Systems Analysis and Policy Planning*, edited by Quade and Boucher, 228–40.

Dugatkin, Lee Alan, and Hudson Kern Reeve. *Game Theory and Animal Behavior*. New York: Oxford University Press, 1998.

Edwards, Paul N. *The Closed World: Computers and the Politics of Discourse in Cold War America*. Cambridge, MA: MIT Press, 1996.

———. *A Vast Machine: Computer Models, Climate Data, and the Politics of Global Warming*. Cambridge MA: MIT Press, 2010.

Edwards, Ward. Review of *Decision Processes*. *American Journal of Psychology* 68, no. 3 (September 1955): 505–7.

Eisenberg, J. F., and Wilton S. Dillon, eds. *Man and Beast: Comparative Social Behavior.* Washington, DC: Smithsonian Institution Press, 1971.

Eisenhart, Churchill. "S.S. Wilks's Princeton Appointment, and Statistics at Princeton before Wilks." In *A Century of Mathematics in America*, vol. 3. Providence, RI: American Mathematical Society, 1989.

Ellsberg, D. "Classic and Current Notions of 'Measurable Utility.'" *Economic Journal* 64, no. 2 55 (1954): 528–56.

Emerson, Alfred E. "Dynamic Homeostasis: A Unifying Principle in Organic, Social, and Ethical Evolution." *Scientific Monthly* 78, no. 2 (1954): 67–85.

Erickson, Paul. "Mathematical Models, Rational Choice, and the Search for Cold War Culture." *Isis* 101, no. 2 (2010): 386–92.

Erickson, Paul, Judy L. Klein, Lorraine Daston, Rebecca Lemov, Thomas Sturm, and Michael D. Gordin. *How Reason Almost Lost Its Mind: The Strange Career of Cold War Rationality.* Chicago: University of Chicago Press, 2013.

Fels, E. M. "Review of Martin Shubik, Strategy and Market Structure." *Weltwirtschaftliches Archiv* 87 (1961): 12–14.

Fisher, Franklin M. "Games Economists Play: A Noncooperative View." *RAND Journal of Economics* 20, no. 1 (1989): 113–24.

Fiske, Marjorie, and Paul F. Lazarsfeld. "The Columbia Office of Radio Research." *Hollywood Quarterly* 1, no. 1 (1945): 51–59.

Fleming, Donald, and Bernard Bailyn, eds. *The Intellectual Migration: Europe and America, 1930–1960.* Cambridge: Belknap Press of Harvard University Press, 1969.

Flood, M. M. "Aerial Bombing Tactics: General Considerations (A World War II Study)." RAND RM-913, September 2, 1952.

———. "Illustrative Example of Application of Koopmans' Transportation Theory to Scheduling Military Tanker Fleet." RAND RM-267, October 17, 1949.

———. "The Objectives of TIMS." *Management Science* 2, no. 2 (1956): 178–84.

———. "Recursive Methods in Business-Cycle Analysis." *Econometrica* 8, no. 4 (1940): 333–53.

———. "Some Experimental Games." RAND RM-789-1, June 20, 1952.

Fontaine, Philippe. "The Homeless Observer: John Harsanyi on Interpersonal Utility Comparisons and Bargaining, 1950–1964." *Journal of the History of Economic Thought* 32, no. 2 (2010): 145–73.

———. "Stabilizing American Society: Kenneth Boulding and the Integration of the Social Sciences, 1943–1980." *Science in Context* 22, no. 2 (2010): 221–65

Fontaine, Philippe, and Robert Leonard. *The Experiment in the History of Economics.* London: Routledge, 2005.

Forman, Paul. "Behind Quantum Electronics." *Historical Studies in the Physical and Biological Sciences* 18 (1987): 149–229.

Fox, Phyllis. "Mina Rees." In *Women of Mathematics: A Bio-Bibliographic*

Sourcebook, edited by Louise Grinstein and Paul Campbell, 175–81. Westport, CT: Greenwood Press, 1987.

Frank, Steve A. "George Price's Contributions to Evolutionary Genetics." *Journal of Theoretical Biology* 175 (1995): 373–88.

Freedman, Lawrence. *The Evolution of Nuclear Strategy*. New York: St. Martin's Press, 1981.

Friedberg, Aaron L. *In the Shadow of the Garrison State: America's Anti-Statism and Its Cold War Grand Strategy*. Princeton Studies in International History and Politics. Princeton, NJ: Princeton University Press, 2000.

Friedman, Milton, and L. J. Savage. "The Expected-Utility Hypothesis and the Measurability of Utility." *Journal of Political Economy* 60, no. 6 (1952): 463–74.

———. "The Utility Analysis of Choices Involving Risk." *Journal of Political Economy* 56, no. 4 (1948): 279–304.

Fry, Thornton C. "Industrial Mathematics." *American Mathematical Monthly* 48, no. 6 (1941): 1–38.

———. "Mathematicians in Industry—the First 75 Years." *Science* 143, no. 3609 (1964): 934–38.

———. "Mathematics as a Profession Today in Industry." *American Mathematical Monthly* 63, no. 2 (1956): 71–80.

Frydenberg, Erica. *Morton Deutsch: A Life and Legacy of Mediation and Conflict Resolution*. Brisbane: Australian Academic Press, 2005.

Fudenberg, D., and J. Tirole. "Noncooperative Game Theory for Industrial Organization: An Introduction and Overview." In *Handbook of Industrial Organization*, vol. 1, edited by R. Schmalensee and R. D. Willig, 259–327. 2 vols. Amsterdam: North-Holland, 1989.

Gale, David. "The Game of Hex and the Brouwer Fixed-Point Theorem." *American Mathematical Monthly* 86, no. 10 (1979): 818–27.

Gale, David, and F. M. Stewart. "Infinite Games with Perfect Information." In *Contributions to the Theory of Games*, vol. 2, ed. Kuhn and Tucker, 193–216.

Galison, Peter. "The Americanization of Unity." *Daedalus* 127, no. 1 (1998): 45–71.

———. *Image and Logic: A Material Culture of Microphysics*. Chicago: University of Chicago Press, 1997.

———. "The Ontology of the Enemy: Norbert Wiener and the Cybernetic Vision." *Critical Inquiry* 21, no. 1 (1994): 228–66.

Gardner, Martin. *The Scientific American Book of Mathematical Puzzles and Diversions*. New York: Simon and Schuster, 1959.

Ghamari-Tabrizi, Sharon. "Simulating the Unthinkable: Gaming Future War in the 1950s and 1960s." *Social Studies of Science* 30, no. 2 (2000): 163–223.

———. *The Worlds of Herman Kahn: The Intuitive Science of Thermonuclear War*. Cambridge, MA: Harvard University Press, 2005.

Gigerenzer, Gerd. "From Tools to Theories: A Heuristic of Discovery in Cognitive Psychology." *Psychological Review* 98, no. 2 (1991): 254–67.

———. "Where Do New Ideas Come From? A Heuristic of Discovery in the Cognitive Sciences." In *Observation and Experiment in the Natural and Social Sciences*, edited by M. C. Galavotti, 99–139. Dordrecht: Kluwer, 2003.

Gigerenzer, Gerd, and Reinhard Selten. "Rethinking Rationality." In *Bounded Rationality: The Adaptive Toolbox*, edited by Gerd Gigerenzer and Reinhard Selten, 1–12. Cambridge, MA: MIT Press, 2001.

Ginsburg, Faye D., and Rayna R. Reiter. *Conceiving the New World Order: The Global Politics of Reproduction*. Berkeley: University of California Press, 1995.

Gintis, Herbert. *The Bounds of Reason: Game Theory and the Unification of the Behavioral Sciences*. Princeton, NJ: Princeton University Press, 2009.

Giocoli, Nicola. *Modeling Rational Agents: From Interwar Economics to Early Modern Game Theory*. Northampton, MA: Edward Elgar, 2003.

———. "Three Alternative (?) Stories on the Late 20th Century Rise of Game Theory." *Studi e Note di Economia* 2 (2009): 187–210.

Girshick, M. A. "A Generalization of the Silent Duel, Two Opponents, One Bullet Each, Arbitrary Accuracy." RAND RM-206, August 1, 1949.

Girshick, M., and D. Blackwell. "A Loud Duel with Equal Accuracy Where Each Duelist has only a Probability of Possessing a Bullet." RAND RM-219, August 19, 1949.

Goeree, Jacob K., and Charles A. Holt. "Stochastic Game Theory: For Playing Games, Not Just for Doing Theory." *Proceedings of the National Academy of Sciences of the United States of America* 96, no. 19 (1999): 10564–67.

Gold, E. "Lewis Fry Richardson. 1881–1953." *Obituary Notices of Fellows of the Royal Society* 9, no. 1 (1954): 216–35.

Goodman, Leo A. "On Methods of Amalgamation." In *Decision Processes*, ed. Thrall, Coombs, and Davis, 39–48.

Green, Bert F. "Latent Structure Analysis and Its Relation to Factor Analysis." *Journal of the American Statistical Association* 47, no. 257 (1952): 71–76.

Green, Donald P., and Ian Shapiro. *Pathologies of Rational Choice Theory: A Critique of Applications in Political Science*. New Haven, CT: Yale University Press, 1994.

Green, Philip. *Deadly Logic: The Theory of Nuclear Deterrence*. Columbus: Ohio State University Press, 1966.

Grier, David Alan. *When Computers Were Human*. Princeton, NJ: Princeton University Press, 2005.

Grimmer-Solem, Erik. *The Rise of Historical Economics and Social Reform in Germany, 1864–1894*. Oxford: Oxford University Press, 2003.

Grinker, Roy R. *Men Under Stress*. Edited by John P Spiegel. Philadelphia: Blakiston, 1945.

———. *Toward a Unified Theory of Human Behavior*. New York: Basic Books, 1956.

Gross, Oliver A. "An Infinite-Dimensional Extension of a Symmetric Blotto Game." RAND RM-718, November 8, 1951.

———. "The Symmetric Blotto Game." RAND RM-424, July 1950.

Gross, Oliver A., and R. A. Wagner. "A Continuous Blotto Game." RAND RM-424, June 1950.

Grossman, Andrew D. *Neither Dead nor Red: Civil Defense and American Political Development during the Early Cold War*. New York: Routledge, 2001.

Guetzkow, Harold S. *Groups, Leadership and Men: Research in Human Relations: Reports on Research Sponsored by the Human Relations and Morale Branch of the Office of Naval Research, 1945–1950*. Pittsburgh, PA: Carnegie Press, 1951.

Gumbel, E. J. "Review of *Theory of Games and Economic Behaviour* by John von Neumann and Oskar Morgenstern." *Annals of the American Academy of Political and Social Science* 239 (1945): 209–10.

Hacking, Ian. *The Emergence of Probability: A Philosophical Study of Early Ideas about Probability, Induction, and Statistical Inference*. Cambridge: Cambridge University Press, 1984.

Halberstam, David. *The Best and the Brightest*. New York: Random House, 1972.

Hamilton, W. D. "The Evolution of Altruistic Behavior." *American Naturalist* 97 (1963): 354–56.

———. "Extraordinary Sex Ratios." *Science* 156 (1967): 477–88.

———. "The Genetical Evolution of Social Behavior, I." *Journal of Theoretical Biology* 7 (1964): 1–26.

———. "The Genetical Evolution of Social Behavior, II." *Journal of Theoretical Biology* 7 (1964): 27–52.

———. "Geometry for the Selfish Herd." *Journal of Theoretical Biology* 31 (1971): 295–311.

———. *Narrow Roads of Gene Land: The Collected Papers of W. D. Hamilton*. New York: W. H. Freeman, 1996.

———. *The Narrow Roads of Gene Land*, vol. 2. Oxford: Oxford University Press, 2001.

———. "Review of K. Mather, Human Diversity." *Population Studies* 19, no. 2 (1965): 203–5

———. "Selection of Selfish and Altruistic Behavior in Some Extreme Models." In *Man and Beast: Comparative Social Behavior*, edited by J. F. Eisenberg and Wilton S. Dillon, 57–91. Washington, DC: Smithsonian Institution Press, 1971.

———. "Selfish and Spiteful Behavior in an Evolutionary Model." *Nature* 228 (1970): 1218–20.

Hammerstein, Peter, and Reinhard Selten. "Game Theory and Evolutionary Biology." In *Handbook of Game Theory with Economic Applications*, vol. 2, edited by Robert J. Aumann and S. Hart, 929–93. Amsterdam: Elsevier, 1994.

Hammond, Debora. *The Science of Synthesis*. Boulder: University Press of Colorado, 2003.

Hands, D. Wade. "On Operationalisms and Economics." *Journal of Economic Issues* 38, no. 4 (2004): 953–68.

Harman, Oren S. *The Price of Altruism: George Price and the Search for the Origins of Kindness*. New York: W. W. Norton, 2010.

Harrison, G. Ainsworth. "The Genetic Structure of Human Populations." In *The Royal Society Population Study Group Abstracts of Proceedings #2*. London: Royal Society, October 21, 1966.

Harrison, Robert W. "Review of *Theory of Games and Economic Behaviour* by John von Neumann and Oskar Morgenstern." *Journal of Farm Economics* 27, no. 3 (1945): 725–26.

Harsanyi, John C. "Bargaining and Conflict Situations in the Light of a New Approach to Game Theory." *American Economic Review* 55, nos. 1/2 (1965): 447–57.

———. "Bargaining in Ignorance of the Opponent's Utility Function." *Journal of Conflict Resolution* 6, no. 1 (1962): 29–38.

———. "A Bargaining Model for the Cooperative *n*-Person Game." In *Contributions to the Theory of Games*, vol. 4, ed. Tucker and Luce, 325–55.

———. "Games with Incomplete Information Played by 'Bayesian' Players, I–III. Part I. The Basic Model." *Management Science* 14, no. 3 (1967): 159–82.

———. "Games with Incomplete Information Played by 'Bayesian' Players, I–III. Part II. Bayesian Equilibrium Points." *Management Science* 14, no. 5 (1968): 320–34.

———. "Games with Incomplete Information Played by 'Bayesian' Players, I–III. Part III. The Basic Probability Distribution of the Game." *Management Science* 14, no. 7 (1968): 486–502.

———. "A Generalized Nash Solution for Two-Person Bargaining Games with Incomplete Information." *Management Science* 18, no. 5 (1972): P80–P106.

———. "Individualistic and Functionalistic Explanations in the Light of Game Theory: The Example of Social Status." In *Problems in the Philosophy of Science*, edited by I. Lakatos and A. Musgrave, 305–48. Amsterdam: North-Holland, 1968.

———. *Rational Behavior and Bargaining Equilibrium in Games and Social Situations*. Cambridge: Cambridge University Press, 1977.

———. "Rational-Choice Models of Political Behavior vs. Functionalist and Conformist Theories." *World Politics* 21, no. 4 (1969): 513–38.

——. "Review of Martin Shubik, Strategy and Market Structure." *Economet-rica* 29, no. 2 (1961): 267–68.

——. "Subjective Probability and the Theory of Games: Comments on Kadane and Larkey's Paper." *Management Science* 28, no. 2 (1982): 120–24.

Hayek, Friedrich. *Capitalism and the Historians.* Chicago: University of Chicago Press, 1954.

——. "The Nature and History of the Problem." In Hayek et al., *Collectivist Economic Planning,* 1–40.

Hayek, Friedrich A. von, N. G. Pierson, Ludwig Von Mises, Georg Halm, and Enrico Barone. *Collectivist Economic Planning: Critical Studies on the Pos-sibilities of Socialism.* London: G. Routledge, 1935.

Heims, Steve J. *Constructing a Social Science for Postwar America: The Cyber-netics Group, 1946–1953.* Cambridge, MA: MIT Press, 1993

——. *The Cybernetics Group.* Cambridge, MA: MIT Press, 1991.

——. *John von Neumann and Norbert Wiener: From Mathematics to the Tech-nologies of Life and Death.* Cambridge, MA: MIT Press, 1980.

Herman, Ellen. *The Romance of American Psychology: Political Culture in the Age of Experts.* Berkeley: University of California Press, 1995.

Heukelom, Floris. "What to Conclude from Psychological Experiments: The Contrasting Cases of Experimental and Behavioral Economics." *History of Political Economy* 43, no. 4 (2011): 649–81.

Hicks, J. R. and R. G. D. Allen. "A Reconsideration of the Theory of Value. Part I." *Economica* 1, no. 1 (1934): 52–76.

——. "A Reconsideration of the Theory of Value. Part II. A Mathematical Theory of Individual Demand Functions." *Economica* 1, no. 2 (May 1934): 196–219.

Hodgson, Geoffrey Martin. "Behind Methodological Individualism." *Cam-bridge Journal of Economics* 10 (1986): 211–24.

——. *How Economics Forgot History: The Problem of Historical Specificity in Social Science.* London: Routledge, 2001.

Hofbauer, Josef, and Karl Sigmund. *The Theory of Evolution and Dynamical Systems: Mathematical Aspects of Selection.* Cambridge: Cambridge Univer-sity Press, 1988.

Hoffman, Paul, Leon Festinger, and Douglas Lawrence. "Tendencies Toward Group Comparability in Competitive Bargaining." In *Decision Processes,* ed. Thrall, Coombs, and Davis, 231–53.

Hölldobler, Bert, and Edward O. Wilson. *The Ants.* Cambridge, MA: Belknap Press of Harvard University Press, 1990.

Horowitz, Irving Louis. *The Rise and Fall of Project Camelot: Studies in the Re-lationship between Social Science and Practical Politics.* Cambridge, MA: MIT Press, 1967.

———. *The War Game: Studies of the New Civilian Militarists.* New York: Ballantine Books, 1963.

Hounshell, David. "The Cold War, RAND, and the Generation of Knowledge, 1946–1962." *Historical Studies in the Physical and Biological Sciences* 27 (1997): 237–67.

Hurwicz, Leonid. "The Design of Mechanisms for Resource Allocation." *American Economic Review* 63, no. 2 (1973): 1–30.

———. "Input-Output Analysis and Economic Structure." *American Economic Review* 45, no. 5 (1955): 945.

———. "Inventing New Institutions: The Design Perspective." *American Journal of Agricultural Economics* 69, no. 2 (1987): 395–402.

———. "On Informationally Decentralized Systems." In *Decision and Organization: A Volume in Honor of Jacob Marschak*, edited by C. B. McGuire and Roy Radner, 297–336. Amsterdam: North-Holland, 1972.

———. "On the Concept and Possibility of Informational Decentralization." *American Economic Review* 59, no. 2 (1969): 513–24.

———. "*The Theory of Games and Economic Behavior*, by John von Neumann; Oskar Morgenstern." *Annals of Mathematical Statistics* 19, no. 3 (1948): 436–37.

Igo, Sarah Elizabeth. *The Averaged American: Surveys, Citizens, and the Making of a Mass Public.* Cambridge, MA: Harvard University Press, 2007.

Ingrao, Bruna, and Giorgio Israel. *The Invisible Hand: Economic Equilibrium in the History of Science.* Cambridge, MA: MIT Press, 1990.

Isaac, Joel. "Epistemic Design: Theory and Data in Harvard'sDepartment of Social Relations." In *Cold War Social Science: Knowledge Production, Liberal Democracy, and Human Nature*, ed. Solovey and Cravens, 79–95.

———. "Tool Shock: Technique and Epistemology in the Postwar Social Sciences." *History of Political Economy* 42 (2010): 133–64.

———. *Working Knowledge: Making the Human Sciences from Parsons to Kuhn.* Cambridge, MA: Harvard University Press, 2012.

Isserman, Maurice. *If I Had A Hammer: The Death of the Old Left and the Birth of the New Left.* Champaign: University of Illinois Press, 1993.

James, Ioan. *History of Topology.* Amsterdam: Elsevier Science, 1999.

Jardini, David. "Out of the Blue Yonder: The RAND Corporation's Diversification into Social Welfare Research, 1946–1968." PhD diss., Carnegie Mellon University, 1996.

———. "Out of the Blue Yonder: The Transfer of Systems Thinking From the Pentagon to the Great Society, 1961–1965." In *Systems, Experts, and Computers: The Systems Approach in Management and Engineering, World War II and After*, edited by Agatha C. Hughes and Thomas P. Hughes, 311–58. Cambridge, MA: MIT Press, 2000.

K., M. G. "Review of 'Theory of Games and Economic Behaviour' by John von

Neumann and Oskar Morgenstern." *Journal of the Royal Statistical Society* 107, nos. 3/4 (1944): 293.

Kahn, Herman. *On Thermonuclear War*. Princeton, NJ: Princeton University Press, 1960.

———. *Thinking about the Unthinkable*. New York: Horizon Press, 1962.

Kahn, Herman, and Irwin Mann. "Game Theory." RAND P-1166 (1957).

———. "Ten Common Pitfalls." RAND RM-1937, Santa Monica, CA: RAND Corporation, RM-1937, July 17, 1957.

———. "Techniques of Systems Analysis." RAND Memorandum RM-1829-1-PR (1957).

———. "War Gaming." RAND P-1167 (1957).

Kahneman, Daniel. "New Challenges to the Rationality Assumption." *Journal of Institutional and Theoretical Economics (JITE) / Zeitschrift für die gesamte Staatswissenschaft* 150, no. 1 (1994): 18–36.

Kaiser, David. *Drawing Theories Apart: The Dispersion of Feynman Diagrams in Postwar Physics*. Chicago: University of Chicago Press, 2005.

———. "A Ψ Is Just a Ψ? Pedagogy, Practice, and the Reconstitution of General Relativity, 1942–1975." *Studies in the History and Philosophy of Modern Physics* 29 (1998): 321–38.

Kakutani, S. "A Generalization of Brouwer's Fixed Point Theorem." *Duke Mathematical Journal* 8, no. 3 (1941): 457–59.

Kalisch, G., J. W. Milnor, J. Nash, and E. D. Nering. "Some Experimental n-Person Games." In *Decision Processes*, ed. Thrall, Coombs, and Davis, 301–28.

Kaplan, Fred M. *The Wizards of Armageddon*. New York: Simon and Schuster, 1983.

Kay, Lily. *Who Wrote the Book of Life? A History of the Genetic Code*. Stanford, CA: Stanford University Press, 2000.

Kaysen, Carl. "A Revolution in Economic Theory?" *Review of Economic Studies* 14, no. 1 (1946): 1–15.

Keller, Evelyn Fox. *The Century of the Gene*. Cambridge, MA: Harvard University Press, 2000.

Kennedy, John L. "The Uses and Limitations of Mathematical Models, Game Theory, and Systems Analysis in Planning and Problem Solution." In *Current Trends: Psychology in the World Emergency*, 97–116. Pittsburgh, PA: University of Pittsburgh Press, 1952.

Kerman, Cynthia Earl. *Creative Tension: The Life and Thought of Kenneth Boulding*. Ann Arbor: University of Michigan Press, 1974.

Kevles, Daniel J. *The Physicists: A History of a Scientific Community in Modern America*. New York: Knopf, 1978.

Kirby, Maurice W. *Operational Research in War and Peace: The British Experience from the 1930s to 1970s*. London: Imperial College Press, 2003.

Kjeldsen, Tinne Hoff. "A Contextualized Historical Analysis of the Kuhn-

Tucker Theorem in Nonlinear Programming: The Impact of World War II." *Historia Mathematica* 27 (2000): 331–61.

———. "The Emergence of Nonlinear Programming: Interactions between Practical Mathematics and Mathematics Proper." *Mathematical Intelligencer* 22, no. 3 (2000): 50–54.

———. "John von Neumann's Conception of the Minimax Theorem: A Journey through Different Mathematical Contexts." *Archive for the History of Exact Sciences* 56 (2001): 39–68.

Klein, Ursula. *Experiments, Models, Paper Tools: Cultures of Organic Chemistry in the Nineteenth Century.* Stanford, CA: Stanford University Press, 2003.

Kleinman, Daniel. *Politics on the Endless Frontier: Postwar Research Policy in the United States.* Durham, NC: Duke University Press, 1995.

Knaster, B., C. Kuratowski, and S. Mazurkiewicz. "Ein Beweis des Fixpunktsatzes für N-dimensionale Simplexe." *Fundamenta Mathematicae* 14 (1929): 132–37.

Kohler, Robert E. "The Ph.D. Machine: Building on the Collegiate Base." *Isis* 81, no. 4 (1990): 638–62.

Koopmans, Tjalling C. *Activity Analysis of Production and Allocation.* New York: Wiley, 1951.

König, Dénes. *Theorie der Endlichen und Unendlichen Graphen.* Leipzig: Teubner, 1936.

———. "Über eine Schlußweise aus dem Endlichen ins Unendliche." *Acta Scientiarum Mathematicarum, Universitatis Szegediensis* 3 (1927): 121–30.

Kraft, Viktor. *The Vienna Circle, the Origin of Neo-Positivism: A Chapter in the History of Recent Philosophy.* New York: Philosophical Library, 1953.

Kreps, D. M. *Game Theory and Economic Modelling.* Oxford: Oxford University Press, 1990.

Kreps, David M., Paul Milgrom, John Roberts, and Robert Wilson. "Rational Cooperation in the Finitely Repeated Prisoners' Dilemma." *Journal of Economic Theory* 27, no. 2 (1982): 245–52.

Kreps, David M., and Robert Wilson. "Reputation and Imperfect Information." *Journal of Economic Theory* 27, no. 2 (1982): 253–79.

Kuhn, H. W. "Extensive Games and the Problem of Information." In *Contributions to the Theory of Games*, vol. 2, ed. Kuhn and Tucker, 193–216.

———. *Lectures on the Theory of Games.* Princeton, NJ: Princeton University Press, 2003.

Kuhn, Harold W., and Albert W. Tucker, eds. *Contributions to the Theory of Games*, vol. 1. Princeton, NJ: Princeton University Press, 1950.

———, eds. *Contributions to the Theory of Games*, vol. 2. Princeton, NJ: Princeton University Press, 1953.

———. "John von Neumann's Work in the Theory of Games and Mathematical Economics." *Bulletin of the American Mathematical Society* 64 (1958): 100–122.

Kurz, Heinz D., and Neri Salvadori. "Sraffa and Von Neumann." *Review of Political Economy* 13, no. 2 (2001): 161–80.

———. "Von Neumann's Growth Model and the 'Classical' Tradition." In *Understanding "Classical" Economics: Studies in Long-Period Theory.* New York: Routledge, 1998.

Kusch, Martin. *Psychologism: A Case Study in the Sociology of Philosophical Knowledge.* London: Routledge, 1995.

Kwa, Chunglin. "Representations of Nature Mediating between Ecology and Science Policy: The Case of the International Biological Programme." *Social Studies of Science* 17, no. 3 (August 1987): 413–42.

Lack, David Lambert. *The Natural Regulation of Animal Numbers.* Oxford: Clarendon Press, 1954.

Lave, Jean. *Cognition in Practice: Mind, Mathematics and Culture in Everyday Life.* New York: Cambridge University Press, 1988.

———. "The Values of Quantification." In *Sociological Review Monograph,* edited by John Law, 88–111. 1986.

Lazarsfeld, Paul F. "The Logical and Mathematical Foundation of Latent Structure Analysis." In *The American Soldier,* vol. 4, *Measurement and Prediction,* ed. Samuel Stouffer et al. Princeton, NJ: Princeton University Press, 1949.

———. *Radio and the Printed Page: An Introduction to the Study of Radio and Its Role in the Communication of Ideas.* New York: Duell, Sloan and Pearce, 1940.

———. "Recent Developments in Latent Structure Analysis." *Sociometry* 18, no. 4 (1955): 391–403.

Lefschetz, Solomon. "Reminiscences of a Mathematical Immigrant in the United States." *American Mathematical Monthly* 77, no. 4 (1970): 344–50.

LeMay, Curtis, and Bill Yenne. *Superfortress: The Boeing B-29 and American Air Power in World War II.* Yardley, PA: Westholme Publishing, 1988.

Lemov, Rebecca. "'Hypothetical Machines': The Science Fiction Dreams of Cold War Social Science." *Isis* 101, no. 2 (2010): 401–11

Lentz, Theodore F. *Towards a Science of Peace: Turning Point in Human Destiny.* New York: Bookman Associates, 1955.

Leonard, Robert J. "'Between Worlds,' or an Imagined Reminiscence by Oskar Morgenstern about Equilibrium and Mathematics in the 1920s." *Journal of the History of Economic Thought* 26, no. 3 (2004): 527–585.

———. "Creating a Context for Game Theory." In *Toward a History of Game Theory,* ed. Weintraub, 29–76.

———. "Ethics and the Excluded Middle: Karl Menger and Social Science in Interwar Vienna." *Isis* 89, no. 1 (1998): 1–26.

———. "From Parlor Games to Social Science: Von Neumann, Morgenstern, and the Creation of Game Theory 1928–1944." *Journal of Economic Literature* 33, no. 2 (1995): 730–61.

———. "Reading Cournot, Reading Nash: The Creation and Stabilisation of the Nash Equilibrium." *Economic Journal* 104, no. 424 (1994): 492–511.

———. *Von Neumann, Morgenstern, and the Creation of Game Theory: From Chess to Social Science, 1900–1960.* New York: Cambridge University Press, 2010.

Leslie, Stuart. *The Cold War and American Science: The Military-Industrial-Academic Complex at MIT and Stanford.* New York: Columbia University Press, 1993.

Levins, Richard. *Evolution in Changing Environments.* Princeton, NJ: Princeton University Press, 1968.

———. "Theory of Fitness in a Heterogeneous Environment." PhD diss., Columbia University, 1965.

Lewin, Kurt. *Field Theory in Social Science.* New York: Harper and Brothers, 1951.

———. "The Research Center for Group Dynamics at Massachusetts Institute of Technology." *Sociometry* 8, no. 2 (1945): 126–36.

Lewin, Kurt, and Martin Gold. *The Complete Social Scientist: A Kurt Lewin Reader.* Washington, DC: American Psychological Association, 1999.

Lewin, Kurt, Fritz Heider, and Grace M. Heider. *Principles of Topological Psychology.* New York: McGraw-Hill, 1936.

Lewin, Kurt, and Ronald Lippitt. "An Experimental Approach to the Study of Autocracy and Democracy: A Preliminary Note." *Sociometry* 1, nos. 3/4 (1938): 292–300.

Lewin, Kurt, Ronald Lippitt, and Ralph K. White. "Patterns of Aggressive Behavior in Experimentally Created 'Social Climates.'" *Journal of Social Psychology* 10 (1939): 271–99.

Lewontin, Richard C. "Evolution and the Theory of Games." *Journal of Theoretical Biology* 1 (1961): 382–403

Light, Jennifer S. *From Warfare to Welfare: Defense Intellectuals and Urban Problems in Cold War America.* Baltimore: Johns Hopkins University Press, 2003.

Lissner, Will. "Mathematical Theory of Poker Is Applied to Business Problems." *New York Times,* March 10, 1946.

Lucas, W. F. "The Proof that a Game May Not Have a Solution." RAND RM-5543-PR, January 1968.

Luce, R. Duncan. "Connectivity and Generalized Cliques in Sociometric Group Structure." *Psychometrika* 15 (1950): 169–90.

———. "A Definition of Stability for *n*-Person Games." *Annals of Mathematics* 59, no. 3 (1954): 357–66.

——, ed. *Developments in Mathematical Psychology: Information, Learning, and Tracking. A Study of the Behavioral Models Project, Bureau of Applied Social Research, Columbia University.* Glencoe, IL: Free Press, 1960.

——. "Dilemma of Research Strategy." *Science* 151, no. 3708 (1966): 318–19.

——. *Individual Choice Behavior: A Theoretical Analysis.* New York: Wiley, 1959.

——. "k-Stability of Symmetric and of Quota Games." *Annals of Mathematics* 62, no. 3 (1955): 517–27.

——. "Networks Satisfying Minimality Conditions." *American Journal of Mathematics* 75, no. 4 (1953): 825–38.

——. "R. Duncan Luce." In *A History of Psychology in Autobiography,* vol. 8, edited by Gardner Lindzey, 245–89. Stanford, CA: Stanford University Press, 1989.

——. "Semiorders and a Theory of Utility Discrimination." *Econometrica* 24, no. 2 (1956): 178–91.

——. "Two Decomposition Theorems for a Class of Finite Oriented Graphs." *American Journal of Mathematics* 74, no. 3 (1952): 701–22.

——. "Ψ-Stability: A New Equilibrium Concept for *n*-Person Game Theory." In *Mathematical Models of Human Behavior: Proceedings of a Symposium,* 32–44. New York: Dunlap and Associates, 1955.

Luce, R. Duncan, and Ward Edwards. "The Derivation of Subjective Scales from Just Noticeable Differences." *Psychological Review* 65, no. 4 (1958): 222–37.

Luce, R. Duncan, D. H. Krantz, Patrick Suppes, and Amos Tversky. *Foundations of Measurement,* vols. 1–3. New York: Academic Press, 1971–89.

Luce, R. Duncan, and Albert D. Perry. "A Method of Matrix Analysis of Group Structure." *Psychometrika* 14, no. 1 (1949): 95–116.

Luce, R. Duncan, and Howard Raiffa. *Games and Decisions.* New York: Wiley, 1957.

MacArthur, Robert H. "Fluctuations of Animal Populations and a Measure of Community Stability." *Ecology* 36 (1955): 533–36.

——. "On the Relative Abundance of Bird Species." *Proceedings of the National Academy of Sciences of the United States of America* 43, no. 3 (1957): 293–95

Macdonald, Dwight. *The Ford Foundation: The Men and the Millions.* New York: Reynal, 1956.

Macrae, Norman. *John von Neumann.* New York: Pantheon Books, 1992.

Macy, Josiah, Jr., Lee S. Christie, and R. Duncan Luce. "Coding Noise in Task-Oriented Groups." *Journal of Abnormal and Social Psychology* 48, no. 3 (1953): 401–9.

Magat, Richard. *The Ford Foundation at Work: Philanthropic Choices, Methods, and Styles.* New York: Plenum Press, 1979.

Magnani, L., and N. J. Nersessian. *Model-Based Reasoning: Science, Technology, Values.* Dordrecht: Kluwer Academic, 2002.

Mancosu, Paolo. *From Hilbert to Brouwer: The Debate on the Foundations of Mathematics in the 1920s.* New York: Oxford University Press, 1988.

Marquis, Donald. "Scientific Methodology in Human Relations." In *Experiments in Social Process,* edited by J. G. Miller, 1–16. New York: McGraw-Hill, 1950.

Marquis, Jefferson P. "The Other Warriors: American Social Science and Nation Building in Vietnam." *Diplomatic History* 24, no. 1 (2000): 79–105.

Marrow, Alfred J. *The Practical Theorist: The Life and Work of Kurt Lewin.* New York: Basic Books, 1969.

Marschak, Jacob. "Neumann's and Morgenstern's New Approach to Static Economics." *Journal of Political Economy* 54, no. 2 (1946): 97–115.

———. "On Mathematics for Economists." *Review of Economics and Statistics* 29, no. 4 (1947): 269–73.

———. "Rational Behavior, Uncertain Prospects, and Measurable Utility." *Econometrica* 18, no. 2 (1950): 111–41.

———. "Towards an Economic Theory of Organization and Information." In *Decision Processes,* ed. Thrall, Coombs, and Davis, 187–220.

———. "Why 'Should' Statisticians and Businessmen Maximize 'Moral Expectation'?" In *Proceedings of the Second Berkeley Symposium on Mathematical Statistics and Probability,* July 31–August 12, 1950. Berkeley: University of California Press, 1951.

———. "Wirtschaftsrechnung und Gemeinwirtschaft." *Archiv für Sozialwissenschaft und Sozialpolitik* 51 (1924).

Mathematica, Inc. *Models of Gradual Reduction of Arms.* Report Prepared for the US Arms Control and Disarmament Agency, Princeton, NJ: Mathematica, September, 1967.

May, Andrew David. "The RAND Corporation and the Dynamics of American Strategic Thought, 1946–1962." PhD diss., Emory University, 1998.

Maynard Smith, John. "The Concept of Information in Biology." *Philosophy of Science* 67, no. 2 (2000): 177–94.

———. "Evolution and the Theory of Games." *American Scientist* 64 (1976): 41–45.

———. *Evolution and the Theory of Games.* New York: Cambridge University Press, 1982.

———. "The Evolution of Alarm Calls." *American Naturalist* 99, no. 904 (1965): 59–63.

———. "Fertility, Mating Behaviour and Sexual Selection in Drosophila Subobscura." *Journal of Genetics* 54 (1956): 261–79.

———. "Game Theory and the Evolution of Behaviour." *Proceedings of the Royal Society of London, Series B, Biological Sciences* 205 (1979): 475–88.

———. "Game Theory without Rationality." *Behavioral and Brain Sciences* 7, no. 1 (1984): 117–22.

———. "Group Selection and Kin Selection." *Nature* 200 (1964): 1145–47.

———. "In Haldane's Footsteps." In *Leaders in the Study of Animal Behavior*, edited by Donald A Dewsbury, 347–56. Lewisburg, PA: Bucknell University Press, 1985.

———. "Natural Selection and the Concept of Protein Space." *Nature* 225 (1970): 563–64.

———. *On Evolution*. Edinburgh: Edinburgh University Press, 1972.

———. "Optimization Theory in Evolution." *Annual Review of Ecology and Systematics* 9 (1978): 31–56.

———. "The Spatial and Temporal Organization of Cells." In *Mathematics and Computer Science in Biology and Medicine*, 247–54. London: Medical Research Council, 1965.

———. "The Theory of Games and the Evolution of Animal Conflicts." *Journal of Theoretical Biology* 47 (1974): 209–21.

Maynard Smith, John, and D. Michie. "Machines That Play Games." *New Scientist* 12 (1964): 367–69.

Maynard Smith, John, and George R. Parker. "The Logic of Asymmetric Contests." *Animal Behavior* 24 (1976): 159–75.

Maynard Smith, John, and George R. Price. "The Logic of Animal Conflict." *Nature* 246 (1973): 15–18.

McArthur, Charles W. *Operations Analysis in the U.S. Army Eighth Air Force in World War II*. Providence, RI: American Mathematical Society, 1990.

McCarthy, Kathleen D. "From Cold War to Cultural Development: The International Cultural Activities of the Ford Foundation, 1950–1980." *Daedalus* 116, no. 1 (1987): 93–117.

McCorduck, Pamela. *Machines Who Think: A Personal Inquiry Into the History and Prospects of Artificial Intelligence*. San Francisco: W. H. Freeman, 1979.

McCulloch, Warren, and Walter Pitts. "A Logical Calculus of Ideas Immanent in Nervous Activity." *Bulletin of Mathematical Biophysics* 5 (1943): 115–33.

McDonald, John. *Strategy in Poker, Business and War*. 1st ed. New York: Norton, 1950.

Mehrtens, Herbert. *Moderne, Sprache, Mathematik: Eine Geschichte des Streits um die Grundlagen der Disziplin und des Subjekts formaler Systeme*. Frankfurt am Main: Suhrkamp, 1990.

Menger, Karl. "Bernoullische Wertlehre Und Petersburger Spiel." *Ergebnisse Eines Mathematischen Kolloquiums* 6 (1934): 26–27.

———. "Ein Satz über Endliche Mengen mit Anwendungen auf die Formale Ethik." *Ergebnisse Eines Mathematischen Kolloquiums* 6 (1934): 23–26.

———. "An Exact Theory of Social Groups and Relations." *American Journal of Sociology* 43, no. 5 (1938): 790–98.

――. "The Formative Years of Abraham Wald and His Work in Geometry." *Annals of Mathematical Statistics* 23, no. 1 (1952): 14–20.

――. *Moral, Wille und Weltgestaltung: Grundlegung zur Logik der Sitten.* Wien: J. Springer, 1934.

――. *Morality, Decision, and Social Organization: Toward a Logic of Ethics.* Dordrecht: D. Reidel Pub. Co., 1974.

――. "The Role of Uncertainty in Economics." In *Essays in Mathematical Economics in Honor of Oskar Morgenstern*, 211–31. Princeton, NJ: Princeton University Press, 1967.

Menger, Karl, and Louise Golland. *Reminiscences of the Vienna Circle and the Mathematical Colloquium.* Dordrecht: Kluwer, 1994.

Mensch, A. *Theory of Games: Techniques and Applications.* New York: American Elsevier, 1966.

Milgrom, Paul, and John Roberts. "Limit Pricing and Entry under Incomplete Information: An Equilibrium Analysis." *Econometrica* 50, no. 2 (1982): 443–59

――. "Predation, Reputation, and Entry Deterrence." *Journal of Economic Theory* 27, no. 2 (1982): 280–312.

Milgrom, Paul, and Nancy Stokey. "Information, Trade and Common Knowledge." *Journal of Economic Theory* 26, no. 1 (1982): 17–27.

Miller, James Grier, ed. *Experiments in Social Process: A Symposium on Social Psychology.* New York: McGraw-Hill, 1950.

――. *Living Systems.* New York: McGraw-Hill, 1978.

――. "Living Systems: Basic Concepts." *Behavioral Science* 10 (1965): 193–237

――. "Living Systems: Cross-Level Hypotheses." *Behavioral Science* 10 (1965): 380–411.

――. "Living Systems: Structure and Process." *Behavioral Science* 10 (1965): 337–79

――. "Living Systems: The Cell." *Currents in Modern Biology* 4 (1971): 78–206.

――. "Living Systems: The Group." *Behavioral Science* 16 (1971): 302–98.

――. "Living Systems: The Organ." *Currents in Modern Biology* 4 (1971): 207–56.

――. "Living Systems: The Organism." *Quarterly Review of Biology* 48, no. 2 (1971): 92–276.

――. "Living Systems: The Society." *Behavioral Science* 20 (1975): 366–535.

――. "Toward a General Theory for the Behavioral Sciences." *American Psychologist* 10 (1955): 513–31.

――. "Toward a General Theory for the Behavioral Sciences." In *The State of the Social Sciences*, edited by Leonard D. White. Chicago: University of Chicago Press, 1956.

Miller, Robert Earnest. "The War that Never Came: Civilian Defense, Mobilization, and Morale during World War II." PhD diss., University of Cincinnati, 1991.

Mirowski, Philip. *Machine Dreams: Economics Becomes a Cyborg Science*. New York: Cambridge University Press, 2002.

———. *More Heat than Light: Economics as Social Physics, Physics as Nature's Economics*. Cambridge: Cambridge University Press, 1989.

———. "Problems in the Paternity of Econometrics: Henry Ludwell Moore." *History of Political Economy* 22, no. 4 (1990): 587–609.

———. "What Were Von Neumann and Morgenstern Trying to Accomplish?" In *Toward a History of Game Theory*, ed. Weintraub, 113–47.

———. "When Games Grow Deadly Serious." In *Economics and National Security: A History of Their Interaction*, edited by Craufurd D. W. Goodwin, 227–60. Durham, NC: Duke University Press, 1991.

Mirowski, Philip, R. Shaw, and Arnold Mandell. "The Rise and Fall of the Concept of Equilibrium in Economic Analysis." *Recherches Économiques de Louvain/Louvain Economic Review* 55, no. 4 (1989): 447–68.

Mitman, Gregg. *The State of Nature: Ecology, Community, and American Social Thought, 1900–1950*. Chicago: University of Chicago Press, 1992.

Moore, Gregory H. *Zermelo's Axiom of Choice: Its Origins, Development, and Influence*. New York: Springer-Verlag, 1982.

Moore, Kelly. *Disrupting Science: Social Movements, American Scientists, and the Politics of the Military, 1945–1975*. Princeton, NJ: Princeton University Press, 2008.

Moreno, J. L. "How Kurt Lewin's 'Research Center for Group Dynamics' Started." *Sociometry* 16, no. 1 (1953): 101–4.

———. "Old and New Trends in Sociometry: Turning Points in Small Group Research." *Sociometry* 17, no. 2 (1954): 179–93.

———. *Who Shall Survive? A New Approach to the Problem of Human Interrelations*. Washington, DC: Nervous and Mental Disease Pub. Co., 1934.

Morgan, Mary S. "Economic Man as Model Man: Ideal Types, Idealization and Caricatures." *Journal of the History of Economic Thought* 28, no. 1 (2006): 1–27.

———. *The World in the Model: How Economists Work and Think*. New York: Cambridge University Press, 2012.

Morgenstern, Oskar. "The Collaboration between Oskar Morgenstern and John von Neumann on the Theory of Games." *Journal of Economic Literature* 14, no. 3 (1976): 805–16.

———. "Das Zeitmoment in der Wertlehre." *Zeitschrift für Nationalökonomie* 5, no. 4 (1934): 433–58.

———. *Die Grenzen der Wirtschaftspolitik*. Wien: Julius Springer, 1934.

———. *The Limits of Economics*. Translated by Vera Smith. London: W. Hodge, 1937.

———. "Logistik und Sozialwissenschaften." *Zeitschrift für Nationalökonomie* 7 (1936): 1–24.

———. "Note on the Formulation of the Study of Logistics." RAND RM-614, May 28, 1951.

———. "Professor Hicks on Value and Capital." *Journal of Political Economy* 49, no. 3 (1941): 361–93.

———. "Prolegomena to a Theory of Organization." RAND RM-734, December 10, 1951.

———. "Review of Rapoport, *Fights, Games, and Debates.*" *Southern Economic Journal* 28, no. 1 (1961): 103–5.

———. "Vollkommene Voraussicht Und Wirtschaftliches Gleichgewicht." *Zeitschrift für Nationalökonomie* 6, no. 3 (1935): 337–57.

———. *Wirtschaftsprognose: Eine Untersuchung ihrer Voraussetzungen und Moglichkeiten.* Wien: Julius Springer 1928.

Morse, Philip M., and George E. Kimball. *Methods of Operations Research.* New York: Wiley, 1951.

Moscati, Ivan. "Early Experiments in Consumer Demand Theory: 1930–1970." *History of Political Economy* 39, no. 3 (2007): 359–401.

Moss, Norman. *Men Who Play God: The Story of the Hydrogen Bomb.* London: Gollancz, 1968.

Mosteller, Frederick. *The Pleasures of Statistics: The Autobiography of Frederick Mosteller.* Edited by Stephen E. Fienberg, David C. Hoaglin, and Judith M. Tanur. New York: Springer Verlag, 2010.

Mosteller, Frederick, and Philip Nogee. "An Experimental Measurement of Utility." *Journal of Political Economy* 59, no. 5 (1951): 371–404.

Motzkin, T. S., H. Raiffa, G. L. Thompson, and R. M. Thrall. "The Double Description Method." In *Contributions to the Theory of Games*, vol. 2, ed. Kuhn and Tucker, 51–74.

Muste, A. J. "Strategy and Conscience: An Exchange." *Liberation* (1965): 17–18.

Myerson, Roger B. "Comments on 'Games with Incomplete Information Played by "Bayesian" Players, I–III': Harsanyi's Games with Incomplete Information." *Management Science* 50, no. 12 (2004): 1818–24.

———. *Game Theory: Analysis of Conflict.* Cambridge, MA: Harvard University Press, 1997.

———. "Learning Game Theory from John Harsanyi." *Games and Economic Behavior* 36, no. 1 (2001): 20–25.

———. "Nash Equilibrium and the History of Economic Theory." *Journal of Economic Literature* 37, no. 3 (1999): 1067–82.

———. "Optimal Auction Design." *Mathematics of Operations Research* 6, no. 1 (1981): 58–73.

———. "Perspectives on Mechanism Design in Economic Theory." *American Economic Review* 98, no. 3 (2008): 586–603.

Nasar, Sylvia. *A Beautiful Mind: A Biography of John Forbes Nash, Jr., Winner of the Nobel Prize in Economics, 1994.* New York: Simon and Schuster, 1998.

Nash, John F. "Equilibrium Points in *n*-Person Games." *Proceedings of the National Academy of Sciences of the United States of America* 36, no. 1 (1950): 48–49.

———. "Non-Cooperative Games." *Annals of Mathematics* 54, no. 2 (1951): 286–95.

———. "Two-Person Cooperative Games." *Econometrica* 21, no. 1 (1953): 128–40.

Newman, M. H. A. "Topology by S. Lefschetz: Analysis Situs by O. Veblen." *Mathematical Gazette* 16, no. 221 (1932): 352–53.

Nti, Kofi O., and Martin Shubik. "Entry in Oligopoly Theory: A Survey." *Eastern Economic Journal* 5, nos. 1/2 (1979): 271–89.

O'Rand, Angela M. "Mathematizing Social Science in the 1950s: The Early Development and Diffusion of Game Theory." In *Toward a History of Game Theory*, ed. Weintraub, 177–204.

Office of Scientific Research and Development. *Analytical Studies in Aerial Warfare*, vol. 2 of *Summary Technical Report of the Applied Mathematics Panel, NDRC*. Edited by Mina Rees. Washington, DC, 1946.

———. *Human Factors in Military Efficiency: Training and Equipment*, vol. 2 of *Summary Technical Report of the Applied Psychology Panel, NDRC*. Washington DC, 1946.

———. *Mathematical Studies Relating to Military Physical Research*, vol. 1 of *Summary Technical Report of the Applied Mathematics Panel, NDRC*. Washington, DC, 1946.

———. *Probability and Statistical Studies in Warfare Analysis*, vol. 3 of *Summary Technical Report of the Applied Mathematics Panel, NDRC*. Washington, DC, 1946.

Oster, George F., and Edward O. Wilson. *Caste and Ecology in the Social Insects*. Princeton, NJ: Princeton University Press, 1978.

Owens, Larry. "Mathematicians at War: Warren Weaver and the Applied Mathematics Panel, 1942–1945." In *History of Modern Mathematics*, edited by David Rowe and John McCleary, 287–305. Boston: Academic Press, 1989.

Palmer, Gregory. *The McNamara Strategy and the Vietnam War: Program Budgeting in the Pentagon, 1960–1968*. Contributions in Political Science 13. Westport, CT: Greenwood Press, 1978.

Peisakoff, M. P. "Continuous Blotto." RAND RM-736, December 4, 1951.

Peltzman, Sam. "The Handbook of Industrial Organization." *Journal of Political Economy* 99, no. 1 (1991): 201–17.

Penrose, L. S. *On the Objective Study of Crowd Behaviour*. London: H. K. Lewis, 1952.

Perrins, Christopher. "Survival of Young Swifts in Relation to Brood-Size." *Nature* 201, no. 4924 (1964): 1147–48.

Peston, M. H. "Review of 'Games and Decisions: Introduction and Critical Survey.'" *Economica* 27, no. 106 (1960): 185–87.

Pickering, Andrew. *The Cybernetic Brain: Sketches of Another Future*. Chicago: University of Chicago Press, 2010.

Pilisuk, Marc, Paul Potter, Anatol Rapoport, and J. Alan Winter. "War Hawks and Peace Doves: Alternate Resolutions of Experimental Conflicts." *Journal of Conflict Resolution* 9, no. 4 (1965): 491–508.

Pilisuk, Marc, and Anatol Rapoport. "Stepwise Disarmament and Sudden Destruction in a Two-Person Game: A Research Tool." *Journal of Conflict Resolution* 8, no. 1 (1964): 36–49.

Pitts, Walter. "Some Observations on the Simple Neuron Circuit." *Bulletin of Mathematical Biophysics* 4, no. 3 (1942): 121–29.

Pooley, Jefferson, and Mark Solovey. "Marginal to the Revolution: The Curious Relationship between Economics and the Behavioral Sciences Movement in Mid-Twentieth-Century America." *History of Political Economy* 42 (2010): 199–233.

Porter, Robert H. "A Review Essay on *Handbook of Industrial Organization*." *Journal of Economic Literature* 29, no. 2 (1991): 553–72.

Poundstone, William. *Prisoner's Dilemma*. New York: Doubleday, 1992.

Price, George R. "Selection and Covariance." *Nature* 227 (1970): 520–21.

Price, Matt. "Roots of Dissent: The Chicago Met Lab and the Origins of the Franck Report." *Isis* 86, no. 2 (1995): 222–44.

Quade, E. S. *Analysis for Military Decisions*. Chicago: Rand McNally, 1964.

Quade, E. S., and Wayne I. Boucher, eds. *Systems Analysis and Policy Planning: Applications in Defense*. New York: Elsevier, 1968.

Quine, W. V. "Commutative Boolean Functions." RAND RM-199, August 10, 1949.

———. "Notes on Information Patterns in Game Theory." RAND RM-216, August 17, 1949.

———. "On Functions of Relations with Especial Reference to Social Welfare." RM-218, August 17, 1949.

———. "A Theorem on Parametric Boolean Functions." RAND RM-196, July 27, 1949.

Raiffa, Howard. "Game Theory at the University of Michigan." In *Toward a History of Game Theory*, ed. Weintraub, 165–75.

———. *Memoir: Analytical Roots of a Decision Scientist*. CreateSpace Independent Publishing Platform, 2011.

Raiffa, Howard, and Stephen E. Fienberg. "The Early Statistical Years, 1947–1967: A Conversation with Howard Raiffa." *Statistical Science* 23, no. 1 (2008): 136–49.

Raiffa, H., H. W. Kuhn, and A. W. Tucker. "Arbitration Schemes for Generalizing Two-Person Games." In *Contributions to the Theory of Games*, vol. 2, ed. Kuhn and Tucker, 361–88.

RAND Corporation. "Conference of Social Scientists, September 14 to 19, 1947–New York." RAND R-106, June 9, 1948.

Rapoport, Anatol. *Certainties and Doubts: A Philosophy of Life.* Toronto: Black Rose Books, 2000.

———. "Chicken à La Kahn: A Critique of Herman Kahn, *On Escalation: Metaphors and Scenarios* (New York: Praeger, 1965)." *Virginia Quarterly Review* 41, no. 3 (1965): 370–89.

———. "Critiques of Game Theory." *Behavioral Science* 4 (1959): 49–66.

———. "Cycle Distributions in Random Nets." *Bulletin of Mathematical Biophysics* 10 (1948): 145–57.

———. "Escape from Paradox." *Scientific American* 217 (1967): 50–56.

———. *Fights, Games, and Debates.* Ann Arbor: University of Michigan Press, 1960.

———. "Formal Games as Probing Tools for Investigating Behavior Motivated by Trust and Suspicion." *Journal of Conflict Resolution* 7, no. 3 (1963): 570–79.

———. "Forms of Output Distribution between Two Individuals Motivated by a Satisfaction Function." *Bulletin of Mathematical Biophysics* 9 (1947): 109–22.

———. "Game Theory without Rationality." *Behavioral and Brain Sciences* 7, no. 1 (1984): 114–15.

———. "Lewis F. Richardson's Mathematical Theory of War." *Journal of Conflict Resolution* 1 (1957): 249–99.

———. "Mathematical Theory of Motivation Interactions of Two Individuals: I." *Bulletin of Mathematical Biophysics* 9 (1947): 17–28.

———. "Mathematical Theory of Motivation Interactions of Two Individuals: II." *Bulletin of Mathematical Biophysics* 9 (1947): 41–61.

———. "Outline of a Probabilistic Approach to Animal Sociology," Parts I–III. *Bulletin of Mathematical Biophysics* 11 (1949): 183–96, 273–81.

———. "Nets with Distance Bias." *Bulletin of Mathematical Biophysics* 13 (1951): 85–91.

———. "New Logic for the Test Ban." *Nation* 192 (1961): 280–84.

———. "Prisoner's Dilemma: Reflections and Recollections." *Simulation and Gaming* 26, no. 4 (1995): 489–502.

———. "Prisoner's Dilemma—Recollections and Observations." In *Game Theory as a Theory of Conflict Resolution*, edited by Anatol Rapoport, 17–34. Dordrecht: Reidel, 1974.

———. "Quantification of Performance on a Logical Task with Uncertainty." In *Symposium on Information Theory in Biology, Gatlinburg, Tennessee, October 29–31, 1956*, edited by Hubert P. Yockey with the assistance of Robert L. Platzman and Henry Quastler, 230–38. New York: Pergamon Press Symposium Publications Division, 1958.

——. "Research for Peace, I." *The Listener,* BBC-TV Review, London, March 31, 1966.

——. "Research for Peace, II." *The Listener,* BBC-TV, London (1966): 504–5, 508.

——. "The Scientific Relevance of C. Wright Mills." In *The New Sociology: Essays in Social Science and Social Theory, in Honor of C. Wright Mills,* edited by Irving Louis Horowitz and C Wright Mills, 94–107. New York: Oxford University Press, 1963.

——. "Some Game-Theoretical Aspects of Parasitism and Symbiosis." *Bulletin of Mathematical Biophysics* 18 (1956): 15–30.

——. "The Sources of Anguish: A Reply to D. G. Brennan's Review of Strategy and Conscience, Liberation, August 1965." *Bulletin of the Atomic Scientists* 21, no. 10 (1965): 25–36.

——. "Spread of Information through a Population with Socio-Structural Bias, I–III." *Bulletin of Mathematical Biophysics* 15 (1953): 523–33, 535–46, and 16 (1954): 75–81.

——. *Strategy and Conscience.* New York: Harper and Row, 1964.

——. "A Study of Disjunctive Reaction Times." *Behavioral Science* 4 (1959): 299–315.

Rapoport, Anatol, and Albert M. Chammah. "The Game of Chicken." *American Behavioral Scientist* 3 (1966): 10–28.

——. *Prisoner's Dilemma: A Study in Conflict and Cooperation.* Ann Arbor: University of Michigan Press, 1965.

Rapoport, Anatol, and C. C. Foster. "Parasitism and Symbiosis in an *n*-Person Non-Constant Sum Game." *Bulletin of Mathematical Biophysics* 18 (1956): 219–31.

Rapoport, Anatol, Melvin Guyer, and David G. Gordon. *The 2×2 Game.* Ann Arbor: University of Michigan Press, 1976.

Rapoport, Anatol, William Hays, and J. David Birch. "The Formation of Hypotheses and Styles in the Process of Solving a Logical Task." In *Decisions, Values, and Groups,* vol. 1, edited by Dorothy Willner, 230–38. New York: Pergamon Press, 1958.

Rapoport, Anatol, Gwen Goodrich Rapoport, and Alfonso Shimbel. "Sanity and the Cold War." *Measure: A Critical Journal* 2 (Spring 1951): 159–74.

Rapoport, Anatol, and Lionel I. Rebhun. "On the Mathematical Theory of Rumor Spread." *Bulletin of Mathematical Biophysics* 14 (1952): 375–83.

Rapoport, Anatol, and Alfonso Shimbel. "Suggested Experimental Procedure for Determining the Satisfaction Function of Animals." *Bulletin of Mathematical Biophysics* 9 (1947): 169–77.

Rapoport, Anatol, and J. David Singer. "The Armers and the Disarmers." *Nation* 196 (1963): 174–77.

———. "Memorandum to the Peace Movement: An Alternative to Slogans." *Nation* 194 (1962): 248–51.

Rashevsky, Nicholas. *Mathematical Theory of Human Relations*. Bloomington, IN: Principia Press, 1947.

———. "A Problem in Mathematical Biophysics of Interaction of Two or More Individuals Which May Be of Interest in Mathematical Sociology." *Bulletin of Mathematical Biophysics* 8 (1947): 9–15.

Rédei, Miklós, and Michael Stöltzner. *John von Neumann and the Foundations of Quantum Physics*. Dordrecht: Kluwer Academic, 2001.

Rees, Mina. "The Computing Program of the Office of Naval Research, 1946–1953." *Annals of the History of Computing* 4, no. 2 (1982): 102–20.

Reingold, Nathan. "Refugee Mathematicians in the United States of America, 1933–1941: Reception and Reflection." *Annals of Science* 38 (1981): 313–38.

Rellstab, Urs. "New Insights Into the Collaboration between John von Neumann and Oskar Morgenstern on the Theory of Games and Economic Behavior." In *Toward a History of Game Theory*, ed. Weintraub, 77–94.

Richardson, L. F. "Could an Arms-Race End without Fighting?" *Nature* 168 (1951): 567–68.

———. "Frequency of Occurrence of Wars and Other Fatal Quarrels." *Nature* 148 (1941): 784.

———. *Generalized Foreign Politics: A Study in Group Psychology*. Cambridge: Cambridge University Press, 1939.

———. "Hints from Physics and Meteorology as to Mental Periodicities." *British Journal of Psychology* 28 (1937): 212–15.

———. *Mathematical Psychology of War*. Oxford: Hunt, 1919.

———. "Mathematical Psychology of War." *Nature* 135 (1935): 830–31.

———. "Thresholds When Sensation Is Regarded as Quantitative." *British Journal of Psychology* 19 (1928): 158–66.

———. "War Moods: I and II." *Psychometrika* 13 (1948): 147–74, 197–232

Richardson, Stephen A. "Lewis Fry Richardson (1881–1953): A Personal Biography." *Conflict Resolution* 1, no. 3 (1957): 300–304.

Rider, Robin E. "Operations Research and Game Theory: Early Connections." In *Toward a History of Game Theory*, ed. Weintraub, 225–39.

Rizvi, S. Abu Turab. "Game Theory to the Rescue?" *Contributions to Political Economy* 13 (1994): 1–28.

———. "The Microfoundations Project in General Equilibrium Theory." *Cambridge Journal of Economics* 18 (1994): 357–77.

———. "The Sonnenschein-Mantel-Debreu Results after Thirty Years." *History of Political Economy* 38, no. suppl 1 (2006): 228–45.

Robertson, Dennis Holme. *Utility and All That: And Other Essays*. London: Allen and Unwin, 1952.

Robertson, Thomas. *The Malthusian Moment: Global Population Growth and the Birth of American Environmentalism*. New Brunswick, NJ: Rutgers University Press, 2012.

Robin, Ron. *The Making of the Cold War Enemy: Culture and Politics in the Military-Intellectual Complex*. Princeton, NJ: Princeton University Press, 2001.

Rosenblueth, Arturo, Norbert Wiener, and Julian Bigelow. "Behavior, Purpose and Teleology." *Philosophy of Science* 10, no. 1 (1943): 18–24.

Roszak, Theodore. *The Making of a Counter Culture: Reflections on the Technocratic Society and Its Youthful Opposition*. Garden City, NY: Doubleday, 1969.

Rubinstein, Ariel. "Comments on the Interpretation of Game Theory." *Econometrica* 59, no. 4 (1991): 909–24.

Runggaldier, Edmund. *Carnap's Early Conventionalism: An Inquiry Into the Historical Background of the Vienna Circle*. Amsterdam: Rodopi, 1984.

Russell, Bertrand. *Common Sense and Nuclear Warfare*. New York: Simon and Schuster, 1959.

Rutherford, Malcolm. *Wesley Clair Mitchell*. London: Pickering and Chatto, 1997.

Saari, Donald. "Mathematical Complexity of Simple Economics." *Notices of the AMS* 42, no. 2 (1995): 222–30.

Sadrieh, Abdolkarim. "Reinhard Selten a Wanderer." In *The Selten School of Behavioral Economics: A Collection of Essays in Honor of Reinhard Selten*, edited by Axel Ockenfels and Abdolkarim Sadrieh, 3–8. New York: Springer, 2010.

Samuelson, Paul A. "Consumption Theory in Terms of Revealed Preference." *Economica* 15, no. 60 (1948): 243–53.

———. "The Empirical Implications of Utility Analysis." *Econometrica* 6, no. 4 (1938): 344–56.

———. "Market Mechanisms and Maximization." RAND P-69, March 28, 1949.

———. "A Note on the Pure Theory of Consumer's Behaviour." *Economica* 5, no. 17 (1938): 61–71.

———. "St. Petersburg Paradoxes: Defanged, Dissected, and Historically Described." *Journal of Economic Literature* 15, no. 1 (1977): 24–55.

Sapolsky, Harvey. *Science and the Navy: The History of the Office of Naval Research*. Princeton, NJ: Princeton University Press, 1990.

Sauermann, Heinz, and Reinhard Selten. "Ein Oligopolexperiment." *Zeitschrift für die gesamte Staatswissenschaft/Journal of Institutional and Theoretical Economics* 115, no. 3 (1959): 427–71.

Scarf, Herbert E. "An Example of an Algorithm for Calculating General Equilibrium Prices." *American Economic Review* 59, no. 4 (1969): 669–77.

———. "Fixed-Point Theorems and Economic Analysis: Mathematical Theorems

Can Be Used to Predict the Probable Effects of Changes in Economic Policy." *American Scientist* 71, no. 3 (1983): 289–96.

Schelling, Thomas C. *Arms and Influence*. New Haven, CT: Yale University Press, 1966.

——. "Bargaining, Communication, and Limited War." *Conflict Resolution* 1, no. 1 (1957): 19–36.

——. "An Essay on Bargaining." *American Economic Review* 46, no. 3 (1956): 281–306.

——. "For the Abandonment of Symmetry in Game Theory." *Review of Economics and Statistics* 41, no. 3 (1959): 213–24.

——. "Nuclear Weapons and Limited War." RAND P-1620, February 20, 1959.

——. "Re-Interpretation of the Solution Concept for 'Non-Cooperative' Games." RAND P-1385, June 2, 1958.

——. "The Reciprocal Fear of Surprise Attack." RAND P-1342, April 16, 1958.

——. "The Retarded Science of International Strategy." *Midwest Journal of Political Science* 4, no. 2 (1960): 107–37.

——. "Review of Strategy and Conscience, by Anatol Rapoport." *American Economic Review* 54, no. 6 (1964): 1082–88.

——. "The Role of Theory in the Study of Conflict." RAND RM-2515, January 1, 1960

——. *The Strategy of Conflict*. Cambridge, MA: Harvard University Press, 1960.

——. "The Strategy of Conflict: Prospectus for a Reorientation of Game Theory." *Journal of Conflict Resolution* 2, no. 3 (1958): 203–64.

——. "Surprise Attack and Disarmament." RAND P-1574, December 10, 1958.

——. "Toward a Theory of Strategy for International Conflict." RAND P-1648, May 8, 1958.

Schlenker, Barry R., and Thomas V. Bonoma. "Fun and Games: The Validity of Games for the Study of Conflict." *Journal of Conflict Resolution* 22, no. 1 (1978): 7–38.

Schlesinger, K. "Über die Produktionsgleichungen der Ökonomischen Wertlehre." *Ergebnisse eines mathematischen Kolloquiums* 6 (1933): 10–11.

Schmidt, Christian. *Game Theory and Economic Analysis: A Quiet Revolution in Economics*. New York: Routledge, 2002.

——. "Game Theory and Economics: An Historical Survey." *Revue D'Économie Politique* 100, no. 5 (1990): 589–618.

Schotter, Andrew. "Core Allocations and Competitive Equilibrium: A Survey." *Zeitschrift für Nationalökonomie* 33 (1973): 281–313.

——. "Oskar Morgenstern's Contribution to the Development of the Theory of Games." In *Toward a History of Game Theory*, ed. Weintraub, 95–112.

Schotter, Andrew, and Gerhard Schwödiauer. "Economics and the Theory of Games: A Survey." *Journal of Economic Literature* 18, no. 2 (1980): 479–527.

Schwalbe, Ulrich, and Paul Walker. "Zermelo and the Early History of Game Theory." *Games and Economic Behavior* 34 (2001): 123–37.

Schwartz, James. "Death of an Altruist." *Lingua Franca* 10, no. 5 (2000): 51–61.

Segerstråle, Ullica. *Defenders of the Truth: The Battle for Science in the Sociobiology Debate and Beyond.* Oxford: Oxford University Press, 2000.

Selten, Reinhard. "Bounded Rationality." *Journal of Institutional and Theoretical Economics* 146, no. 4 (1990): 649–58.

———. "The Chain Store Paradox." *Theory and Decision* 9, no. 2 (1978): 127–60.

———. "New Challenges to the Rationality Assumption: Comment." *Journal of Institutional and Theoretical Economics* 150, no. 1 (1994): 42–44.

———. "A Note on Evolutionary Stable Strategies in Asymmetric Animal Conflicts." *Journal of Theoretical Biology* 84 (1980): 91–101.

———. "Reexamination of the Perfectness Concept for Equilibrium Points in Extensive Games." *International Journal of Game Theory* 4, no. 1 (1975): 25–55.

———. "A Simple Model of Imperfect Competition, Where 4 Are Few and 6 Are Many." *International Journal of Game Theory* 2 (1973): 141–201.

———. "Spieltheoretische Behandlung Eines Oligopolmodells Mit Nachfrageträgheit." *Journal of Institutional and Theoretical Economics* 131, no. 2 (1975): 374.

———. "Spieltheoretische Behandlung eines Oligopolmodells mit Nachfrageträgheit: Teil I: Bestimmung des Dynamischen Preisgleichgewichts." *Journal of Institutional and Theoretical Economics* 121, no. 2 (1965): 301–24.

———. "Spieltheoretische Behandlung eines Oligopolmodells mit Nachfrageträgheit: Teil II: Eigenschaften des Dynamischen Preisgleichgewichts." *Journal of Institutional and Theoretical Economics* 121, no. 4 (1965): 667–89.

Sent, Esther-Mirjam. "Behavioral Economics: How Psychology Made Its (Limited) Way Back Into Economics." *History of Political Economy* 36, no. 4 (2005): 735–60.

Servos, John W. "Mathematics and the Physical Sciences in America, 1880–1930." *Isis* 77, no. 4 (1986): 611–29.

Shannon, C. E. "A Mathematical Theory of Communication." *Bell System Technical Journal* 27 (1948): 379–423, 623–56.

Shapiro, Ian. *The Flight from Reality in the Human Sciences.* Princeton, NJ: Princeton University Press, 2005.

Shapley, Lloyd S. "Additive and Non-Additive Set Functions." PhD diss., Princeton University, 1953.

———. "Note on a Duel Game." RAND RM-82, December 28, 1948.

———. "Note on Duels with Continuous Firing." RAND RM-118, March 11, 1949.

———. "Notes on the n-Person Game—II: The Value of an *n*-Person Game." RAND RM-670, August 21, 1951.

——. "The St. Petersburg Paradox: A Con Game?" *Journal of Economic Theory* 14 (1977): 439–42.

——. "A Tactical Reconnaissance Model." RAND RM-205, August 1949.

——. "A Value for *n*-Person Games." In *Contributions to the Theory of Games*, vol. 2, ed. Kuhn and Tucker, 307–17.

Sharpless, John. "Population Science, Private Foundations, and Development: The Transformation of Demographic Knowledge in the United States, 1945–1965." In *International Development and the Social Sciences: Essays in the History and Politics of Knowledge*, edited by Frederick Cooper and Randall Packard, 176–200. Berkeley: University of California Press.

Sherman, S. "Total Reconnaissance with Total Countermeasures; Simplified Model." P-106, August 5, 1949.

Shionoya, Yuichi, ed. *The German Historical School: The Historical and Ethical Approach to Economics*. London: Routledge, 2001.

Shubik, Martin. "Edgeworth Market Games." In *Contributions to the Theory of Games*, vol. 4, ed. Luce and Tucker, 267–79.

——. "Game Theory: Some Observations." *Yale School of Management Working Papers, Series B* 132 (2000).

——. "Game Theory and Operations Research: Some Musings 50 Years Later." *Operations Research* 50, no. 1 (2002): 192–96.

——. "Game Theory at Princeton, 1949–1955: A Personal Reminiscence." In *Toward a History of Game Theory*, ed. Weintraub, 151–63.

——. *Game Theory in the Social Sciences: Concepts and Solutions*. Cambridge, MA: MIT Press, 1982.

——. "Information, Theories of Competition, and the Theory of Games." *Journal of Political Economy* 60, no. 2 (1952): 145–50.

——. "Oskar Morgenstern: Mentor and Friend." *International Journal of Game Theory* 7, nos. 3/4 (1978): 131–35.

——. "Review of Schelling, the Strategy of Conflict, and Rapoport, Fights, Games and Debates." *Journal of Political Economy* 69, no. 5 (1961): 501–3.

——. "The Role of Game Theory in Economics." *Kyklos* 7, no. 2 (1953): 21–34.

——. *Strategy and Market Structure: Competition, Oligopoly, and the Theory of Games*. New York: Wiley, 1959.

——. "What Is an Application and When Is Theory a Waste of Time?" *Management Science* 33, no. 12 (1987): 1511–22.

Shubik, Martin, ed. *Essays in Mathematical Economics in Honor of Oskar Morgenstern*. Princeton, NJ: Princeton University Press, 1967.

Sills, David L. "Paul F. Lazarsfeld." In *National Academy of Sciences Biographical Memoirs*, vol. 56. Washington, DC: National Academy Press, 1987.

Simon, Herbert A. "A Behavioral Model of Rational Choice." *Quarterly Journal of Economics* 69, no. 1 (1955): 99–118.

———. "Review of *Games and Decisions: Introduction and Critical Survey.*" *American Sociological Review* 23, no. 3 (1958): 342–43.

———. "Review of *Theory of Games and Economic Behaviour* by John von Neumann and Oskar Morgenstern." *American Journal of Sociology* 50, no. 6 (1945): 558–60.

Simpson, Christopher. *Universities and Empire: Money and Politics in the Social Sciences during the Cold War.* New York: W. W. Norton, 1998.

Singer, J. David. "Cosmopolitan Attitudes and International Relations Courses: Some Tentative Conclusions." *Journal of Politics* (1964): 318–38

———. *Human Behavior and International Politics: Contributions from the Social-Psychological Sciences.* Chicago: Rand McNally, 1965.

———. "International Politics and the Behavioral Sciences." *International Review of Political Sciences* 1 (1964): 103–14.

———. "Inter-Nation Influence: A Formal Model." *American Political Science Review* 2 (1963): 420–30.

———. "Soviet and American Foreign Policy Attitudes: A Content Analysis of Elite Articulations." *Journal of Conflict Resolution* 8, no. 4 (1964): 424–85.

Sismondo, Sergio. "Modeling Strategies: Creating Autonomy for Biology's Theory of Games." *History and Philosophy of the Life Sciences* 19 (1997): 147–61.

Slobodkin, Lawrence B. "The Strategy of Evolution." *American Scientist* 52, no. 3 (1964): 342–57.

Smith, Bruce L. R. *The Rand Corporation: Case Study of a Nonprofit Advisory Corporation.* Cambridge, MA: Harvard University Press, 1966.

Smith, C. A. B. *Life, Form, and Number: An Inaugural Lecture Delivered at University College London 17 May 1965.* London: H. K. Lewis, 1966.

———. "Review of *Theory of Games and Economic Behaviour* by John von Neumann and Oskar Morgenstern." *Mathematical Gazette* 29, no. 285 (1945): 131–33.

Smoker, Paul. "On Mathematical Models in Arms Races." *Journal of Peace Research* 2, no. 1 (1965): 94–95.

Snyder, Glenn H. "'Prisoner's Dilemma' and 'Chicken' Models in International Politics." *International Studies Quarterly* 15, no. 1 (1971): 66–103.

Sober, Elliott. "Evolution, Population Thinking, and Essentialism." *Philosophy of Science* 47, no. 3 (1980): 350–83.

Sober, Elliott, and David Sloan Wilson. *Unto Others: The Evolution and Psychology of Unselfish Behavior.* Cambridge, MA: Harvard University Press, 1998.

Solovey, Mark. "Project Camelot and the 1960s Epistemological Revolution: Rethinking the Politics-Patronage-Social Science Nexus." *Social Studies of Science* 31, no. 2 (2001): 171–206.

———. "Riding Natural Scientists' Coattails Onto the Endless Frontier: The

SSRC and the Quest for Scientific Legitimacy." *Journal of the History of the Behavioral Sciences* 40, no. 4 (2004): 393–422.

———. *Shaky Foundations: The Politics-Patronage-Social Science Nexus in Cold War America*. New Brunswick, NJ: Rutgers University Press, 2013.

Solovey, Mark, and Hamilton Cravens, eds. *Cold War Social Science: Knowledge Production, Liberal Democracy, and Human Nature*. New York: Palgrave Macmillan, 2012.

Sonnenschein, Hugo. "Market Excess Demand Functions." *Econometrica* 40, no. 3 (1972): 549–63.

Specht, R. D. "RAND: A Personal View of Its History." RAND P-1601, October 23, 1958.

Steigerwald, David. *The Sixties and the End of Modern America*. New York: St. Martin's Press, 1995.

Stevens, S. S. "On the Theory of Scales of Measurement." *Science*, n.s. 103, no. 2684 (1946): 677–80.

———. "The Operational Basis of Psychology." *American Journal of Psychology* 47, no. 2 (1935): 323–30.

———. "The Operational Definition of Psychological Concepts." *Psychological Review* 42, no. 6 (1935): 517.

Stigler, George J. "The Development of Utility Theory, I." *Journal of Political Economy* 58, no. 4 (August 1950): 307–27.

———. "The Development of Utility Theory, II." *Journal of Political Economy* 58, no. 5 (October 1950): 373–96.

———. *The Organization of Industry*. Homewood, IL: R. D. Irwin, 1968.

Stone, Richard. "The Theory of Games." *Economic Journal* 58, no. 230 (1948): 185–201.

Streissler, Erich, and Karl Milford. "Theoretical and Methodological Positions of German Economists in the Middle of the Nineteenth Century." *History of Economic Ideas* 1, no. 3/2 (1993): 66–71.

Strodtbeck, Fred L., and A. Paul Hare. "Bibliography of Small Group Research (From 1900 Through 1953)." *Sociometry* 17, no. 2 (1954): 107–78.

Strotz, Robert H. "Cardinal Utility." *American Economic Review* 43, no. 2 (1953): 384–97.

Sugden, Robert. "Ken Binmore's Evolutionary Social Theory." *Economic Journal* 111, no. 469 (2001): F213–F243.

———. "Review of *A General Theory of Equilibrium Selection in Games*." *Economica* 56, no. 224 (1989): 530.

Sutton, Francis X. "The Ford Foundation: The Early Years." *Daedalus* 116, no. 1 (1987): 41–91.

———. "The Ford Foundation's Development Program in Africa." *African Studies Bulletin* 3, no. 4 (1960): 1–7.

Szaniawski, Klemens. *The Vienna Circle and the Lvov-Warsaw School.* Dordrecht: Kluwer Academic Publishers, 1989.

Taylor, Peter. "Technocratic Optimism: H. T. Odum, and the Partial Transformation of Ecological Metaphor after World War II." *Journal of the History of Biology* 21 (1988): 213–44.

Thomas, William. *Rational Action: The Sciences of Policy in Britain and America, 1940–1960* (Cambridge, MA: MIT Press, 2015).

Thrall, R. M., C. H. Coombs, and R. L. Davis, eds. *Decision Processes.* New York: John Wiley and Sons, 1954.

Thurstone, L. L. "An Experiment in the Prediction of Choice." In *Proceedings of the Fourth Research Conference.* Chicago: Council on Research, American Meat Institute, 1952.

———. "The Indifference Function." *Journal of Social Psychology* 3 (1931): 139–67.

———. "The Method of Paired Comparisons for Social Values." *Journal of Abnormal and Social Psychology* 21 (1927): 384–400.

———. "Methods of Food-Tasting Experiments." In *Proceedings of the Second Conference on Research.* Chicago: American Meat Institute, March 1950.

Todt, Horst. "Reinhard Selten and the Scientific Climate in Frankfurt during the Fifties." In *The Selten School of Behavioral Economics: A Collection of Essays in Honor of Reinhard Selten*, edited by Axel Ockenfels and Abdolkarim Sadrieh, 29–32. New York: Springer, 2010.

Tolon, Kaya. "Futures Studies: A New Social Science Rooted in Cold War Strategic Thinking." In *Cold War Social Science: Knowledge Production, Liberal Democracy, and Human Nature*, ed. Solovey and Cravens, 45–62.

Trachtenberg, Marc. *History and Strategy.* Princeton, NJ: Princeton University Press, 1991.

Tribe, Keith. "Die Vernunft des List: National Economy and the Critique of Cosmopolitical Economy." In *Strategies of Economic Order: German Economic Discourse, 1750–1950.* Cambridge: Cambridge University Press, 1995.

———. *Governing Economy: The Reformation of German Economic Discourse, 1750–1840.* New York: Cambridge University Press, 1988.

Trist, Eric, and Hugh Murray. *The Social Engagement of Social Science: A Tavistock Anthology.* 3 vols. Philadelphia: University of Pennsylvania Press, 1990.

Trivers, Robert. "The Evolution of Reciprocal Altruism." *Quarterly Review of Biology* 46 (1971): 35–57.

Tucker, Albert W. *Game Theory and Programming.* Stillwater: Dept. of Mathematics, Oklahoma Agricultural and Mechanical College, 1955.

———. "Solomon Lefschetz: A Reminiscence." *Two-Year College Mathematics Journal* 14, no. 3 (1983): 225–27.

Tucker, A. W., and R. D. Luce, eds. *Contributions to the Theory of Games*, vol. 4. Princeton, NJ: Princeton University Press, 1959.

Tversky, Amos. "Clyde Hamilton Coombs, 1912–1988." *National Academy of Sciences Biographical Memoirs* 61 (1992): 59–77.

Uebel, Thomas E. *Rediscovering the Forgotten Vienna Circle: Austrian Studies on Otto Neurath and the Vienna Circle.* Dordrecht: Kluwer Academic Publishers, 1991.

Ulam, Stanislaw M. *Adventures of a Mathematician.* New York: Scribner, 1976.

———. "John von Neumann, 1903–1957." *Annals of the History of Computing* 4, no. 2 (1982): 1–49.

Ulam, S., H. W. Kuhn, A. W. Tucker, and Claude E. Shannon. "John von Neumann, 1903–1957." In *The Intellectual Migration: Europe and America, 1930–1960,* edited by Donald Fleming and Bernard Bailyn, 235–69. Cambridge, MA: Harvard University Press, 1968.

Urton, Gary. *The Social Life of Numbers: A Quechua Ontology of Numbers and Philosophy of Arithmetic.* Austin: University of Texas Press, 1997.

Vajda, S. *An Introduction to Linear Programming and the Theory of Games.* New York: Wiley, 1960.

van Dalen, Dirk. *Mystic, Geometer, and Intuitionist: The Life of L.E.J. Brouwer.* New York: Oxford University Press, 1999.

van Damme, Eric. "On the Contributions of John C. Harsanyi, John F. Nash, and Reinhard Selten." *International Journal of Game Theory* 24 (1995): 3–11.

———. *Refinements of the Nash Equilibrium Concept.* Berlin: Springer-Verlag, 1983.

———. *Stability and Perfection of Nash Equilibria.* Princeton, NJ: Princeton University Press, 1991.

Van Heijenoort, Jean. *From Frege to Gödel: A Source Book in Mathematical Logic, 1879–1931.* Cambridge, MA: Harvard University Press, 1967.

Verran, Helen. *Science and an African Logic.* Chicago: University of Chicago Press, 2001.

Vickrey, William. "Counterspeculation, Auctions, and Competitive Sealed Tenders." *Journal of Finance* 16, no. 1 (1961): 8–37.

Viner, Jacob. "The Utility Concept in Value Theory and Its Critics, I." *Journal of Political Economy* 33, no. 4 (August 1925): 369–87.

———. "The Utility Concept in Value Theory and Its Critics, II." *Journal of Political Economy* 33, no. 6 (December 1925): 638–59.

von Mises, Ludwig. *Socialism: An Economic and Sociological Analysis.* Indianapolis, IN: Liberty Fund, 1981.

von Neumann, John. "A Certain Zero-Sum Two-Person Game Equivalent to the Optimal Assignment Problem." In *Contributions to the Theory of Games,* vol. 2, ed. Kuhn and Tucker, 5–12.

———. *The Computer and the Brain.* New Haven, CT: Yale University Press, 1958.

———. "Eine Axiomatizierung der Mengenlehre." *Journal für die Reine und Angewandte Mathematick* 154 (1925): 219–40.

——. *Mathematische Grundlagen der Quantenmechanik.* Berlin: Springer, 1932.

——. "A Model of General Economic Equilibrium." *Review of Economic Studies* 13, no. 33 (1945): 1–9.

——. "A Numerical Method to Determine Optimum Strategy." *Naval Research Logistics Quarterly* 1 (1954): 109–15.

——. "On the Theory of Games of Strategy." In *Contributions to the Theory of Games*, vol. 4, ed. Tucker and Luce, 13–42. Translated by Sonya Bargmann.

——. "Über ein Ökonomisches Gleichungssystem und eine Verallgemeinerung des Brouwerschen Fixpunktsatzes." *Ergebnisse eines mathematischen Kolloquiums* 8 (1937): 73–83.

——. "Zur Theorie der Gesellschaftsspiele." *Mathematische Annalen* 100 (1928): 295–320.

von Neumann, John, and M. Fréchet. "Communication on the Borel Notes." *Econometrica* 21, no. 1 (1953): 124–27.

von Neumann, John, D. Hilbert, and L. Nordheim. "Über die Grundlagen der Quantenmechanic." *Mathematische Annalen* 98 (1927): 1–30.

von Neumann, John, and Oskar Morgenstern. *Theory of Games and Economic Behavior.* Princeton, NJ: Princeton University Press, 1944; 2nd ed., 1947.

Vonneumann, Nicholas A. *John von Neumann—As Seen by His Brother.* Meadowbrook, PA: Private Printing, 1987.

Waddington, C. H. *The Strategy of the Genes: A Discussion of Some Aspects of Theoretical Biology.* London: Allen and Unwin, 1957.

Wald, Abraham. "Generalization of a Theorem by v. Neumann Concerning Zero Sum Two Person Games." *Annals of Mathematics* 46, no. 2 (1945): 281–86.

——. "Statistical Decision Functions Which Minimize the Maximum Risk." *Annals of Mathematics* 46, no. 2 (1945): 265–80.

——. "Über die Eindeutige Positive Lösbarkeit der Neuen Produktionsgleichungen." *Ergebnisse eines mathematischen Kolloquiums* 6 (1933): 12–20.

Walras, Léon. *Éléments d'économie politique pure: Ou, théorie de la richesse sociale.* 4th ed. Lausanne: F. Rouge, 1900.

Wang, Jessica. *American Science in an Age of Anxiety: Scientists, Anticommunism, and the Cold War.* Chapel Hill: University of North Carolina Press, 1999.

Warwick, Andrew. "Cambridge Mathematics and Cavendish Physics: Cunningham, Campbell, and Einstein's Relativity, 1905–1911. Part I: The Uses of Theory." *Studies in the History and Philosophy of Science* 23 (1992): 625–56.

——. *Masters of Theory: The Pursuit of Mathematical Physics in Victorian Cambridge.* Cambridge: Cambridge University Press, 2003.

Weaver, Warren. "Science and Complexity." *American Scientist* 36, no. 4 (1948): 536–44.

Weintraub, E. Roy. *General Equilibrium Analysis: Studies in Appraisal.* Cambridge: Cambridge University Press, 1985.

———. *How Economics Became a Mathematical Science.* Durham, NC: Duke University Press, 2002.

———, ed. *Toward a History of Game Theory.* Durham, NC: Duke University Press, 1992.

Weintraub, E. Roy, and Philip Mirowski. "The Pure and the Applied: Bourbakism Comes to Mathematical Economics." *Science in Context* 7, no. 2 (1994): 245–72.

Weintraub, Sidney. "Oligopoly and Game Theory: A Review of Martin Shubik's *Strategy and Market Structure.*" *Kyklos* 13, no. 3 (1960): 400–406.

Weisbord, Marvin Ross. *Productive Workplaces Revisited: Dignity, Meaning, and Community in the 21st Century.* San Francisco: Jossey-Bass, 2004.

Weisner, Louis. "Review of *Theory of Games and Economic Behavior* by John von Neumann and Oskar Morgenstern." *Science and Society* 9 (1945): 366–69.

Wells, Tom. *Wild Man: The Life and Times of Daniel Ellsberg.* New York: Palgrave, 2001.

Wiener, Norbert. *Cybernetics: Or, Control and Communication in the Animal and the Machine.* New York: Wiley and Sons, 1948.

Wigner, Eugene Paul. *Symmetries and Reflections: Scientific Essays of Eugene P. Wigner.* Bloomington: Indiana University Press, 1967.

Williams, George C. *Adaptation and Natural Selection: A Critique of Some Current Evolutionary Thought.* Princeton, NJ: Princeton University Press, 1966.

Wilson, Edward O. "The Ergonomics of Caste in the Social Insects." *American Naturalist* 102, no. 923 (1968): 41–66.

———. *Naturalist.* Washington, DC: Island Press, 1994.

———. "The Social Biology of Ants." *Annual Review of Entomology* 8 (1963): 345–68.

———. *Sociobiology: The New Synthesis.* Cambridge, MA: Belknap Press of Harvard University Press, 1975.

Wohlstetter, Albert. "The Delicate Balance of Terror." *Foreign Affairs* 37 (1959): 211–34.

Wright, Quincy. "Criteria for Judging the Relevance of Researches on the Problems of Peace." In *Research for Peace: Essays*, edited by Quincy Wright. Amsterdam: North-Holland, 1954.

Wynne-Edwards, V. C. *Animal Dispersion in Relation to Social Behavior.* Edinburgh: Oliver and Boyd, 1962.

———. "Ecology and the Evolution of Social Ethics." In *Biology and the Human Sciences: The Herbert S. Spencer Lectures 1970*, edited by J. W. S. Pringle, 49–69. Oxford: Clarendon Press, 1972.

———. "The Overfishing Principle Applied to Natural Populations and Their Food-Resources: And a Theory of Natural Conservation." In *Proceedings of the XIIth International Ornithological Congress, Helsinki 1958*, 790–94. Helsinki: Tilgmannin Kirjapaino, 1960.

———. "Population, Affluence, and the Environment." *Environmental Education* 3 (1973): 10–18.

———. "Survival of Young Swifts in Relation to Brood-Size." *Nature* 201, no. 4924 (1964): 1148–49.

Zermelo, Ernst. "Über eine Anwendung der Mengenlehre auf die Theorie des Schachspiels." In *Proceedings of the 5th International Congress of Mathematicians, Cambridge, 22–28 August 1912*, vol. 2, edited by E. W. Hobson and A. E. H. Love, 501–4. Cambridge: Cambridge University Press, 1913.

———. "Untersuchungen über die Grundlagen der Mengenlehre. I." *Mathematische Annalen* 65 (1908): 261–81.

Zuckerman, Solly. "Judgment and Control in Modern Warfare." *Foreign Affairs* 40, no. 2 (1962): 196–212.

Index